STUDY SMARTER

CHAPTER

Test Prep
VIDEO CD

Step-by-step solutions on video for all
chapter test exercises from the text

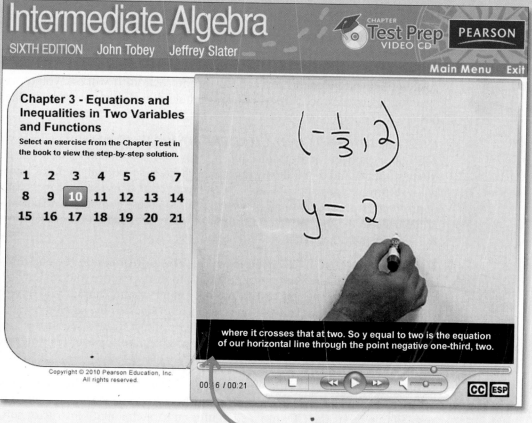

Intermediate Algebra

SIXTH EDITION John Tobey Jeffrey Slater

CHAPTER Test Prep VIDEO CD PEARSON

Main Menu   Exit

**Chapter 3 - Equations and
Inequalities in Two Variables
and Functions**

Select an exercise from the Chapter Test in
the book to view the step-by-step solution.

1	2	3	4	5	6	7
8	9	**10**	11	12	13	14
15	16	17	18	19	20	21

$$\left(-\frac{1}{3}, 2\right)$$

$$y = 2$$

where it crosses that at two. So y equal to two is the equation
of our horizontal line through the point negative one-third, two.

00:16 / 00:21

CC ESP

English and Spanish Subtitles Available

INCLUDED WITH EVERY NEW COPY OF THIS TEXTBOOK!

To the Student

With this edition of the Tobey/Slater/Blair Developmental Mathematics series, we are committed to helping you get the most out of your learning experience by showing you how important math can be in your daily lives. For example, the new **Use Math to Save Money** feature presents practical examples of how math can help you save money, cut costs, and spend less. One of the best ways you can save money is to pass your course the first time you take it. This text and its features and MyMathLab can help you do that.

The best place to start is the **How Am I Doing? Guide to Math Success.** This clear path for you to follow is based upon how our successful students have utilized the textbook in the past. Here is how it works:

EXAMPLES and PRACTICE PROBLEMS: When you study an Example, you should immediately do the Practice Problem that follows to make sure you understand each step in solving a particular problem. The worked-out solution to every Practice Problem can be found in the back of the text, starting at page SP-1, so you can check your work and receive immediate guidance in case you need to review.

EXERCISE SETS—Practice, Practice, Practice: You learn math by *doing* math. The best way to learn math is to *practice, practice, practice*. Be sure that you complete every exercise your instructor assigns as homework. In addition, check your answers to the odd-numbered exercises in the back of the text to see whether you have correctly solved each problem.

QUICK QUIZ: After every exercise set, be sure to do the problems in the Quick Quiz. This will tell you immediately if you have understood the key points of the homework exercises.

CONCEPT CHECK: At the end of the Quick Quiz is a Concept Check. This will test your understanding of the key concept of the section. It will ask you to explain in your own words how a procedure works. It will help you clarify your knowledge of the methods of the section.

HOW AM I DOING? MID-CHAPTER REVIEW: This feature allows you to check if you understand the important concepts covered to that point in a particular chapter. Many students find that halfway through a chapter is a crucial point for review because so many different types of problems have been covered. This review covers each of the types of problems from the first half of the chapter. Do these problems and check your answers at the back of the text. If you need to review any of these problems, simply refer back to the section and objective indicated next to the answer.

HOW AM I DOING? CHAPTER TEST: This test (found at the end of every chapter) provides you with an excellent opportunity to both practice and review for any test you will take in class. Take this test to see how much of the chapter you have mastered. By checking your answers, you can once again refer back to the section and the objective of any exercise you want to review further. This allows you to see at once what has been learned and what still needs more study as you prepare for your test or exam.

HOW AM I DOING? CHAPTER TEST PREP VIDEO CD: If you need to review any of the exercises from the *How Am I Doing? Chapter Test*, this video CD found at the front of the text provides a clear explanation of how to do each step of every problem on the test. Simply insert the CD into a computer and watch a math instructor solve each of the Chapter Test exercises in detail. By reviewing these problems, you can study through any points of difficulty and better prepare yourself for your upcoming test or exam.

These steps provide a clear path you can follow in order to successfully complete your math course. More importantly, the **How Am I Doing? Guide to Math Success** is a tool to help you achieve an understanding of mathematics. We encourage you to take advantage of this new feature.

John Tobey and Jeffrey Slater
North Shore Community College

6th Edition

Intermediate Algebra

John Tobey

North Shore Community College
Danvers, Massachusetts

Jeffrey Slater

North Shore Community College
Danvers, Massachusetts

With Contributions from Jennifer Crawford

Prentice Hall
is an imprint of

Upper Saddle River, NJ 07458

Library of Congress Cataloging-in-Publication Data

Tobey, John
   Intermediate algebra / John Tobey, Jeffrey Slater.—6th ed.
     p. cm.
   ISBN 978-0-321-57829-7 (student ed.) — ISBN 978-0-321-57841-9 (annotated instructor's ed.)
  1. Algebra—Textbooks. I. Slater, Jeffrey. II. Title.
   QA154.3.T64 2010
   512.9—dc22   2008031087

Editorial Director, Mathematics: *Christine Hoag*
Editor in Chief: *Paul Murphy*
Executive Project Manager: *Kari Heen*
Senior Project Editor: *Lauren Morse*
Assistant Editors: *Georgina Brown and Christine Whitlock*
Production Management: *Elm Street Publishing Services*
Senior Managing Editor: *Linda Mihatov Behrens*
Operations Specialist: *Ilene Kahn*
Senior Operations Supervisor: *Diane Peirano*
Marketing Manager: *Marlana Voerster*
Marketing Assistant: *Nathaniel Koven*
Art Director: *Heather Scott*
Interior/Cover Designer: *Tamara Newnam*
AV Project Manager: *Thomas Benfatti*
Executive Manager, Course Production: *Peter Silvia*
Media Producer: *Audra J. Walsh*
Associate Producer: *Emilia Yeh*
Manager, Content Development: *Rebecca Williams*
QA Manager: *Marty Wright*
Senior Content Developer: *Mary Durnwald*
Photo Research Development Manager: *Elaine Soares*
Image Permission Coordinator: *Kathy Gavilanes*
Photo Researcher: *Sheila Norman*
Manager, Cover Visual Research and Permissions: *Karen Sanatar*
Cover Image: *Ryan McVay/Photodisc/Getty Images, Inc.*
Compositor: *Macmillan Publishing Solutions*
Art Studios: *Scientific Illustrators and Laserwords*

Photo credits appear on page P-1, which constitutes a continuation of the copyright page.

Prentice Hall
is an imprint of

PEARSON

© 2010, 2006, 2002, 1998, 1995, 1991 by Pearson Education, Inc.
Pearson Prentice Hall
Pearson Education, Inc.
Upper Saddle River, New Jersey 07458

Printed in the United States of America

10 9 8 7 6 5 4 3 2 1

ISBN 10: 0-321-57829-5
ISBN 13: 978-0-321-57829-7

Pearson Education Ltd., London
Pearson Education Singapore, Pte. Ltd.
Pearson Education Canada, Inc.
Pearson Education, Japan
Pearson Education Australia PTY, Limited

Pearson Education North Asia, Ltd., Hong Kong
Pearson Educación de Mexico, S.A. de C.V.
Pearson Education Malaysia, Pte. Ltd.
Pearson Education Upper Saddle River,
  New Jersey

This book is dedicated to John Tobey III, Marcia Tobey Salzman, and Melissa Tobey LaBelle. They are three college graduates who would make any parent proud. They have each worked hard to achieve a master of arts degree and have proved that they all can do two things at once and do well at both.

Contents

How Am I Doing? Guide to Math Success   (inside front cover)

To the Student   ii

Preface   xiii

Acknowledgments   xvii

Diagnostic Pretest   xx

CHAPTER 1

Basic Concepts   1

1.1   The Real Number System   2
1.2   Operations with Real Numbers   9
1.3   Powers, Square Roots, and the Order of Operations   17
      How Am I Doing? Sections 1.1–1.3   25
1.4   Integer Exponents and Scientific Notation   26
1.5   Operations with Variables and Grouping Symbols   37
1.6   Evaluating Variable Expressions and Formulas   43
      Putting Your Skills to Work: Use Math to Save Money   51
      Chapter 1 Organizer   52
      Chapter 1 Review Problems   54
      How Am I Doing? Chapter 1 Test   56

CHAPTER 2

Linear Equations and Inequalities   57

2.1   First-Degree Equations with One Unknown   58
2.2   Literal Equations and Formulas   66
2.3   Absolute Value Equations   72
2.4   Using Equations to Solve Word Problems   78
      How Am I Doing? Sections 2.1–2.4   84
2.5   Solving More-Involved Word Problems   85
2.6   Linear Inequalities   94
2.7   Compound Inequalities   103
2.8   Absolute Value Inequalities   109
      Putting Your Skills to Work: Use Math to Save Money 116
      Chapter 2 Organizer   117
      Chapter 2 Review Problems   119
      How Am I Doing? Chapter 2 Test   123
      Cumulative Test for Chapters 1–2   124

CHAPTER 3

Equations and Inequalities in Two Variables and Functions

		125
3.1	Graphing Linear Equations with Two Unknowns	126
3.2	Slope of a Line	135
3.3	Graphs and the Equations of a Line	143
	How Am I Doing? Sections 3.1–3.3	153
3.4	Linear Inequalities in Two Variables	154
3.5	Concept of a Function	160
3.6	Graphing Functions from Equations and Tables of Data	169
	Putting Your Skills to Work: Use Math to Save Money	183
	Chapter 3 Organizer	184
	Chapter 3 Review Problems	186
	How Am I Doing? Chapter 3 Test	191
	Cumulative Test for Chapters 1–3	193

CHAPTER 4

Systems of Linear Equations and Inequalities

		195
4.1	Systems of Linear Equations in Two Variables	196
4.2	Systems of Linear Equations in Three Variables	210
	How Am I Doing? Sections 4.1–4.2	217
4.3	Applications of Systems of Linear Equations	218
4.4	Systems of Linear Inequalities	229
	Putting Your Skills to Work: Use Math to Save Money	237
	Chapter 4 Organizer	238
	Chapter 4 Review Problems	240
	How Am I Doing? Chapter 4 Test	244
	Cumulative Test for Chapters 1–4	245

CHAPTER 5

Polynomials   247

5.1	Introduction to Polynomials and Polynomial Functions: Adding, Subtracting, and Multiplying	248
5.2	Dividing Polynomials	255
5.3	Synthetic Division	260
5.4	Removing Common Factors; Factoring by Grouping	264
5.5	Factoring Trinomials	270
	How Am I Doing? Sections 5.1–5.5	277
5.6	Special Cases of Factoring	278
5.7	Factoring a Polynomial Completely	285
5.8	Solving Equations and Applications Using Polynomials	289
	Putting Your Skills to Work: Use Math to Save Money	297
	Chapter 5 Organizer	298
	Chapter 5 Review Problems	300
	How Am I Doing? Chapter 5 Test	303
	Cumulative Test for Chapters 1–5	305

CHAPTER 6

Rational Expressions and Equations   307

6.1	Rational Expressions and Functions: Simplifying, Multiplying, and Dividing	308
6.2	Adding and Subtracting Rational Expressions	316
6.3	Complex Rational Expressions	323
	How Am I Doing? Sections 6.1–6.3	329
6.4	Rational Equations	330
6.5	Applications: Formulas and Advanced Ratio Exercises	335
	Putting Your Skills to Work: Use Math to Save Money	343
	Chapter 6 Organizer	344
	Chapter 6 Review Problems	346
	How Am I Doing? Chapter 6 Test	350
	Cumulative Test for Chapters 1–6	351

CHAPTER 7

Rational Exponents and Radicals

Rational Exponents and Radicals   353

7.1   Rational Exponents   354
7.2   Radical Expressions and Functions   361
7.3   Simplifying, Adding, and Subtracting Radicals   369
7.4   Multiplying and Dividing Radicals   375
      How Am I Doing? Sections 7.1–7.4   385
7.5   Radical Equations   386
7.6   Complex Numbers   392
7.7   Variation   400
      Putting Your Skills to Work: Use Math to Save Money   407
      Chapter 7 Organizer   408
      Chapter 7 Review Problems   411
      How Am I Doing? Chapter 7 Test   415
      Cumulative Test for Chapters 1–7   417

CHAPTER 8

Quadratic Equations and Inequalities

Quadratic Equations and Inequalities   419

8.1   Quadratic Equations   420
8.2   The Quadratic Formula and Solutions to Quadratic Equations   427
8.3   Equations That Can Be Transformed into Quadratic Form   436
      How Am I Doing? Sections 8.1–8.3   443
8.4   Formulas and Applications   444
8.5   Quadratic Functions   455
8.6   Quadratic Inequalities in One Variable   464
      Putting Your Skills to Work: Use Math to Save Money   471
      Chapter 8 Organizer   472
      Chapter 8 Review Problems   475
      How Am I Doing? Chapter 8 Test   479
      Cumulative Test for Chapters 1–8   481

CHAPTER 9

The Conic Sections

483

9.1	The Distance Formula and the Circle	484
9.2	The Parabola	492
9.3	The Ellipse	503
	How Am I Doing? Sections 9.1–9.3	511
9.4	The Hyperbola	512
9.5	Nonlinear Systems of Equations	520
	Putting Your Skills to Work: Use Math to Save Money	526
	Chapter 9 Organizer	527
	Chapter 9 Review Problems	530
	How Am I Doing? Chapter 9 Test	533
	Cumulative Test for Chapters 1–9	535

CHAPTER 10

Additional Properties of Functions

537

10.1	Function Notation	538
10.2	General Graphing Procedures for Functions	545
	How Am I Doing? Sections 10.1–10.2	553
10.3	Algebraic Operations on Functions	554
10.4	Inverse of a Function	561
	Putting Your Skills to Work: Use Math to Save Money	570
	Chapter 10 Organizer	571
	Chapter 10 Review Problems	573
	How Am I Doing? Chapter 10 Test	578
	Cumulative Test for Chapters 1–10	580

CHAPTER 11

Logarithmic and Exponential Functions   582

11.1 The Exponential Function   583

11.2 The Logarithmic Function   592

11.3 Properties of Logarithms   599

How Am I Doing? Sections 11.1–11.3   606

11.4 Common Logarithms, Natural Logarithms, and Change of Base of Logarithms   607

11.5 Exponential and Logarithmic Equations   615

Putting Your Skills to Work: Use Math to Save Money   626

Chapter 11 Organizer   627

Chapter 11 Review Problems   629

How Am I Doing? Chapter 11 Test   631

Practice Final Examination   632

Appendix A: Tables   A-1

Table A-1: Table of Square Roots   A-1

Table A-2: Table of Exponential Values   A-2

Appendix B: Determinants and Cramer's Rule   A-3

Appendix C: Solving Systems of Linear Equations Using Matrices   A-11

Solutions to Practice Problems   SP-1

Answers to Selected Exercises   SA-1

Glossary   G-1

Subject Index   I-1

Applications Index   I-7

Photo Credits   P-1

TO THE INSTRUCTOR

One of the hallmark characteristics of *Intermediate Algebra* that makes the text easy to learn and teach from is the building-block organization. Each section is written to stand on its own, and every homework set is completely self-testing. Exercises are paired and graded and are of varying levels and types to ensure that all skills and concepts are covered. As a result, the text offers students an effective and proven learning program suitable for a variety of course formats—including lecture-based classes; discussion-oriented classes; distance learning classes; modular, self-paced courses; mathematics laboratories; and computer-supported centers. The book has been written to be especially helpful in online courses. The authors usually teach at least one course online each semester.

Intermediate Algebra is part of a series that includes the following:

Tobey/Slater/Blair, *Basic College Mathematics,* Sixth Edition

Tobey/Slater/Blair, *Essentials of Basic College Mathematics,* Second Edition

Blair/Tobey/Slater, *Prealgebra,* Fourth Edition

Tobey/Slater, *Beginning Algebra,* Seventh Edition

Tobey/Slater/Blair, *Beginning Algebra: Early Graphing,* Second Edition

Tobey/Slater, *Intermediate Algebra,* Sixth Edition

Tobey/Slater/Blair, *Beginning and Intermediate Algebra,* Third Edition

We have visited and listened to teachers across the country and have incorporated a number of suggestions into this edition to help you with the particular learning delivery system at your school. The following pages describe the key continuing features and changes in the sixth edition.

NEW! FEATURES IN THE SIXTH EDITION

Quick Quiz

At the end of each problem section, there is a **Quick Quiz** for the student. The quiz contains three problems that cover all of the essential content of that entire section of the book. If a student can do those three problems, the student has mastered the mathematics skills of that section. If a student cannot do those three problems, the student is made aware that further study is needed to obtain mastery. At the end of the Quick Quiz is a **Concept Check** question. This question stresses a mastery of the concepts of each section of the book. The question asks the student to explain how and why a solution method actually works. It forces the student to analyze problems and reflect on the mathematical concepts that have been learned. The student is asked to explain in his or her own words the mathematical procedures that have been practiced in a given section.

Classroom Quiz

Adjacent to each Quick Quiz for the student, the Annotated Instructor's Edition of the book contains a **Classroom Quiz.** This quiz allows the instructor to give a short quiz in class that covers the essential content of each section of the book. Immediately, an instructor can find out if the students have mastered the material in this section or not. No more will an instructor have to rush to pick out the right balance of problems that test a student's knowledge of a section. The Classroom Quiz is instantly available to assist the instructor in assessing student knowledge.

Putting Your Skills to Work: Use Math to Save Money

Each chapter of the book presents a simple, down-to-earth, practical example of how to save money. Students are given a straightforward, realistic way to cut costs and spend less. They are shown logical ways to get out of debt. They are given motivating examples of how other students saved money. Students are very motivated to read these articles and soon begin to see how the course will actually help them in everyday life. Many of these activities were contributed by professors, and our thanks go out to Mary Pearce, Suellen Robinson, Mike Yarbrough, Betty Ludlum, Connie Buller, Armando Perez, and Maria Luisa Mendez for their insightful contributions.

Teaching Examples

Having in-class practice problems readily available is extremely helpful to both new and experienced instructors. These instructor examples, called **Teaching Examples,** are included in the margins of the Annotated Instructor's Edition for practice in class.

KEY FEATURES IN THE SIXTH EDITION

The **How Am I Doing? Guide to Math Success** shows students how they can effectively use this textbook to succeed in their mathematics course. This clear path for them to follow is based upon how students have successfully utilized this textbook in the past.

The text has been designed so the **Examples and Practice Problems** are clearly connected in a cohesive unit. This encourages students to try the Practice Problem associated with each Example to ensure they understand each step in solving a particular problem. The worked-out solution to every Practice Problem can be found in the back of the text so students can check their work. **New!** to this edition, the answers are now included next to each Practice Problem in the Annotated Instructor's Edition.

Each **exercise set** progresses from easy to medium to challenging problems, with appropriate quantities of each, and has paired even and odd problems. All concepts are fully represented, with every Example from the section covered by a group of exercises. Exercise sets include **Mixed Practice** problems, which require students to identify the type of problem and the best method they should use to solve it, as well as **Verbal and Writing Skills** exercises, which allow students to explain new concepts fully in their own words. Throughout the text the application exercises have been updated. These **Applications** relate to everyday life, global issues beyond the borders of the United States, and other academic disciplines. Roughly 25 percent of the applications have been contributed by actual students based on scenarios they have encountered in their home or work lives.

Many students find that halfway through a chapter is a crucial point for review because so many different types of problems have been covered. The **How Am I Doing? Mid-Chapter Review** covers each of the types of problems from the first half of the chapter and allows instructors to check if students understand the important concepts covered to that point. Specific section and objective references are provided with each answer to indicate where a student should look for further review.

Developing Problem-Solving Abilities

As authors, we are committed to producing a textbook that emphasizes mathematical reasoning and problem-solving techniques as recommended by AMATYC, NCTM, AMS, NADE, MAA, and other bodies. To this end, the problem sets are built on a wealth of real-life and real-data applications. Unique problems have been developed and incorporated into the exercise sets that help train students in data interpretation, mental mathematics, estimation, geometry and graphing, number sense, critical thinking, and decision making.

The successful **Mathematics Blueprint for Problem Solving** strengthens problem-solving skills by providing a consistent and interactive outline to help students organize their approach to problem solving. Once students fill in the blueprint, they can refer back to their plan as they do what is needed to solve the problem. Because of its flexibility, this feature can be used with single-step problems, multi-step problems, applications, and nonroutine problems that require problem-solving strategies.

The **Developing Your Study Skills** boxes are integrated throughout the text to provide students with techniques for improving their study skills and succeeding in math courses.

Integration and Emphasis on Geometry Due to the emphasis on geometry on many statewide exams, geometry problems are integrated throughout the text. Examples and exercises that incorporate a principle of geometry are marked with a triangle icon for easy identification.

When students encounter mathematics in real-world publications, they often encounter data represented in a **graph, chart, or table** and are asked to make a reasonable conclusion based on the data presented. This emphasis on graphical interpretation is a continuing trend with today's expanding technology. In this text, students are asked to make simple interpretations, to solve medium-level problems, and to investigate challenging applied problems based on the data shown in a chart, graph, or table.

Mastering Mathematical Concepts

Text features that develop the mastery of concepts include the following:

Concise **Learning Objectives** listed at the beginning of each section allow students to preview the goals of that section.

To Think About questions extend the concept being taught, providing the opportunity for all students to stretch their minds, to look for patterns, and to make conclusions based on their previous experience. These critical-thinking questions may follow Examples in the text and appear in the exercise sets.

Almost every exercise set concludes with a section of **Cumulative Review** problems. These problems review topics previously covered, and are designed to assist students in retaining the material.

Calculator boxes are placed in the margin of the text to alert students to a scientific calculator application. In the exercise section a scientific calculator icon is used to indicate problems that are designed for solving with a calculator. There is also instruction on how to use a scientific calculator in an appendix.

Reviewing Mathematical Concepts

At the end of each chapter, we have included problems and tests to provide your students with several different formats to help them review and reinforce the ideas that they have learned. This assists them not only with that specific chapter, but reviews previously covered topics as well.

The concepts and mathematical procedures covered are reviewed at the end of each chapter in a unique **Chapter Organizer.** It lists concepts and methods, and provides a completely worked-out example for each type of problem.

Chapter Review Problems are grouped by section as a quick refresher at the end of the chapter. They can also be used by the student as a quiz of the chapter material.

Found at the end of the chapter, the **How Am I Doing? Chapter Test** is a representative review of the material from that particular chapter that simulates an actual testing format. This provides the students with a gauge of their preparedness for the actual examination.

At the end of each chapter is a **Cumulative Test.** One-half of the content of each cumulative test is based on the math skills learned in previous chapters. By completing these tests for each chapter, the students build confidence that they have mastered not only the contents of the chapter but those of previous chapters as well.

RESOURCES FOR THE STUDENT

Student Solutions Manual
(ISBNs: 0-321-57838-4, 978-0-321-57838-9)

- Solutions to all odd-numbered section exercises
- Solutions to every exercise (even and odd) in the Quick Quiz, mid-chapter reviews, chapter reviews, chapter tests, and cumulative reviews

Worksheets for Classroom or Lab Practice
(ISBNs: 0-321-59418-5, 978-0-321-59418-1)

- Extra practice exercises for every section of the text with ample space for students to show their work

Chapter Test Prep Video CD

Provides step-by-step video solutions to each problem in each How Am I Doing? Chapter Test in the textbook. Automatically included with every new copy of the text, inside the front cover.

Lecture Series on DVD
(ISBNs: 0-321-59413-4, 978-0-321-59413-6)

- Organized by section, contain problem-solving techniques and examples from the textbook
- Step-by-step solutions to selected exercises from each textbook section

MathXL® Tutorials on CD
(ISBNs: 0-321-59419-3, 978-0-321-59419-8)
This interactive tutorial CD-ROM provides:

- Algorithmically generated practice exercises correlated at the objective level
- Practice exercises accompanied by an example and a guided solution
- Tutorial video clips within the exercise to help students visualize concepts
- Easy-to-use tracking of student activity and scores and printed summaries of students' progress

RESOURCES FOR THE INSTRUCTOR

Annotated Instructor's Edition
(ISBNs: 0-321-57841-4, 978-0-321-57841-9)

- Complete student text with answers to all practice problems, section exercises, mid-chapter reviews, chapter reviews, chapter tests, cumulative tests, and practice final exam.
- Teaching Tips placed in the margin at key points where students historically need extra help
- **New!** Teaching Examples provide in-class practice problems and are placed in the margins accompanying each example.

Instructor's Solutions Manual
(ISBNs: 0-321-57839-2, 978-0-321-57839-6)

- Detailed step-by-step solutions to the even-numbered section exercises
- Solutions to every exercise (odd and even) in the Classroom Quiz, mid-chapter reviews, chapter reviews, chapter tests, cumulative tests, and practice final

Instructor's Resource Manual with Tests and Mini-Lectures
(ISBNs: 0-321-57839-2, 978-0-321-57839-6)

- For each section there is one Mini-Lecture with key learning objectives, classroom examples, and teaching notes.

- Two short group activities per chapter are provided in a convenient ready-to-use handout format.
- Three forms of additional practice exercises that help instructors support students of different ability and skill levels.
- Answers are included for all items.
- Alternate test forms with answers:
  - Two Chapter Pretests per chapter (1 free response, 1 multiple choice)
  - Six Chapter Tests per chapter (3 free response, 3 multiple choice)
  - Two Cumulative Tests per even-numbered chapter (1 free response, 1 multiple choice)
  - Two Final Exams (1 free response, 1 multiple choice)

TestGen®

- Enables instructors to build, edit, print, and administer tests.
- Features a computerized bank of questions developed to cover all text objectives.
- Creates multiple but equivalent versions of the same question or test with the click of a button.
- Instructors can modify questions or add new questions.
- Tests can be printed or administered online.

The software and testbank are available for download from Pearson Education's online catalog.

Pearson Adjunct Support Center

The Pearson Adjunct Support Center is staffed by qualified mathematics instructors with more than 50 years of combined experience at both the community college and university level. Assistance is provided for faculty in the following areas:

- Suggested syllabus consultation
- Tips on using materials packed with your book
- Book-specific content assistance
- Teaching suggestions including advice on classroom strategies

MEDIA RESOURCES

MathXL® www.mathxl.com

MathXL is a powerful online homework, tutorial, and assessment system that accompanies Pearson Education textbooks in mathematics and statistics. With MathXL, instructors can create, edit, and assign online homework and tests using algorithmically generated exercises correlated at the objective level to the textbook. They can also create and assign their own online exercises and import TestGen tests for added flexibility. All student work is tracked in MathXL's online gradebook. Students can take chapter tests in MathXL and receive personalized study plans based on their test results. The study plan diagnoses weaknesses and links students directly to tutorial exercises for the objectives they need to study and retest. Students can also access supplemental animations and video clips directly from selected exercises. MathXL is available to qualified adopters. For more information, visit our Web site at www.mathxl.com, or contact your sales representative.

MyMathLab® www.mymathlab.com

MyMathLab is a series of text-specific, easily customizable online courses for Pearson Education textbooks in mathematics and statistics. Powered by CourseCompass™ (our online teaching and learning environment) and MathXL® (our online homework, tutorial, and assessment system), MyMathLab gives instructors the tools they need to deliver all or a portion of their course online, whether students are in a lab or working at home or elsewhere. MyMathLab provides a rich and flexible set of course materials, featuring free-response exercises that are algorithmically generated for unlimited practice and mastery. Students can also use online tools, such as video lectures, animations, and a multimedia textbook, to independently improve their understanding and performance. Instructors can use MyMathLab's homework and test managers to select and assign online exercises correlated directly to the textbook, and they can create and assign their own online exercises and import TestGen tests for added flexibility. MyMathLab's online gradebook—designed specifically for mathematics and statistics—automatically tracks students' homework and test results and gives the instructor control over how to calculate final grades. Instructors can also add offline (paper-and-pencil) grades to the gradebook. MyMathLab also includes access to the Pearson Tutor Center, which provides students with tutoring via toll-free phone, fax, e-mail, and interactive Web sessions. MyMathLab is available to qualified adopters. For more information, visit our Web site at www.mymathlab.com, or contact your sales representative.

ACKNOWLEDGMENTS

This book is the product of many years of work and many contributions from faculty and students across the country. We would like to thank the many reviewers and participants in focus groups and special meetings with the authors in preparation of previous editions. Our deep appreciation to each of the following:

Khadija Ahmed, *Monroe County Community College*

George J. Apostolopoulos, *DeVry Institute of Technology*

Sohrab Bakhtyari, *St. Petersburg College*

Katherine Barringer, *Central Virginia Community College*

Rita Beaver, *Valencia Community College*

Jamie Blair, *Orange Coast College*

Larry Blevins, *Tyler Junior College*

Brenda Callis, *Rappahannock Community College*

Joan P. Capps, *Raritan Valley Community College*

Judy Carlson, *Indiana University-Purdue University Indianapolis*

Judy Carter, *North Shore Community College*

Robert Christie, *Miami-Dade Community College*

Nelson Collins, *Joliet Junior College*

Mike Contino, *California State University at Heyward*

Pamela Cox, *East Mississippi Community College*

Judy Dechene, *Fitchburg State University*

Floyd L. Downs, *Arizona State University*

Sherilyn Duesing, *Lake Superior State University*

Barbara Edwards, *Portland State University*

Lucy Edwards, *Las Positas College*

Janice F. Gahan-Rech, *University of Nebraska at Omaha*

Colin Godfrey, *University of Massachusetts, Boston*

Mary Beth Headlee, *Manatee Community College*

Laura Kaufmann, *Orange Coast College*

Jeffrey Kroll, *Brazosport College*

Doug Mace, *Baker College*

Carl Mancuso, *William Paterson College*

James Matovina, *Community College of Southern Nevada*

Janet McLaughlin, *Montclair State College*

Anna Maria Mendiola, *Laredo Community College*

Beverly Meyers, *Jefferson College*

Gloria Mills, *Tarrant County Junior College*

Norman Mittman, *Northeastern Illinois University*

Jim Osborn, *Baker College*

Linda Padilla, *Joliet Junior College*

Cathy Panik, *Manatee Community College*

Elizabeth A. Polen, *County College of Morris*

Joel Rappaport, *Miami-Dade Community College*

Rosalie Reiter, *North Central Texas College*

James Rewalt, *Albuquerque Technical-Vocational Institute*

Richard Riggs, *New Jersey City University*

Graciela Rodriguez, *Laredo Community College*

Ronald Ruemmler, *Middlesex County College*

Dennis Runde, *Manatee Community College*

Sally Search, *Tallahassee Community College*

Ara B. Sullenberger, *Tarrant County Junior College*

Katalin Szucs, *East Carolina University*

Margie Thrall, *Manatee Community College*

Michael Trappuzanno, *Arizona State University*

Dennis Vargo, *Albuquerque Technical-Vocational Institute*

Cora S. West, *Florida Community College at Jacksonville*

Joseph Willett, *Vermilion Community College*

Jerry Wisnieski, *Des Moines Community College*

In addition, we want to thank the following individuals for providing splendid insight and suggestions for this new edition:

Suzanne Battista, *St. Petersburg Junior College, Clearwater*

Karen Bingham, *Clarion College*

Nadine Branco, *Western Nevada Community College*

Connie Buller, *Metropolitan Community College*

John Close, *Salt Lake Community College*

Colin Godfrey, *University of Massachusetts, Boston*

Shanna Goff, *Grand Rapids Community College*

Edna Greenwood, *Tarrant County College, Northwest*

Peter Kaslik, *Pierce College*

Joyce Keenan, *Horry-Georgetown Technical College*

Nam Lee, *Griffin Technical Institute*

Tanya Lee, *Career Technical College*

Betty Ludlum, *Austin Community College*

Mary Marlin, *West Virginia Northern Community College*

Carolyn T. McIntyre, *Horry-Georgetown Technical College*

Maria Luisa Mendez, *Laredo Community College*

Steven J. Meyer, *Erie Community College*

Marcia Mollé, *Metropolitan Community College*

Jay L. Novello, *Horry-Georgetown Technical College*

Sandra Lee Orr, *West Virginia State University*

Mary Pearce, *Wake Technical Community College*

Armando Perez, *Laredo Community College*

Regina Pierce, *Davenport University*

Anne Praderas, *Austin Community College*

Suellen Robinson, *North Shore Community College*

Kathy Ruggieri, *Lansdale School of Business*

Randy Smith, *Des Moines Area Community College*

Dina Spain, *Horry-Georgetown Technical College*

Lori Welder, *PACE Institute*

Michael Yarbrough, *Cosumnes River College*

We have been greatly helped by a supportive group of colleagues who not only teach at North Shore Community College but have also provided a number of ideas as well as extensive help on all of our mathematics books. Our special best wishes to our colleague Bob Campbell, who recently retired. He has given us a friendly smile and encouraging ideas for 35 years! Also, a special word of thanks to Wally Hersey, Judy Carter, Rick Ponticelli, Lora Connelly, Sharyn Sharaf, Donna Stefano,

Nancy Tufo, Elizabeth Lucas, Anne O'Shea, Marsha Pease, Walter Stone, Evangeline Cornwall, Rumiya Masagutova, Charles Peterson, and Neha Jain.

Jenny Crawford provided major contributions to this revision. She provided new problems, new ideas, and great mental energy. She greatly assisted us during the production process. She made helpful decisions. Her excellent help was much appreciated. She has become an essential part of our team as we work to provide the best possible textbook.

A special word of thanks goes to Cindy Trimble and Associates and Beverly Fusfield for their excellent work in accuracy checking manuscript and reviewing page proofs.

Each textbook is a combination of ideas, writing, and revisions from the authors and wise editorial direction and assistance from the editors. We especially want to thank our editor at Pearson Education—Paul Murphy. He has a true vision of how authors and editors can work together as partners, and it has been a rewarding experience to create and revise textbooks together with him. We especially want to thank our Project Manager Lauren Morse for patiently answering questions and solving many daily problems. We also want to thank our entire team at Pearson Education—Marlana Voerster, Nathaniel Koven, Christine Whitlock, Georgina Brown, Linda Behrens, Tom Benfatti, Heather Scott, Ilene Kahn, Audra Walsh, and MiMi Yeh—as well as Allison Campbell and Karin Kipp at Elm Street Publishing Services for their assistance during the production process.

Nancy Tobey served as our administrative assistant. Daily she was involved with mailing, photocopying, collating, and taping. A special thanks goes to Nancy. We could not have finished the book without you.

Book writing is impossible for us without the loyal support of our families. Our deepest thanks and love to Nancy, Johnny, Melissa, Marcia, Shelley, Rusty, and Abby. Your understanding, your love and help, and your patience have been a source of great encouragement. Finally, we thank God for the strength and energy to write and the opportunity to help others through this textbook.

We have spent more than 37 years teaching mathematics. Each teaching day, we find that our greatest joy is helping students learn. We take a personal interest in ensuring that each student has a good learning experience in taking this course. If you have some personal comments, suggestions, or ideas for future editions of this textbook, please write to us at:

Prof. John Tobey and Prof. Jeffrey Slater
Pearson Education
Office of the College Mathematics Editor
75 Arlington Street, Suite 300
Boston, MA 02116

or e-mail us at

jtobey@northshore.edu

We wish you success in this course and in your future life!

John Tobey
Jeffrey Slater

Diagnostic Pretest: Intermediate Algebra

1. _____

2. _____

3. _____

4. _____

5. _____

6. _____

7. _____

8. _____

9. _____

10. _____

11. _____

12. _____

13. _____

14. _____

15. _____

Chapter 1

1. Evaluate: $3 - (-4)^2 + 16 \div (-2)$

2. Simplify: $(3xy^{-2})^3(2x^2y)$

3. Simplify: $3x - 4x[x - 2(3 - x)]$

4. Evaluate $F = \dfrac{9}{5}C + 32$ when $C = -35°$.

Chapter 2

Solve the following:

5. $-8 + 2(3x + 1) = -3(x - 4)$

6. When ice floats in water, approximately $\frac{8}{9}$ of the height of the ice lies under water. If the tip of an iceberg is 23 feet above the water, how deep is the iceberg below the waterline?

7. $4 + 5(x + 3) \geq x - 1$

8. $\left| 3\left(\dfrac{2}{3}x - 4 \right) \right| \leq 12$

Chapter 3

9. Find the slope and the y-intercept of $3x - 5y = -7$.

10. Find the equation of a line that passes through $(-5, 6)$ and $(2, -3)$.

11. Is this relation also a function? $\{(5, 6), (-6, 5), (6, 5), (-5, 6)\}$

12. If $f(x) = -2x^2 - 6x + 1$, find $f(-3)$.

Chapter 4

Solve the following:

13. $3x + 5y = 30$
$5x + 3y = 34$

14. $2x - y + 2z = 8$
$x + y + z = 7$
$4x + y - 3z = -6$

15. A speedboat can travel 90 miles with the current in 2 hours. It can travel upstream 105 miles against the current in 3 hours. How fast is the boat in still water? How fast is the current?

16. Graph the system.

$$x - y \le -4$$
$$2x + y \le 0$$

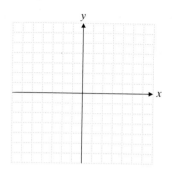

Chapter 5

17. Multiply. $(3x - 4)(2x^2 - x + 3)$

18. Divide. $(2x^3 + 7x^2 - 4x - 21) \div (x + 3)$

19. Factor: $125x^3 - 8y^3$

20. Solve: $2x^2 - 7x - 4 = 0$

Chapter 6

21. Simplify. $\dfrac{10x - 5y}{12x + 36y} \cdot \dfrac{8x + 24y}{20x - 10y}$

22. Combine: $2x - 1 + \dfrac{2}{x + 2}$

23. Simplify. $\dfrac{\dfrac{1}{x + h} - \dfrac{1}{x}}{h}$

24. Solve for x. $\dfrac{x}{x - 2} + \dfrac{3x}{x + 4} = \dfrac{6}{x^2 + 2x - 8}$

Chapter 7

Assume that all expressions under radicals represent nonnegative numbers.

25. Multiply and simplify. $\left(\sqrt{3} + \sqrt{2x}\right)\left(\sqrt{7} - \sqrt{2x^3}\right)$

26. Rationalize the denominator. $\dfrac{3\sqrt{x} + \sqrt{y}}{\sqrt{x} - \sqrt{y}}$

27. Solve and check your solutions. $2\sqrt{x - 1} = x - 4$

Chapter 8

28. Solve: $x^2 - 2x - 4 = 0$

29. $x^4 - 12x^2 + 20 = 0$

16. _____

17. _____

18. _____

19. _____

20. _____

21. _____

22. _____

23. _____

24. _____

25. _____

26. _____

27. _____

28. _____

29. _____

30. _____

31. _____

32. _____

33. _____

34. _____

35. _____

36. _____

37. _____

38. _____

39. _____

40. _____

41. _____

30. Graph $f(x) = (x - 2)^2 + 3$. Label the vertex.

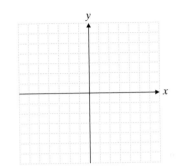

Chapter 9

31. Write in standard form the equation of a circle with center at $(5, -2)$ and a radius of 6.

32. Write in standard form the equation of an ellipse whose center is at $(0, 0)$ and whose intercepts are at $(3, 0)$, $(-3, 0)$, $(0, 4)$ and $(0, -4)$.

33. Solve the following nonlinear system of equations.
$$x^2 + 4y^2 = 9$$
$$x + 2y = 3$$

Chapter 10

34. If $f(x) = 2x^2 - 3x + 4$, find $f(a + 2)$.

35. Graph on one axis $f(x) = |x + 3|$ and $g(x) = |x + 3| - 3$.

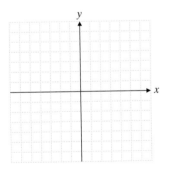

36. If $f(x) = \dfrac{3}{x + 2}$ and $g(x) = 3x^2 - 1$, find $g[f(x)]$.

37. If $f(x) = -\dfrac{1}{2}x - 5$, find $f^{-1}(x)$.

Chapter 11

38. Find y if $\log_5 125 = y$.

39. Find b if $\log_b 4 = \dfrac{2}{3}$.

40. What is $\log 10{,}000$?

41. Solve for x: $\log_6(5 + x) + \log_6 x = 2$

CHAPTER

1

Mathematicians and scientists are able to use formulas to predict a number of things. This is a photograph of the famous Pantheon Pendulum located in the Pantheon building in Paris, France. Here in 1851, the physicist Jean Foucault demonstrated the rotation of Earth by constructing a 67-meter pendulum suspended from the dome. We will study the use of pendulums in this chapter.

Basic Concepts

1.1   THE REAL NUMBER SYSTEM   2

1.2   OPERATIONS WITH REAL NUMBERS   9

1.3   POWERS, SQUARE ROOTS, AND THE ORDER OF OPERATIONS   17

     HOW AM I DOING? SECTIONS 1.1–1.3   25

1.4   INTEGER EXPONENTS AND SCIENTIFIC NOTATION   26

1.5   OPERATIONS WITH VARIABLES AND GROUPING SYMBOLS   37

1.6   EVALUATING VARIABLE EXPRESSIONS AND FORMULAS   43

     CHAPTER 1 ORGANIZER   52

     CHAPTER 1 REVIEW PROBLEMS   54

     HOW AM I DOING? CHAPTER 1 TEST   56

1 Identifying Subsets of the Real Numbers

A **set** is a collection of objects called **elements.** A set of numbers is simply a listing, within braces { }, of the numbers (elements) in the set. For example,

$$S = \{1, 3, 5, \dots\}$$

is the set of positive odd integers. The three dots (called an *ellipsis*) here mean that the set is **infinite;** in other words, we haven't written all the possible elements of the set. A set that contains no elements is called the **empty set** and is symbolized by $\varnothing$ or { }.

Some important sets of numbers that we will study are the following:

- Natural numbers
- Whole numbers
- Integers

- Rational numbers
- Irrational numbers
- Real numbers

DEFINITION

The **natural numbers** N (also called the *positive integers*) are the counting numbers:

$$N = \{1, 2, 3, \dots\}.$$

The **whole numbers** W are the natural numbers plus 0:

$$W = \{0, 1, 2, 3, \dots\}.$$

The **integers** I are the whole numbers plus the *negatives* of all natural numbers:

$$I = \{\dots, -3, -2, -1, 0, 1, 2, 3, \dots\}.$$

The **rational numbers** Q include the integers and all *quotients* of integers (but division by zero is not allowed):

$$Q = \left\{ \frac{a}{b} \,\middle|\, a \text{ and } b \text{ are integers but } b \neq 0 \right\}.$$

The last expression means "the set of all fractions a divided by b, such that a and b are integers but b is not equal to zero." (The | is read "such that.") The letters a and b are **variables;** that is, they can represent different numbers.

A rational number can be written as a **terminating decimal,** $\frac{1}{8} = 0.125$, or as a **repeating decimal,** $\frac{2}{3} = 0.6666\dots$. For repeating decimals we often use a bar over the repeating digits. Thus, we write $0.232323\dots$ as $0.\overline{23}$ to show that the digits 23 repeat indefinitely. A terminating decimal has a finite number of digits. A repeating decimal goes on forever, but the digits repeat in a definite pattern.

Some numbers in decimal notation are nonterminating and nonrepeating. In other words, we can't write them as quotients of integers. Such numbers are called **irrational numbers.**

The **irrational numbers** are numbers whose decimal forms are nonterminating and nonrepeating.

For example, $\sqrt{3} = 1.7320508\dots$ can be carried out to an infinite number of decimal places with no repeating pattern of digits. In Chapter 7 we will study irrational numbers extensively. We can now describe the set of real numbers.

The set R of **real numbers** is the set of all numbers that are rational or irrational.

The following figure will help you see the relationships among the sets of numbers that we have described.

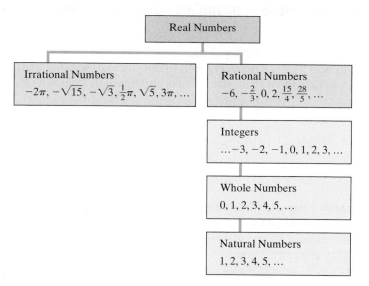

The figure shows that the natural numbers are contained within the set of whole numbers (or we could say that the set of whole numbers contains the set of natural numbers). If all the members of one set are also members of another set, then the first set is a **subset** of the second set. The natural numbers are a subset of the whole numbers. The natural numbers are also a subset of the integers, rational numbers, and real numbers.

EXAMPLE 1 Name the sets to which each of the following numbers belongs.

(a) 5   **(b)** 0.2666...   **(c)** 1.4371826138526...

(d) 0   **(e)** $\dfrac{1}{9}$

Solution

(a) 5 is a natural number. Thus, we can say that it is also a whole number, an integer, a rational number, and a real number.

(b) 0.2666... is a repeating decimal. Thus, it is a rational number and a real number.

(c) 1.4371826138526... doesn't have any repeating pattern. Therefore, it is an irrational number and a real number.

(d) 0 is a whole number, an integer, a rational number, and a real number.

(e) $\dfrac{1}{9}$ is a quotient of integers and can be written as 0.111... (or $0.\overline{1}$), so it is a rational number and a real number.

Practice Problem 1 Name the sets to which the following numbers belong.

(a) 1.26   **(b)** 3   **(c)** $\dfrac{3}{7}$ **(d)** −2   **(e)** 5.182671...

② Identifying the Properties of Real Numbers

A property of something is a characteristic that we know to be true. For example, a property of road surfaces is that if they are not treated with chemicals, water on them will form a skin of ice when the temperature drops to 32°F.

Real numbers also have properties. There are some characteristics of real numbers that we have found to be true. For example, we know that we can add any two

real numbers and the order in which we add the numbers does not affect the sum: $6 + 8 = 8 + 6$. Mathematicians call this property the **commutative property.** Since all real numbers have this characteristic, we can write the property using letters:

$$a + b = b + a,$$

where a and b are any real numbers.

We will look at another property with which you may be familiar, the **associative property.** It states that the way in which you group numbers does not affect their sum:

$$(3 + 8) + 4 = 3 + (8 + 4)$$
$$11 + 4 = 3 + 12$$
$$15 = 15.$$

We can write this property using letters:

$$(a + b) + c = a + (b + c).$$

The **identity property** of addition states that if we add a unique real number (called an **identity element**) to any other real number, that number is not changed. Unique means one and only one such number has this property. For addition, zero (0) is the identity element. Thus,

$$99 + 0 = 99,$$

and, in general, $a + 0 = a$.

The last property we will describe is the additive **inverse property.** For any real number a there is a unique real number $-a$ such that if we add them we obtain the identity element. For example,

$$8 + (-8) = 0 \quad \text{and} \quad -1.5 + 1.5 = 0.$$

In general, $a + (-a) = 0$.

These are the properties of real numbers for addition. These properties apply to multiplication as well. Multiplication of real numbers is commutative: $3 \cdot 8 = 8 \cdot 3$. It is also associative: $3 \cdot (5 \cdot 6) = (3 \cdot 5) \cdot 6$. Note that the dot $\cdot$ indicates multiplication. Thus, $3 \cdot 5$ means 3 times 5; $a \cdot b$ means a times b. These properties as well as a few others are summarized in the following table.

Properties of Real Numbers

Addition	Property	Multiplication
	For all real numbers a, b, c:	
$a + b = b + a$	Commutative properties	$a \cdot b = b \cdot a$
$a + (b + c) = (a + b) + c$	Associative properties	$a \cdot (b \cdot c) = (a \cdot b) \cdot c$
$a + 0 = a$	Identity properties	$a \cdot 1 = a$
$a + (-a) = 0$	Inverse properties	$a \cdot \dfrac{1}{a} = 1$ when $a \neq 0$
	Distributive property of multiplication over addition	
	$a(b + c) = a \cdot b + a \cdot c$	

Note that the additive inverse of a number is the **opposite** of the number. The additive inverse of 6 is -6. The additive inverse of $-\frac{2}{3}$ is $\frac{2}{3}$. This idea is different from the idea of a multiplicative inverse. The multiplicative inverse of a number is the **reciprocal** of a number. The multiplicative inverse of 7 is $\frac{1}{7}$. The multiplicative inverse of $-\frac{8}{9}$ is $-\frac{9}{8}$.

TO THINK ABOUT: Multiplication Properties What is the identity element of multiplication? Why? Give a numerical example of the inverse property of multiplication. What is $\dfrac{1}{a}$ if a is $\dfrac{2}{5}$?

We will use each of these properties to simplify expressions and to solve equations. It will be helpful for you to become familiar with these properties and to recognize what these properties can do.

EXAMPLE 2 State the name of the property that justifies each statement.

(a) $5 + (7 + 1) = (5 + 7) + 1$ **(b)** $x + 0.6 = 0.6 + x$

Solution

(a) Associative property of addition   **(b)** Commutative property of addition

Practice Problem 2 State the name of the property that justifies each statement.

(a) $9 + 8 = 8 + 9$ **(b)** $17 + 0 = 17$

NOTE TO STUDENT: Fully worked-out solutions to all of the Practice Problems can be found at the back of the text starting at page SP-1

The distributive property links multiplication with addition. It states that

$$a(b + c) = (a \cdot b) + (a \cdot c).$$

We can illustrate this property with a numerical example.

$$5(2 + 7) = (5 \cdot 2) + (5 \cdot 7)$$
$$5(9) = 10 + 35$$
$$45 = 45$$

EXAMPLE 3 State the name of the property that justifies each statement.

(a) $5 \cdot y = y \cdot 5$ **(b)** $6 \cdot \dfrac{1}{6} = 1$ **(c)** $5(9 + 3) = 5 \cdot 9 + 5 \cdot 3$

Solution

(a) Commutative property of multiplication
(b) Inverse property of multiplication
(c) Distributive property of multiplication over addition

Practice Problem 3 State the name of the property that justifies each statement.

(a) $6 \cdot (2 \cdot w) = (6 \cdot 2) \cdot w$ **(b)** $4 \cdot \dfrac{1}{4} = 1$ **(c)** $6(8 + 7) = 6 \cdot 8 + 6 \cdot 7$

A word about percents: Throughout this book the topic of percent will be used. If you remember percents but need a brief refresher, here are three basic facts.

1. A **percent** means parts per hundred. So 30% means 30 parts per hundred. We write this as 30% = 0.30.

2. To find a percent of a number, change the percent to a decimal and multiply the number by the decimal. To find 30% of 67, we calculate $0.30 \times 67 = 20.1$.

3. To answer such questions as "what percent is 36 out of 48?" we write the fractional form of "36 out of 48," which is $\frac{36}{48}$. We then change the fraction to a decimal by calculating $36 \div 48 = 0.75$. Finally, we change the decimal to percent form, $0.75 = 75\%$. Thus, 36 out of 48 is 75%.

If this brief review is not sufficient for you to recall the basic ideas of percent, you should refer immediately to a book on beginning algebra or basic college mathematics and study these topics carefully. You will frequently see exercises involving percent in the section exercises.

Verbal and Writing Skills

1. How do integers differ from whole numbers?

2. How do rational numbers differ from integers?

3. What is a terminating decimal?

4. What is a repeating decimal?

Check the column(s) to which the number belongs.	Natural Numbers	Whole Numbers	Integers	Rational Numbers	Irrational Numbers	Real Numbers
5. 13,001						
6. 0						
7. −42						
8. $-\dfrac{144}{4}$			✔	✔		✔
9. −6.1313						
10. $10.\overline{59}$						
11. $-\dfrac{8}{7}$						
12. $-5\dfrac{1}{2}$						
13. $\dfrac{\pi}{5}$						
14. $\sqrt{7}$					✔	✔
15. 0.79						
16. 1.314278619... (no discernible pattern)						
17. 7.040040004... (pattern of increasing numbers of zeros between the 4s)						
18. 54.989898...						

Exercises 19–26 refer to the set $\left\{ -25, -\dfrac{28}{7}, -\dfrac{18}{5}, -\pi, -0.763, -0.333\ldots, 0, \dfrac{1}{10}, \dfrac{2}{7}, \dfrac{\pi}{4}, \sqrt{3}, 9, 52.8, \dfrac{283}{5} \right\}.$

19. List the negative integers.

20. List the rational numbers.

21. List the negative real numbers.

22. List the irrational numbers.

23. List the positive rational numbers.

24. List the negative rational numbers.

25. List the negative rational numbers that are not integers.

26. List the positive rational numbers that are integers.

List all the elements of each set.

27. $\{a \mid a$ is a natural number less than $8\}$

28. $\{b \mid b$ is a positive even integer$\}$

Name the property that justifies each statement. All variables represent real numbers.

29. $8 + (-5) = (-5) + 8$

30. $(6 + 1) + 3 = 6 + (1 + 3)$

31. $\frac{1}{14} \cdot 14 = 1$

32. $4\left(\frac{1}{3} + 6\right) = 4 \cdot \frac{1}{3} + 4 \cdot 6$

33. $5.6 + 0 = 5.6$

34. $5.6 + (-5.6) = 0$

35. $-\frac{1}{6} \cdot \frac{5}{9} = \frac{5}{9} \cdot \left(-\frac{1}{6}\right)$

✓36. $(-2.6) + 7 = 7 + (-2.6)$
commutative

37. $\frac{3}{7} \cdot 1 = \frac{3}{7}$

38. $\left(-\frac{1}{5}\right) + \left(\frac{1}{5}\right) = 0$

39. $-6(4.5 + 3) = -6(4.5) + (-6)(3)$

40. $0 + \pi = \pi$

41. $2.4 + (3 + \sqrt{5}) = (2.4 + 3) + \sqrt{5}$

42. $3\left(-\frac{x}{2}\right) = -\frac{x}{2}(3)$

43. $4 \cdot (3 \cdot x) = (4 \cdot 3) \cdot x$

44. $w \cdot 1 = 1 \cdot w$

45. $-\sqrt{3}\left(5 \cdot \frac{1}{2}\right) = \left(-\sqrt{3} \cdot 5\right) \cdot \frac{1}{2}$

✓46. $y \cdot \frac{1}{y} = 1$ *Inverse*

47. What is the additive inverse of -12?

48. What is the additive inverse of -1.3?

49. What is the multiplicative inverse of $-\frac{5}{3}$?

50. What is the multiplicative inverse of -10?

To Think About *Using real numbers to visualize an applied situation*

51. Siberian Tiger The Siberian tiger is the largest living cat in the world, an endangered subspecies of the tiger family. It is estimated that there are no more than 210 of these animals left in the wild, and approximately the same number in captivity. Of the eight subspecies of tiger, three are extinct, and the remaining five are endangered.

(a) What is the percentage of subspecies of tiger that have become extinct?
(b) What is the percentage of subspecies of tiger that are endangered?
(c) How many subspecies of tiger are not extinct and are not endangered?

52. Mountain Height According to the world's geologists, the tallest mountain in the world, measured from its base, is the island of Hawaii. The entire island is one big mountain. The tallest peak is Mauna Kea, 13,784 feet above sea level and rising from a sea bottom that is 18,000 feet below sea level. Mount Everest's peak is 29,022 feet above sea level, but it rises from a plateau 12,000 feet high. How high is each mountain measured from its base to its peak? (*Source:* U.S. Department of the Interior.) How many of these mountains measure more than 40,000 feet from base to peak?

Game Show Winnings In a game show the incremental winnings are $100, $200, $300, $500, $1000, $2000, $4000, $8000, $16,000, $32,000, $64,000, $125,000, $250,000, $500,000, and $1,000,000.

53. Where does the pattern not follow a logical progression?

54. What number would you obtain if you followed the original progression of the first three numbers—$100, $200, $300—and followed that pattern twenty times after the $1 million mark?

55. Suppose a new television game show is established where the top prize is $1,000,000. However, the incremental winnings (listed in reverse order) are $500,000, $250,000, $125,000, $62,500, $31,250, etc. Would the set of all numbers that are possible winnings be a set of whole numbers?

56. *Pass Completions* During three games of the 2007 NFL season, Tom Brady of the New England Patriots completed 64 out of 102 pass attempts. What percent of his pass attempts were completed? Round to the nearest whole percent.

57. *Fire Deaths* The regional fire marshall said that in the past 20 years, smoke detectors have reduced deaths in home fires by about 60%. Twenty years ago 154 people were killed in home fires. If the fire marshall is correct, how many people per year are killed in house fires?

58. *Gas Mileage* In 2007, President Bush signed into law a requirement that by the year 2020, cars manufactured in the United States must achieve an average gas mileage of 35 miles per gallon. This will be about a 27% increase over the average gas mileage in 2007. What was the average gas mileage in 2007? Round to the nearest tenth of a percent.

Quick Quiz 1.1 Consider the set of numbers $\{-2\pi, -0.5333\ldots, -\frac{3}{11}, 0, 2, \sqrt{7}, \frac{55}{7}, 23.5, 77.222\}$.

1. List the irrational numbers.

2. List the negative real numbers.

3. Name the property that justifies this statement.
$3(-3.56 + 9) = 3(-3.56) + 3(9)$

4. **Concept Check** Explain what properties would be needed to justify this statement.
$(4)(3.5 + 9.3) = (9.3 + 3.5)(4)$

1.2 OPERATIONS WITH REAL NUMBERS

1 Finding the Absolute Value of a Real Number

We can think of real numbers as points on a line, called the **real number line,** where positive numbers lie to the right of zero and negative numbers lie to the left of zero.

A number line helps us to understand the important concept of absolute value and lets us see how to add and subtract real numbers.

Positive

$$-5 \quad -4 \quad -3 \quad -2 \quad -1 \quad 0 \quad 1 \quad 2 \quad 3 \quad 4 \quad 5$$

All real numbers can be placed in order on a number line. For any two numbers graphed on a number line, the number to the left is less than the number to the right. This means that the number to the right is greater than the number on the left.

The symbol $<$ means **is less than.** Since -5 is to the left of -3 on the number line, we can write this as $-5 < -3$. The symbol $>$ means **is greater than.** Since -3 is to the right of -5 on the number line, we can write this as $-3 > -5$.

Larger numbers

$$-6 \quad -5 \quad -4 \quad -3 \quad -2 \quad -1 \quad 0 \quad 1 \quad 2 \quad 3 \quad 4$$

If we are considering some number x that is greater than 4, we would say that x is greater than 4 and write $x > 4$. But if we wanted to write some number that could be 4 or greater than 4, the approach is a little different. We would say that x is greater than or equal to 4 and write $x \geq 4$. In a similar way, we can say "x is less than or equal to 4."

We will discuss inequalities in great detail in Section 2.6. We provide this brief introduction so that we can refer to inequalities while discussing the concept of absolute value.

The **absolute value** of a number x, written $|x|$, can be visualized as its distance from 0 on a number line.

For example, the absolute value of 5 is 5, because 5 is located 5 units from 0 on a number line. The absolute value of -5 is also 5 because -5 is located 5 units from 0 on a number line. We illustrate this concept on the number line below.

$$|-5| = 5 \qquad |5| = 5$$
$$-5 \quad -4 \quad -3 \quad -2 \quad -1 \quad 0 \quad 1 \quad 2 \quad 3 \quad 4 \quad 5$$

Even though -5 and 5 are located on opposite sides of 0, we are concerned only with distance, which is always positive. (We never say, for instance, that we traveled -10 miles.) Thus, we see that the absolute value of any number is always positive or zero. We formally define absolute value as follows:

DEFINITION OF ABSOLUTE VALUE

Absolute value of x: $\qquad |x| = \begin{cases} x, & \text{if } x \geq 0 \\ -x, & \text{if } x < 0 \end{cases}$

EXAMPLE 1 Evaluate.

(a) $|6|$ **(b)** $|-8|$ **(c)** $|0|$ **(d)** $|5 - 3|$ **(e)** $|-1.9|$

Solution

(a) $|6| = 6$ **(b)** $|-8| = 8$ **(c)** $|0| = 0$

(d) $|5 - 3| = |2| = 2$ **(e)** $|-1.9| = 1.9$

Practice Problem 1 Evaluate.

(a) $|-4|$ **(b)** $|3.16|$ **(c)** $|8 - 8|$ **(d)** $|12 - 7|$ **(e)** $\left| -2\frac{1}{3} \right|$

Student Learning Objectives

After studying this section, you will be able to:

1 Find the absolute value of any real number.

2 Add, subtract, multiply, and divide real numbers.

3 Perform mixed operations of addition, subtraction, multiplication, and division in the proper order.

NOTE TO STUDENT: Fully worked-out solutions to all of the Practice Problems can be found at the back of the text starting at page SP-1

2 Adding, Subtracting, Multiplying, and Dividing Real Numbers

ADDING REAL NUMBERS

Addition of real numbers can be pictured on a number line. For example, to add $+6$ and $+5$, we start at 6 on a number line and move 5 units in the positive direction (to the right).

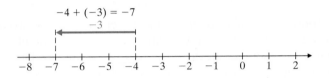

$$6 + 5 = 11$$

To add two negative numbers, say -4 and -3, we start at -4 and move 3 units in the negative direction (to the left). In other words, we add their absolute values and assign their common sign to the sum. We state this procedure as a rule.

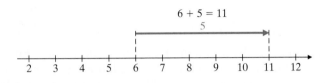

$$-4 + (-3) = -7$$

RULE 1.1

To add two real numbers with the *same* sign, add their absolute values. The sum takes the common sign.

EXAMPLE 2 Add.

(a) $-5 + (-0.6)$ **(b)** $-\dfrac{1}{2} + \left(-\dfrac{1}{3}\right)$ **(c)** $\dfrac{2}{5} + \dfrac{3}{7}$

Solution   We apply Rule 1.1 to all three problems.

(a) $-5 + (-0.6) = -5.6$

(b) $-\dfrac{1}{2} + \left(-\dfrac{1}{3}\right) = -\dfrac{3}{6} + \left(-\dfrac{2}{6}\right) = -\dfrac{5}{6}$

> Obtain a common denominator before adding.

(c) $\dfrac{2}{5} + \dfrac{3}{7} = \dfrac{14}{35} + \dfrac{15}{35} = \dfrac{29}{35}$

Practice Problem 2 Add.

(a) $3.4 + 2.6$ **(b)** $-\dfrac{3}{4} + \left(-\dfrac{1}{6}\right)$ **(c)** $-5 + (-37)$

$-12°\,F$

One place where we routinely see positive and negative numbers is a thermometer. If the temperature is $-12°F$, and then there is a drop of $-1.5°F$, the new temperature is $-12 + (-1.5) = -13.5°F$.

What do we do if the numbers have different signs? Let's add −4 and 3. Again using a number line, we start at −4 and move 3 units in the positive direction.

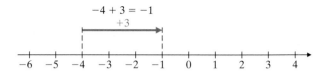

$$-4 + 3 = -1$$

RULE 1.2

To add two real numbers with different signs, find the difference between their absolute values. The answer takes the sign of the number with the larger absolute value.

EXAMPLE 3   Add.

(a) $12 + (-5)$

(b) $-\dfrac{1}{3} + \dfrac{1}{4}$

Solution

(a) $12 + (-5) = 7$

(b) $-\dfrac{1}{3} + \dfrac{1}{4} = -\dfrac{4}{12} + \dfrac{3}{12} = -\dfrac{1}{12}$

Practice Problem 3   Add.

(a) $24 + (-30)$

(b) $-\dfrac{1}{5} + \dfrac{2}{3}$

Notice that if you add two numbers that are opposites, the answer will *always* be zero. The **opposite** of a number is the number with the same absolute value but a different sign. The opposite of −6 is 6. The opposite of $\frac{2}{3}$ is $-\frac{2}{3}$.

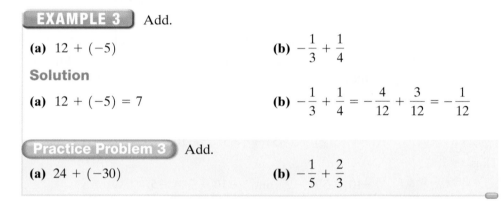

The opposite of a number is also called the **additive inverse.**

TO THINK ABOUT: Additive Inverse   Why is the opposite of a number called the additive inverse of the number?

SUBTRACTING REAL NUMBERS

We define subtraction in terms of addition and use our rules for adding real numbers. So actually, you already know how to subtract if you understand how to add real numbers.

RULE 1.3

To subtract b from a, add the opposite (additive inverse) of b to a. Thus, $a - b = a + (-b)$.

Calculator

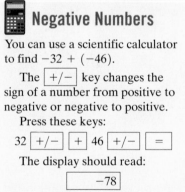

Negative Numbers

You can use a scientific calculator to find $-32 + (-46)$.

The $\boxed{+/-}$ key changes the sign of a number from positive to negative or negative to positive.

Press these keys:

$32 \boxed{+/-} \boxed{+} 46 \boxed{+/-} \boxed{=}$

The display should read:

$$\boxed{-78}$$

On a graphing calculator, use the $\boxed{(-)}$ key to indicate a negative number. Press this before you enter the number. Note that this is not the same as pressing the subtraction key $\boxed{-}$. In general, operations on a graphing calculator are entered in the same order as written.

Try the following:

(a) $-256 + 184$

(b) $94 + (-51)$

(c) $-18 - (-24)$

(d) $-6 + (-10) - (-15)$

EXAMPLE 4   Use Rule 1.3 to subtract.

(a) $5 - 7$ **(b)** $-12 - (-3)$ **(c)** $-0.06 - 0.55$

Solution

(a) $5 - 7 = 5 + (-7) = -2$ **(b)** $-12 - (-3) = -12 + 3 = -9$
(c) $-0.06 - 0.55 = -0.06 + (-0.55) = -0.61$

Practice Problem 4   Use Rule 1.3 to subtract.

(a) $-8 - (-3)$ **(b)** $\dfrac{1}{2} - \left(-\dfrac{1}{4}\right)$ **(c)** $-0.35 - 0.67$

MULTIPLYING AND DIVIDING REAL NUMBERS

Recall that the product or quotient of two positive numbers is positive.

$$(+2)(+5) = (2)(5) = 2 + 2 + 2 + 2 + 2 = 10$$

Thus,

$$(+10) \div (+5) = +2 \quad \text{because} \quad (+2)(+5) = +10.$$

Suppose that the signs of the numbers we want to multiply or divide are different. What will be the sign of the answer? We again use repeated addition for multiplication.

$$-2(5) = (-2) + (-2) + (-2) + (-2) + (-2) = -10$$

Similarly,

$$-5(2) = (-5) + (-5) = -10.$$

Thus,

$$-10 \div 5 = -2 \quad \text{and} \quad -10 \div 2 = -5.$$

Notice that multiplying or dividing a negative number by a positive number gives a negative answer.

RULE 1.4

When you multiply or divide two real numbers with different signs, the answer is a negative number.

If b is any real number, then $b \cdot 0 = 0$. Also, $0 \cdot b = 0$. This is sometimes called the **multiplication property of zero**.

EXAMPLE 5   Evaluate.

(a) $5(-8)$ **(b)** $(-5)\left(\dfrac{2}{17}\right)$ **(c)** $\dfrac{1.6}{-0.08}$ **(d)** $\dfrac{-50}{10}$

Solution

(a) $5(-8) = -40$ **(b)** $(-5)\left(\dfrac{2}{17}\right) = \left(-\dfrac{5}{1}\right)\left(\dfrac{2}{17}\right) = -\dfrac{10}{17}$

(c) $\dfrac{1.6}{-0.08} = -20$ **(d)** $\dfrac{-50}{10} = -5$

Practice Problem 5   Evaluate.

(a) $\left(-\dfrac{2}{5}\right)\left(\dfrac{3}{4}\right)$ **(b)** $\dfrac{150}{-30}$ **(c)** $\dfrac{0.27}{-0.003}$ **(d)** $\dfrac{-12}{24}$

Now let us look at multiplication and division of two negative numbers. We begin by letting a and b be two positive numbers. Then we have the following:

$$-a(0) = 0 \qquad \text{The multiplication property of zero states that any number times 0 is zero.}$$

$$-a(-b + b) = 0 \qquad \text{Rewrite 0 as } -b + b \text{ using the additive inverse property.}$$

$$(-a)(-b) + (-a)(b) = 0 \qquad \text{Use the distributive property.}$$

$$ab \quad + \quad -ab \quad = 0 \qquad \text{By the additive inverse property.}$$

We know that $(-a)(b) = -ab$, a negative number. For the last statement above to be true, $(-a)(-b)$ must be the positive number ab. Thus, we state that $(-a)(-b) = ab$.

RULE 1.5

When you multiply or divide two real numbers with like signs, the answer is a positive number.

EXAMPLE 6 Evaluate.

(a) $\dfrac{-3}{5} \cdot \dfrac{-2}{11}$ 　　　　 **(b)** $\dfrac{-15}{-3}$ 　　　　 **(c)** $\dfrac{2}{3} \div \dfrac{5}{7}$

Solution

(a) $\dfrac{-3}{5} \cdot \dfrac{-2}{11} = \dfrac{(-3)(-2)}{(5)(11)} = \dfrac{6}{55}$ 　　　　 **(b)** $\dfrac{-15}{-3} = 5$

(c) $\dfrac{2}{3} \div \dfrac{5}{7} = \dfrac{2}{3} \cdot \dfrac{7}{5} = \dfrac{(2)(7)}{(3)(5)} = \dfrac{14}{15}$

Practice Problem 6 Evaluate.

(a) $-\dfrac{2}{7} \cdot \left(-\dfrac{3}{5}\right)$ 　　　　 **(b)** $-60 \div (-5)$ 　　　　 **(c)** $\dfrac{3}{7} \div \dfrac{1}{5}$

Before we leave this topic, we take one more look at division by zero.

$$-\frac{5}{3} \div 0 = ? \quad \text{means} \quad -\frac{5}{3} = ? \times 0.$$

In words, "What times 0 is equal to $-\frac{5}{3}$?" By the multiplication property of zero, any number times zero is zero. Thus, our division problem is impossible to solve. We say that division by 0 is undefined.

Division by 0 is undefined.

③ Performing Mixed Operations

It is very important to perform all mathematical operations in the proper order. Otherwise, you will get wrong answers.

If addition, subtraction, multiplication, and division are written horizontally, do the operations in the following order.

1. Do all multiplications and divisions from left to right.

2. Do all additions and subtractions from left to right.

EXAMPLE 7   Evaluate. $20 \div (-4) \times 3 + 2 + 6 \times 5$

Solution

1. Beginning at the left, we do multiplication and division as we encounter it. Here we encounter division first and then multiplication.

$$20 \div (-4) \times 3 + 2 + 6 \times 5 = -5 \times 3 + 2 + 6 \times 5$$
$$= -15 + 2 + 6 \times 5$$
$$= -15 + 2 + 30$$

2. Next we add and subtract from left to right.

$$-15 + 2 + 30 = -13 + 30$$
$$= 17$$

Practice Problem 7   Evaluate. $5 + 7(-2) - (-3) + 50 \div (-2)$

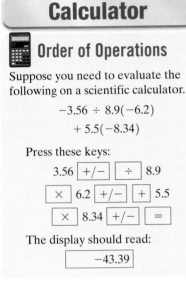

Calculator

Order of Operations

Suppose you need to evaluate the following on a scientific calculator.

$$-3.56 \div 8.9(-6.2)$$
$$+ 5.5(-8.34)$$

Press these keys:

3.56 [+/−] [÷] 8.9

[×] 6.2 [+/−] [+] 5.5

[×] 8.34 [+/−] [=]

The display should read:

| −43.39 |

You may encounter a fraction with several operations in the numerator or denominator or both. In such cases complete the operations in the numerator and then complete the operations in the denominator before simplifying the fraction.

EXAMPLE 8   Evaluate.

(a) $-20 \div (-5)(3) + 6 - 5(-2)$ **(b)** $\dfrac{13 - (-3)}{5(-2) - 6(-3)}$

Solution

(a) $-20 \div (-5)(3) + 6 - 5(-2)$

$= 4(3) + 6 - 5(-2)$

$= 12 + 6 + 10$

$= 28$

(b) $\dfrac{13 - (-3)}{5(-2) - 6(-3)} = \dfrac{13 + 3}{-10 + 18} = \dfrac{16}{8} = 2$

Practice Problem 8   Evaluate.

(a) $6(-2) + (-20) \div (2)(3)$ **(b)** $\dfrac{7 + 2 - 12 - (-1)}{(-5)(-6) + 4(-8)}$

Developing Your Study Skills

Why Is Homework Necessary?

Mathematics is a set of skills that you learn by practicing, not by watching someone else do it. Your instructor may make solving a mathematics exercise look very easy, but for you to learn the necessary skills, you must practice them over and over again, just as your instructor once had to do. There is no other way. Learning mathematics is like learning to play a musical instrument, to type, or to play a sport. No matter how much you watch someone else do it, how many "how to" books you read, or how easy it may seem to be, the key to success is to practice on a regular basis.

Homework provides this practice. The amount of practice needed varies for each individual, but usually students should do most or all of the exercises provided at the end of each section in the text. The more exercises you do, the better you will get. Some exercises in a set are more difficult than others, and they often stress different concepts. Only by working all the exercises will you understand the full range of concepts and techniques.

Verbal and Writing Skills

1. Explain in your own words how to add two real numbers with the same sign or with different signs.

2. Explain in your own words how to multiply or divide two real numbers with the same sign or with different signs.

Evaluate. Assume a and b are greater than 0.

3. $\left|-\dfrac{2}{3}\right|$ **4.** $|-27|$ **5.** $|8.3|$ **6.** $\left|3\dfrac{1}{2}\right|$

7. $|9-14|$ **8.** $|2-6|$ **9.** $|-b|$ **10.** $|a|$

Perform the operations indicated. Write your answer in simplest form.

11. $-6+(-12)$ **12.** $-17+(-3)$ **13.** $-3-(-5)$ **14.** $-8-(+3)$

15. $5\left(-\dfrac{1}{3}\right)$ **16.** $(-16)(-2)$ **17.** $\dfrac{-18}{-2}$ **18.** $(-42)\div 7$

19. $(-0.3)(0.1)$ **20.** $(1.2)(-5)$ **21.** $-4.9+10.5$ **22.** $1.4-(-3.6)$

23. $-\dfrac{5}{12}+\dfrac{7}{18}$ **24.** $-\dfrac{7}{9}+\dfrac{1}{2}$ **25.** $(-2.4)\div 6$ **26.** $3.6\div(-3)$

27. $\left(-\dfrac{2}{3}\right)\left(-\dfrac{7}{4}\right)$ **28.** $\dfrac{4}{5}\left(-\dfrac{15}{11}\right)$ $\dfrac{-60}{55} = \dfrac{-12}{11}$ **29.** $(-4)\left(\dfrac{1}{2}\right)+(-7)(-3)$ **30.** $(6)\left(-\dfrac{2}{3}\right)+(-8)(2)$

Perform each of the following operations, if possible.

31. $12+(-12)$ **32.** $-5.9+5.9$ **33.** $\dfrac{-5}{0}$ **34.** $\dfrac{-12}{0}$

35. $\dfrac{0}{-3}$ **36.** $\dfrac{0}{4}$ **37.** $\dfrac{3-3}{-8}$ **38.** $\dfrac{-5+5}{6}$

39. $\dfrac{-7+(-7)}{-14}$ **40.** $\dfrac{-4+(-4)}{-20}$

Mixed Practice *Perform the following operations in the proper order. Write your answer in simplest form.*

41. $\dfrac{17}{18}+\left(-\dfrac{5}{6}\right)$ $= \dfrac{85}{108}$ **42.** $\dfrac{19}{22}+\left(-\dfrac{1}{2}\right)$ $\dfrac{19}{44}$

43. $\dfrac{5}{6}\div\left(-\dfrac{3}{4}\right)$ **44.** $-\dfrac{4}{5}\div\dfrac{7}{10}$

45. $5 + 6 - (-3) - 8 + 4 - 3$

46. $12 - 3 - (-4) + 6 - 5 - 8$

47. $\dfrac{9(-3) + 7}{3 - 7}$

48. $\dfrac{10 - 2(5)}{4 - 5}$

49. $6(-2) + 3 - 5(-3) - 4$

50. $-7(-3) - 10 + 3(-1) + 8$

51. $15 + 20 \div 2 - 4(3)$

52. $8(0.5) + 9 \div (-0.2)$

53. $\dfrac{6 - 2(7)}{5 - 6}$

54. $\dfrac{2(6) + 3}{5(-1) - 7}$

55. $\dfrac{1 + 49 \div (-7) - (-3)}{-1 - 2}$

56. $\dfrac{72 \div (-4) + 3(-4)}{5 - (-5)}$

57. $4\left(-\dfrac{1}{2}\right) + \dfrac{2}{3}(9)$

58. $-3.5(6) - 2.1(3)$

59. $\dfrac{1.63482 - 2.48561}{(16.05436)(0.07814)}$

60. $(1.783)(2.5725) - (1.0526)(-5.9812)$

To Think About

61. Three numbers are multiplied: $a \cdot b \cdot c = d$. The value of d is a negative number. What are the possible signs of a, b, and c?

62. Three numbers are multiplied: $a \cdot b \cdot c = d$. The value of d is a positive number. What are the possible signs of a, b, and c?

Cumulative Review *Name the property illustrated by each equation.*

63. **[1.1.2]** $5 + 17 = 17 + 5$

64. **[1.1.2]** $4 \cdot (3 \cdot 6) = (4 \cdot 3) \cdot 6$

Refer to the set of numbers $\left\{-16, -\frac{1}{2}\pi, 0, \sqrt{3}, 9.36, \frac{19}{2}, 10.\overline{5}\right\}$.

65. **[1.1.1]** List the irrational numbers.

66. **[1.1.1]** List the rational numbers.

Quick Quiz 1.2 Perform the following operations in the proper order. Write your answer in simplest form.

1. $-8 + (3)(-4)$

2. $5.6 - (-3.8)$

3. $-12 + 15 \div (-5) + 3(6)$

4. **Concept Check** Explain what steps need to be taken in what order to perform the following operations.

$$\dfrac{3(-2) + 8}{5 - 9}$$

① Raising a Number to a Power

Exponents or powers are used to indicate repeated multiplication. For example, we can write $6 \cdot 6 \cdot 6 \cdot 6$ as 6^4. In the expression 6^4, 4 is the **exponent** or **power** that tells us how many times the **base,** 6, appears as a factor. This is called **exponential notation.**

$$\text{exponent} \longrightarrow \overset{\text{4 factors}}{6^4 = \overbrace{6 \cdot 6 \cdot 6 \cdot 6}}$$
$$\underset{\text{base}}{\uparrow}$$

> **EXPONENTIAL NOTATION**
>
> If x is a real number and n is a positive integer, then
>
> $$x^n = \underbrace{x \cdot x \cdot x \cdot x \cdots}_{n \text{ factors}}$$

EXAMPLE 1 Write in exponential notation.

(a) $(-2)(-2)(-2)(-2)(-2)$ **(b)** $x \cdot x \cdot x \cdot x \cdot x \cdot x \cdot x \cdot x$

Solution

(a) $(-2)(-2)(-2)(-2)(-2) = (-2)^5$ **(b)** $x \cdot x \cdot x \cdot x \cdot x \cdot x \cdot x \cdot x = x^8$

Practice Problem 1 Write in exponential notation.

(a) $(-4)(-4)(-4)(-4)$ **(b)** $z \cdot z \cdot z \cdot z \cdot z \cdot z \cdot z$

EXAMPLE 2 Evaluate.

(a) $(-2)^4$ **(b)** -2^4 **(c)** 3^5 **(d)** $(-5)^3$ **(e)** $\left(\dfrac{1}{3}\right)^3$

Solution

(a) $(-2)^4 = (-2)(-2)(-2)(-2) = (4)(-2)(-2) = (-8)(-2) = 16$

    Notice that we are raising -2 to the fourth power. That is, the base is -2. We use parentheses to clearly indicate that the base is negative.

(b) $-2^4 = -(2 \cdot 2 \cdot 2 \cdot 2) = -16$

    Here the base is 2. The base is not -2. We wish to find the negative of 2 raised to the fourth power.

(c) $3^5 = 3 \cdot 3 \cdot 3 \cdot 3 \cdot 3 = 243$

(d) $(-5)^3 = (-5)(-5)(-5) = (25)(-5) = -125$

(e) $\left(\dfrac{1}{3}\right)^3 = \left(\dfrac{1}{3}\right)\left(\dfrac{1}{3}\right)\left(\dfrac{1}{3}\right) = \dfrac{1}{27}$

Practice Problem 2 Evaluate.

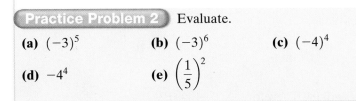

(a) $(-3)^5$ **(b)** $(-3)^6$ **(c)** $(-4)^4$

(d) -4^4 **(e)** $\left(\dfrac{1}{5}\right)^2$

TO THINK ABOUT: Raising Negative Numbers to a Power   Look at Practice Problem 2. What do you notice about raising a negative number to an even power? To an odd power? Will this always be true? Why?

② Finding Square Roots

We say that a square root of 16 is 4 because $4 \cdot 4 = 16$. You will note that, since $(-4)(-4) = 16$, another square root of 16 is -4. For practical purposes, we are usually interested in the nonnegative square root. We call this the **principal square root.** $\sqrt{}$ is the principal square root symbol and is called a **radical.**

$$\text{radical} \rightarrow \sqrt{9} = 3$$
$$\text{radicand} \underset{\big|}{} \quad \text{principal square root}$$

The number or expression under the radical sign is called the **radicand.** Both 16 and 9 are called *perfect squares* because their square roots are integers.

> If x is an integer and a is a positive real number such that $a = x^2$, then x is a **square root** of a, and a is a **perfect square.**

EXAMPLE 3   Find the square roots of 25. What is the principal square root?

Solution

Since $(-5)^2 = 25$ and $5^2 = 25$, the square roots of 25 are 5 and -5. The principal square root is 5.

Practice Problem 3   What are the square roots of 49? What is the principal square root of 49?

Calculator

Square Roots

You can use a scientific calculator to evaluate $\sqrt{625}$.

Press these keys:

625 $\boxed{\sqrt{}}$

The display should read:

$\boxed{25}$

On a graphing calculator, the square root operation is usually accessed by pressing a $\boxed{\text{SHIFT}}$ or $\boxed{\text{2nd}}$ key followed by pressing the $\boxed{x^2}$ key. This should precede the entry of the number you are taking the square root of.

EXAMPLE 4   Evaluate.

(a) $\sqrt{81}$ **(b)** $\sqrt{0}$ **(c)** $-\sqrt{49}$

Solution

(a) $\sqrt{81} = 9$ because $9^2 = 81$.   **(b)** $\sqrt{0} = 0$ because $0^2 = 0$.
(c) $-\sqrt{49} = -\left(\sqrt{49}\right) = -(7) = -7$

Remember: A principal square root is *positive* or *zero.*

Practice Problem 4   Evaluate.

(a) $\sqrt{100}$ **(b)** $\sqrt{1}$ **(c)** $-\sqrt{36}$

We can find the square root of a positive number. However, there is no real number for $\sqrt{-4}$ or $\sqrt{-9}$. The square root of a negative number is not a real number.

EXAMPLE 5   Evaluate.

(a) $\sqrt{0.04}$ **(b)** $\sqrt{\dfrac{25}{36}}$ **(c)** $\sqrt{-16}$

Solution

(a) $(0.2)^2 = (0.2)(0.2) = 0.04$. Therefore, $\sqrt{0.04} = 0.2$.

(b) We can write $\sqrt{\dfrac{25}{36}}$ as $\dfrac{\sqrt{25}}{\sqrt{36}}$, and $\dfrac{\sqrt{25}}{\sqrt{36}} = \dfrac{5}{6}$. Thus, $\sqrt{\dfrac{25}{36}} = \dfrac{5}{6}$.

(c) This is not a real number.

Practice Problem 5 Evaluate.

(a) $\sqrt{0.09}$ **(b)** $\sqrt{\dfrac{4}{81}}$ **(c)** $\sqrt{-25}$

The square roots of some numbers are irrational numbers. For example, $\sqrt{3}$ and $\sqrt{7}$ are irrational numbers. We often use rational numbers to *approximate* square roots that are irrational. They can be found using a calculator with a square root key. To approximate $\sqrt{3}$ on most calculators, we enter the 3 and then press the $\boxed{\sqrt{}}$ key. Using a calculator, we might get 1.7320508 as our approximation. If you do not have a calculator, you can use the square root table at the back of the book. From the table we have $\sqrt{3} \approx 1.732$.

③ The Order of Operations of Real Numbers

Parentheses are used in numerical and algebraic expressions to group numbers and variables. When evaluating an expression containing parentheses, evaluate the numerical expressions inside the parentheses first. When we need more than one set of parentheses, we may also use brackets. To evaluate such an expression, work from the inside out.

EXAMPLE 6 Evaluate.

(a) $2(8 + 7) - 12$ **(b)** $12 - 3[7 + 5(6 - 9)]$

Solution

(a)
$$
\begin{aligned}
2(8 + 7) - 12 &= 2(15) - 12 && \text{Work inside the parentheses first.}\\
&= 30 - 12 && \text{Multiply.}\\
&= 18 && \text{Subtract.}
\end{aligned}
$$

(b)
$$
\begin{aligned}
12 - 3[7 + 5(6 - 9)] &= 12 - 3[7 + 5(-3)] && \text{Begin with the innermost grouping symbols.}\\
&= 12 - 3[7 + (-15)] && \text{Multiply inside the grouping symbols.}\\
&= 12 - 3(-8) && \text{Add inside the grouping symbols.}\\
&= 12 + 24 && \text{Multiply.}\\
&= 36 && \text{Add.}
\end{aligned}
$$

Practice Problem 6 Evaluate.

(a) $6(12 - 8) + 4$ **(b)** $5[6 - 3(7 - 9)] - 8$

A fraction bar acts like a grouping symbol. We must evaluate the expressions above and below a fraction bar before we divide.

Calculator

Parentheses and Exponents

A scientific calculator will perform calculations in the right order for expressions involving exponents and parentheses. Let us evaluate

$$(4.89 - 8.34)^3 - 7(2.87 + 9.05) + 12.78.$$

On most scientific calculators we use the following keystrokes:

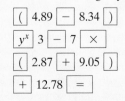

The display will show the following value:

$$\boxed{-111.723625}$$

EXAMPLE 7 Evaluate. $\dfrac{(5)(-2)(-3)}{6 - 8 + 4}$

Solution $\dfrac{(5)(-2)(-3)}{6 - 8 + 4} = \dfrac{30}{2} = 15$

Practice Problem 7 Evaluate.

$$\dfrac{(-6)(3)(2)}{5 - 12 + 3}$$

A radical or absolute value bars group the quantities within them. Thus, we simplify the numerical expressions within the grouping symbols before we find the square root or the absolute value.

EXAMPLE 8 Evaluate.

(a) $\sqrt{(-3)^2 + (4)^2}$ **(b)** $|5 - 8 + 7 - 13|$

Solution

(a) $\sqrt{(-3)^2 + (4)^2} = \sqrt{9 + 16} = \sqrt{25} = 5$

(b) $|5 - 8 + 7 - 13| = |-9| = 9$

Practice Problem 8 Evaluate.

(a) $\sqrt{(-5)^2 + 12^2}$ **(b)** $|-3 - 7 + 2 - (-4)|$

When many arithmetic operations or grouping symbols are used, we use the following order of operations.

> **ORDER OF OPERATIONS FOR CALCULATIONS**
> 1. Combine numbers inside grouping symbols.
> 2. Raise numbers to their indicated powers and take any indicated roots.
> 3. Multiply and divide numbers from left to right.
> 4. Add and subtract numbers from left to right.

EXAMPLE 9 Evaluate. $(4 - 6)^3 + 5(-4) + 3$

Solution

$$\begin{aligned}
(4 - 6)^3 + 5(-4) + 3 &= (-2)^3 + 5(-4) + 3 &&\text{Combine } 4 - 6 \text{ in parentheses.} \\
&= -8 + 5(-4) + 3 &&\text{Cube } -2. \ (-2)^3 = -8. \\
&= -8 - 20 + 3 &&\text{Multiply } 5(-4). \\
&= -25 &&\text{Combine } -8 - 20 + 3.
\end{aligned}$$

Practice Problem 9 Evaluate. $-7 - 2(-3) + (4 - 5)^3$

EXAMPLE 10 Evaluate. $2 + 66 \div 11 \cdot 3 + 2\sqrt{36}$

Solution

$$
\begin{aligned}
2 + 66 \div 11 \cdot 3 + 2\sqrt{36} &= 2 + 66 \div 11 \cdot 3 + 2 \cdot 6 \quad \text{Evaluate } \sqrt{36}. \\
&= 2 + 6 \cdot 3 + 2 \cdot 6 \qquad\quad \text{Divide } 66 \div 11. \\
&= 2 + 18 + 12 \qquad\qquad \text{Multiply } 6 \cdot 3 \text{ and } 2 \cdot 6. \\
&= 32 \qquad\qquad\qquad\quad \text{Add } 2, 18, \text{ and } 12.
\end{aligned}
$$

Students often find that they make errors in exercises like Example 10 when they try to omit steps. Therefore, we recommend that when doing exercises in this section, you should make a separate step for each of the four priorities listed in the preceding box.

Practice Problem 10 Evaluate. $5 + 6 \cdot 2 - 12 \div (-2) + 3\sqrt{4}$

EXAMPLE 11 Evaluate. $\dfrac{2 \cdot 6^2 - 12 \div 3}{4 - 8}$

Solution We evaluate the numerator first.

$$
\begin{aligned}
2 \cdot 6^2 - 12 \div 3 &= 2 \cdot 36 - 12 \div 3 \quad \text{Raise to a power.} \\
&= 72 - 4 \qquad\qquad\quad \text{Multiply and divide from left to right.} \\
&= 68 \qquad\qquad\qquad\; \text{Subtract.}
\end{aligned}
$$

Next we evaluate the denominator.

$$4 - 8 = -4$$

Thus,

$$\frac{2 \cdot 6^2 - 12 \div 3}{4 - 8} = \frac{68}{-4} = -17$$

Practice Problem 11 Evaluate.

$$\frac{2(3) + 5(-2)}{1 + 2 \cdot 3^2 + 5(-3)}$$

As we become a more technologically-oriented society, it is becoming more necessary to understand in what order a computer or a scientific calculator performs arithmetic operations. Be sure you take the time to master this procedure. If you have a scientific calculator, be sure to become familiar with how it works for these types of problems.

Developing Your Study Skills

Why Study Mathematics?

Students often question the value of mathematics, particularly algebra. They see little real use for algebra in their everyday lives. This is understandable at the beginning or intermediate levels of algebra because applications of algebra may not be obvious.

The extensive usefulness of mathematics becomes clear as you take higher-level courses, such as college algebra, statistics, trigonometry, and calculus. You may not be planning to take these higher-level courses now, but your college major may require you to do so.

In our present-day technological world, it is easy to see mathematics at work. Many vocational and professional areas—such as the fields of business, statistics, economics, psychology, finance, computer science, chemistry, physics, engineering, electronics, nuclear energy, banking, quality control, and teaching—require a certain level of expertise in mathematics. Those who want to work in these fields must be able to function at a given mathematical level. Those who cannot will not succeed.

So if your field of study requires you to take higher-level mathematics courses, it is very important that you master the basics of this course. Then you will be ready to advance with your career plans.

Verbal and Writing Skills

1. In the expression a^3, identify the base and the exponent.

2. When a negative number is raised to an odd power, is the result positive or negative?

3. When a negative number is raised to an even power, is the result positive or negative?

4. Will $-a^n$ always be negative? Why or why not?

5. What are the square roots of 121? Why are there two answers?

6. What is the principal square root?

Write in exponential form.

7. $9 \cdot 9 \cdot 9 \cdot 9$

8. $12 \cdot 12 \cdot 12 \cdot 12 \cdot 12 \cdot 12 \cdot 12$

9. $(-6)(-6)(-6)(-6)(-6)$

10. $(-8)(-8)(-8)(-8)$

11. $x \cdot x \cdot x \cdot x \cdot x \cdot x \cdot y \cdot y \cdot y$

12. $a \cdot a \cdot b \cdot b \cdot b \cdot b \cdot b$

Evaluate.

13. 2^5

14. 7^3

15. $(-5)^2$

16. $(-4)^3$

17. -6^2

18. -3^4

19. -1^4

20. $(-3)^2$

21. $\left(\dfrac{2}{3}\right)^2$

22. $\left(-\dfrac{1}{5}\right)^3$

23. $\left(-\dfrac{1}{4}\right)^4$ $\quad -\dfrac{1}{284}$

24. $\left(\dfrac{2}{3}\right)^4$

25. $(0.7)^2$

26. $(-0.5)^2$

27. $(0.04)^3$

28. $(0.03)^3$

Find each expression containing the principal square root.

29. $\sqrt{81}$

30. $\sqrt{121}$

31. $-\sqrt{16}$

32. $-\sqrt{64}$

33. $\sqrt{\dfrac{4}{9}}$

34. $\sqrt{\dfrac{1}{36}}$

35. $\sqrt{0.09}$

36. $\sqrt{0.25}$

37. $\sqrt{9 + 7}$

38. $\sqrt{10 + 15}$

39. $\sqrt{3599 + 1}$

40. $\sqrt{445 - 4}$

41. $\sqrt{\dfrac{5}{36} + \dfrac{31}{36}}$

42. $\sqrt{\dfrac{1}{9} + \dfrac{3}{9}}$

43. $\sqrt{-36}$

44. $\sqrt{-49}$

45. $-\sqrt{-0.36}$

46. $-\sqrt{-0.49}$

Follow the proper order of operations to evaluate each of the following.

47. $5(3 - 9) + 7$

48. $4(3 - 5) + 9$

49. $-15 \div 3 + 7(-4)$

50. $16 \div (-8) - 6(-2)$

51. $(-2)(-10) + (-4)^2$

✓

52. $(-8)(-5) + (7)^2$

53. $(5 + 2 - 8)^3 - (-7)$

54. $(8 - 6 - 7)^2 \div 5 - 6$

55. $(-2)^3 + (-4)^2 - 3$

56. $-5(-10) + (-4)^3 - (-20)$

57. $-5^2 + 3(1 - 8)$

58. $-8^2 - 4(1 - 12)$

Mixed Practice *Follow the proper order of operations to evaluate each of the following.*

59. $5[(1.2 - 0.4) - 0.8]$

60. $-2[(3.6 + 0.3) - 0.9]$

61. $4(-6) - 3^2 + \sqrt{25}$

62. $9(-2) - 2^4 + \sqrt{81}$

63. $\dfrac{7 + 2(-4) + 5}{8 - 6}$

✓**64.** $\dfrac{15 + 5^2 - 10}{3 + 2}$

65. $\dfrac{-3(2^3 - 1)}{3 - 10}$

66. $\dfrac{4 + 2(3^2 - 12)}{4 - 6}$

67. $\dfrac{|2^2 - 5| - 3^2}{-5 + 3}$

68. $\dfrac{-2 + |2^3 - 10|}{3 - 4}$

69. $\dfrac{\sqrt{(-5)^2 - 3 + 14}}{|19 - 6 + 3 - 25|}$

70. $\dfrac{\sqrt{(-2)^2 - 3} + 3}{6 - |3 \cdot 2 - 8|}$

71. $\dfrac{\sqrt{6^2 - 3^2 - 2}}{(-3)^2 - 4}$

72. $\dfrac{\sqrt{4 \cdot 7 + 2^3}}{3^2 - 5}$

🖩 **73.** $(5.986)^5$

🖩 **74.** $(0.325)^4$

To Think About

Coin Toss *When a coin is tossed n times, there are 2^n possible results. For example, if a coin is tossed three times, there are $2^3 = 8$ possible results. If we record the outcome of a head as H and the outcome of a tail as T, we can list the eight results as: HHH, HHT, HTH, THH, HTT, TTH, THT, and TTT.*

75. How many more different results are there when a coin is tossed eight times than when it is tossed six times?

76. How many more different results are there when a coin is tossed twelve times than when it is tossed ten times?

Cumulative Review *State the property illustrated by each equation.*

77. [1.1.2] $a \cdot \dfrac{1}{a} = 1$

78. [1.1.2] $b + (-b) = 0$

79. [1.2.2] *Polar Bears* Due to conservation projects and regulated hunting, the polar bear population has risen from five thousand to forty thousand. What is the percentage of increase in the polar bear population? (*Hint:* The percent of increase is obtained by dividing the amount of increase by the original amount.)

80. [1.2.2] *Astronomy* The pressure at the center of the planet Jupiter is 81,000 tons per square inch, and the pressure at the center of Earth is 27,000 tons per square inch. How many times greater is the pressure at the center of Jupiter than the pressure at the center of Earth?

81. [1.2.2] *Commission Sales* Myles works as a salesman on commission. His income increased by 10% in April. It increased by another 10% in May. By what percent did his income increase from March to May?

82. [1.2.2] *Staff Reduction* The number of employees at Fleet Bank decreased last year by 10%. This year it is expected to decrease by 20%. By what percent will the number of employees decrease over the two-year period?

Quick Quiz 1.3

1. Evaluate. $\left(\dfrac{2}{3}\right)^{5}$

Follow the proper order of operations to evaluate each of the following.

2. $\dfrac{8 + 3(-2) + 4}{9 - 12}$

3. $(8 - 10)^3 - 30 \div (-3) + \sqrt{17 + 8}$

4. Concept Check   Explain what operations need to be done in what order to evaluate the following.

$$\dfrac{\sqrt{(-3)^3 - 6(-2) + 15}}{|3 - 5|}$$

How are you doing with your homework assignments in Sections 1.1 to 1.3? Do you feel you have mastered the material so far? Do you understand the concepts you have covered? Before you go further in the textbook, take some time to do each of the following problems.

1.1

Exercises 1 and 2 refer to the set $\left\{\pi, \sqrt{9}, \sqrt{7}, -5, 3, \frac{6}{2}, 0, \frac{1}{2}, 0.666\ldots\right\}$.

1. List the irrational real numbers.

2. List the integers.

3. What number sets does $\sqrt{3}$ belong to?

4. What is the property of real numbers that justifies the following equation?

$$(x + y) + z = x + (y + z)$$

5. What is the property of real numbers that justifies the following equation?

$$12\left(\frac{1}{12}\right) = 1$$

1.2

Simplify.

6. $30 \div (-6) + 3 - 2(-5)$

7. $6\left(-\dfrac{2}{3}\right) + (-5)(-2)$

8. $\dfrac{15 - (4)(3)}{2 - 8}$

9. $\dfrac{-5 + (-5)}{-15}$

10. $-9 + 6(-2) - (-3)$

1.3

Evaluate.

11. $\sqrt{\dfrac{16}{49}}$

12. $\sqrt{0.81}$

13. 4^4

14. $12 - \sqrt{3^3 + 6(-3)}$

15. $(-4)^3 + 2(3^2 - 2^2)$

16. $\dfrac{4 - 6^2}{3 + \sqrt{16 + 9}}$

17. $|2^2 - 5 - 6|$

18. $\dfrac{\sqrt{(-2)^2 + 5}}{|12 - 15|}$

Now turn to page SA-1 for the answers to each of these problems. Each answer also includes a reference to the objective in which the problem is first taught. If you missed any of these problems, you should stop and review the Examples and Practice Problems in the referenced objective. A little review now will help you master the material in the upcoming sections of the text.

1. _____

2. _____

3. _____

4. _____

5. _____

6. _____

7. _____

8. _____

9. _____

10. _____

11. _____

12. _____

13. _____

14. _____

15. _____

16. _____

17. _____

18. _____

Student Learning Objectives

After studying this section, you will be able to:

 Rewrite expressions with negative exponents as expressions with positive exponents.

2 Use the product rule of exponents.

3 Use the quotient rule of exponents.

4 Use the power rules of exponents.

5 Express numbers in scientific notation.

① Rewriting Expressions with Negative Exponents as Expressions with Positive Exponents

Before we formally define the meaning of a negative exponent, let us look for a pattern.

On this side we decrease each exponent by 1 to obtain the expression on the next line.

$3^4 = 81$
$3^3 = 27$
$3^2 = 9$
$3^1 = 3$
$3^0 = 1$
$3^{-1} = ?$
$3^{-2} = ?$

On this side we divide each number by 3 to obtain the number on the next line.

What results would you expect on the last two lines? $3^{-1} = \frac{1}{3}$? Then $3^{-2} = \frac{1}{3^2} = \frac{1}{9}$. Do you see the pattern? Then we would have

$$3^{-3} = \frac{1}{3^3} = \frac{1}{27} \quad \text{and} \quad 3^{-4} = \frac{1}{3^4} = \frac{1}{81}.$$

Now we are ready to make a formal definition of a negative exponent.

DEFINITION OF NEGATIVE EXPONENTS

If x is any nonzero real number and n is an integer,

$$x^{-n} = \frac{1}{x^n}.$$

EXAMPLE 1 Simplify. Do not leave negative exponents in your answers.

(a) 2^{-5} **(b)** w^{-6}

Solution

(a) $2^{-5} = \frac{1}{2^5} = \frac{1}{32}$ **(b)** $w^{-6} = \frac{1}{w^6}$

Practice Problem 1 Simplify. Do not leave negative exponents in your answers.

(a) 3^{-2} **(b)** z^{-8}

EXAMPLE 2 Simplify. $\left(\frac{2}{3}\right)^{-4}$

Solution $\left(\frac{2}{3}\right)^{-4} = \frac{1}{\left(\frac{2}{3}\right)^4} = \frac{1}{\frac{16}{81}} = (1)\left(\frac{81}{16}\right) = \frac{81}{16}$

Practice Problem 2 Simplify.

$$\left(\frac{3}{4}\right)^{-2}$$

② The Product Rule of Exponents

Numbers and variables with exponents can be multiplied quite simply if *the base is the same*. For example, we know that

$$(x^3)(x^2) = (x \cdot x \cdot x)(x \cdot x).$$

Since the factor x appears five times, it must be true that

$$x^3 \cdot x^2 = x^5.$$

Hence we can state a general rule.

RULE 1.6 PRODUCT RULE OF EXPONENTS

If x is a real number and n and m are integers, then

$$x^m \cdot x^n = x^{m+n}.$$

Remember that we don't usually write an exponent of 1. Thus, $3 = 3^1$ and $x = x^1$.

EXAMPLE 3   Multiply. Leave your answers in exponential form.

(a) $4^3 \cdot 4^{10}$ **(b)** $y \cdot y^6 \cdot y^3$ **(c)** $(a + b)^2(a + b)^3$

Solution

(a) $4^3 \cdot 4^{10} = 4^{3+10} = 4^{13}$

(b) $y \cdot y^6 \cdot y^3 = y^{1+6+3} = y^{10}$

(c) $(a + b)^2(a + b)^3 = (a + b)^{2+3} = (a + b)^5$ (The base is $a + b$.)

Practice Problem 3   Multiply. Leave your answers in exponential form.

(a) $2^8 \cdot 2^{15}$ **(b)** $x^2 \cdot x^8 \cdot x^6$ **(c)** $(x + 2y)^4(x + 2y)^{10}$

EXAMPLE 4   Multiply.

(a) $(3x^2)(5x^6)$ **(b)** $(5x^2y)(-2xy^3)$

Solution

(a) $(3x^2)(5x^6) = (3 \cdot 5)(x^2 \cdot x^6) = 15x^8$

(b) $(5x^2y)(-2xy^3) = (5)(-2)(x^2 \cdot x^1)(y^1 \cdot y^3) = -10x^3y^4$

Practice Problem 4   Multiply.

(a) $(7w^3)(2w)$ **(b)** $(-5xy)(-2x^2y^3)$

USING THE PRODUCT RULE WITH NEGATIVE EXPONENTS

Rule 1.6 says that the exponents are integers. Thus, they can be negative.

EXAMPLE 5 Multiply, then simplify. Do not leave negative exponents in your answer. $(8a^{-3}b^{-8})(2a^5b^5)$

Solution
$$(8a^{-3}b^{-8})(2a^5b^5) = 16a^{-3+5}b^{-8+5}$$
$$= 16a^2b^{-3}$$
$$= 16a^2\left(\frac{1}{b^3}\right) = \frac{16a^2}{b^3}$$

NOTE TO STUDENT: Fully worked-out solutions to all of the Practice Problems can be found at the back of the text starting at page SP-1

Practice Problem 5 Multiply, then simplify. Do not leave negative exponents in your answer.
$$(7xy^{-2})(2x^{-5}y^{-6})$$

③ The Quotient Rule of Exponents

We now develop the rule for dividing numbers with exponents. We know that
$$\frac{x^5}{x^3} = \frac{x \cdot x \cdot \cancel{x} \cdot \cancel{x} \cdot \cancel{x}}{\cancel{x} \cdot \cancel{x} \cdot \cancel{x}} = x \cdot x = x^2.$$

Note that $x^{5-3} = x^2$. This leads us to the following general rule.

RULE 1.7 QUOTIENT RULE OF EXPONENTS

If x is a nonzero real number and m and n are integers,
$$\frac{x^m}{x^n} = x^{m-n}.$$

EXAMPLE 6 Divide. Leave your answers in exponential form with no negative exponents.

(a) $\frac{x^{12}}{x^3}$ (b) $\frac{5^{16}}{5^7}$ (c) $\frac{y^3}{y^{20}}$ (d) $\frac{2^{20}}{2^{30}}$

Solution

(a) $\frac{x^{12}}{x^3} = x^{12-3} = x^9$ (b) $\frac{5^{16}}{5^7} = 5^{16-7} = 5^9$

(c) $\frac{y^3}{y^{20}} = y^{3-20} = y^{-17} = \frac{1}{y^{17}}$ (d) $\frac{2^{20}}{2^{30}} = 2^{20-30} = 2^{-10} = \frac{1}{2^{10}}$

Practice Problem 6 Divide. Leave your answers in exponential form with no negative exponents.

(a) $\frac{w^8}{w^6}$ (b) $\frac{3^7}{3^3}$ (c) $\frac{x^5}{x^{16}}$ (d) $\frac{4^5}{4^8}$

Our quotient rule leads us to an interesting situation if $m = n$.
$$\frac{x^m}{x^m} = x^{m-m} = x^0$$

But what exactly is x^0? Whenever we divide any nonzero value by itself we always get 1, so we would therefore expect that $x^0 = 1$. But can we prove that? Yes.

$$\text{Since} \quad x^{-n} = \frac{1}{x^n},$$

$$\text{Then} \quad x^{-n} \cdot x^n = 1$$
$$x^{-n+n} = 1 \quad \text{Using the product rule.}$$
$$x^0 = 1 \quad \text{Since } -n + n = 0.$$

RULE 1.8 RAISING A NUMBER TO THE ZERO POWER

For any nonzero real number x, $x^0 = 1$.

EXAMPLE 7   Simplify. Do not leave negative exponents in your answers.

(a) $3x^0$ **(b)** $(3x)^0$ **(c)** $(-2x^{-5})(y^3)^0$

Solution

(a) $3x^0 = 3(1)$ Since $x^0 = 1$.   **(b)** $(3x)^0 = 1$ Note that the entire expression
        $= 3$ is raised to the zero power.

(c) $(-2x^{-5})(y^3)^0 = (-2x^{-5})(1)$

$$= (-2)\left(\frac{1}{x^5}\right) = \frac{-2}{x^5} \quad \text{or} \quad -\frac{2}{x^5}$$

Practice Problem 7   Simplify. Do not leave negative exponents in your answers.

(a) $6y^0$ **(b)** $(3xy)^0$ **(c)** $(5^{-3})(2a)^0$

EXAMPLE 8   Divide. Then simplify your answers. Do not leave negative exponents in your answers.

(a) $\dfrac{26x^3y^4}{-13xy^8}$ **(b)** $\dfrac{-150a^3b^4c^2}{-300abc^2}$

Solution

(a) $\dfrac{26x^3y^4}{-13xy^8} = \dfrac{26}{-13} \cdot \dfrac{x^3}{x} \cdot \dfrac{y^4}{y^8} = -2x^2y^{-4} = -\dfrac{2x^2}{y^4}$

(b) $\dfrac{-150a^3b^4c^2}{-300abc^2} = \dfrac{-150}{-300} \cdot \dfrac{a^3}{a} \cdot \dfrac{b^4}{b} \cdot \dfrac{c^2}{c^2} = \dfrac{1}{2} \cdot a^2 \cdot b^3 \cdot c^0 = \dfrac{a^2b^3}{2}$

Practice Problem 8   Divide. Then simplify your answers. Do not leave negative exponents in your answers.

(a) $\dfrac{30x^6y^5}{20x^3y^2}$ **(b)** $\dfrac{-15a^3b^4c^4}{3a^5b^4c^2}$

For the remainder of this chapter, we will assume that for all exercises involving exponents, a simplified answer should not contain negative exponents.

EXAMPLE 9 Divide, then simplify your answer. $\dfrac{3x^{-5}y^{-6}}{27x^2y^{-8}}$

Solution

$$\frac{3x^{-5}y^{-6}}{27x^2y^{-8}} = \frac{1}{9}x^{-5-2}y^{-6-(-8)} = \frac{1}{9}x^{-5-2}y^{-6+8} = \frac{1}{9}x^{-7}y^2 = \frac{y^2}{9x^7}$$

Practice Problem 9 Divide, then simplify your answer.

$$\frac{2x^{-3}y}{4x^{-2}y^5}$$

NOTE TO STUDENT: Fully worked-out solutions to all of the Practice Problems can be found at the back of the text starting at page SP-1

4 The Power Rules of Exponents

Note that $(x^4)^3 = x^4 \cdot x^4 \cdot x^4 = x^{4+4+4} = x^{4 \cdot 3} = x^{12}$. In the same way we can show

$$(xy)^3 = x^3y^3$$

$$\text{and} \left(\frac{x}{y}\right)^3 = \frac{x^3}{y^3} \qquad (y \neq 0).$$

Therefore, we have the following rules.

RULE 1.9 POWER RULES OF EXPONENTS

If x and y are any real numbers and n and m are integers,

$$(x^m)^n = x^{mn}, \quad (xy)^n = x^ny^n, \text{ and}$$

$$\left(\frac{x}{y}\right)^n = \frac{x^n}{y^n}, \qquad \text{if } y \neq 0.$$

EXAMPLE 10 Use the power rules of exponents to simplify.

(a) $(x^6)^5$ **(b)** $(2^8)^4$ **(c)** $[(a + b)^2]^4$

Solution

(a) $(x^6)^5 = x^{6 \cdot 5} = x^{30}$

(b) $(2^8)^4 = 2^{32}$ Careful. Don't change the base of 2.

(c) $[(a + b)^2]^4 = (a + b)^8$ The base is $a + b$.

Practice Problem 10 Use the power rules of exponents to simplify.

(a) $(w^3)^8$ **(b)** $(5^2)^5$ **(c)** $[(x - 2y)^3]^3$

EXAMPLE 11 Simplify.

(a) $(3xy^2)^4$ **(b)** $\left(\dfrac{2a^2b^3}{3ab^4}\right)^3$ **(c)** $(2a^2b^{-3}c^0)^{-4}$

Solution

(a) $(3xy^2)^4 = 3^4x^4y^8 = 81x^4y^8$ **(b)** $\left(\dfrac{2a^2b^3}{3ab^4}\right)^3 = \dfrac{2^3a^6b^9}{3^3a^3b^{12}} = \dfrac{8a^3}{27b^3}$

(c) $(2a^2b^{-3}c^0)^{-4} = 2^{-4}a^{-8}b^{12} = \dfrac{b^{12}}{2^4a^8} = \dfrac{b^{12}}{16a^8}$

Practice Problem 11 Simplify.

(a) $(4x^3y^4)^2$

(b) $\left(\dfrac{4xy}{3x^5y^6}\right)^3$

(c) $(3xy^2)^{-2}$

We need to derive one more rule. You should be able to follow the steps.

$$\frac{x^{-m}}{y^{-n}} = \frac{\dfrac{1}{x^m}}{\dfrac{1}{y^n}} = \frac{1}{x^m} \cdot \frac{y^n}{1} = \frac{y^n}{x^m}$$

RULE 1.10 RULE OF NEGATIVE EXPONENTS

If n and m are positive integers and x and y are nonzero real numbers, then

$$\frac{x^{-m}}{y^{-n}} = \frac{y^n}{x^m}.$$

For example, $\dfrac{x^{-5}}{y^{-6}} = \dfrac{y^6}{x^5}$ and $\dfrac{2^{-3}}{x^{-4}} = \dfrac{x^4}{2^3} = \dfrac{x^4}{8}$.

EXAMPLE 12 Simplify.

(a) $\dfrac{3x^{-2}y^3z^{-1}}{4x^3y^{-5}z^{-2}}$

(b) $\left(\dfrac{5xy^{-3}}{2x^{-4}yz^{-3}}\right)^{-2}$

Solution

(a) First remove all negative exponents.

$$\frac{3x^{-2}y^3z^{-1}}{4x^3y^{-5}z^{-2}} = \frac{3y^3y^5z^2}{4x^3x^2z^1} \quad \begin{array}{l}\text{Only variables with negative exponents}\\\text{will change their position.}\end{array}$$

$$= \frac{3y^8z^2}{4x^5z^1}$$

$$= \frac{3y^8z}{4x^5}$$

(b) First remove the parentheses by using the power rules of exponents.

$$\left(\frac{5xy^{-3}}{2x^{-4}yz^{-3}}\right)^{-2} = \frac{5^{-2}x^{-2}y^6}{2^{-2}x^8y^{-2}z^6}$$

$$= \frac{2^2y^6y^2}{5^2x^8x^2z^6}$$

$$= \frac{4y^8}{25x^{10}z^6}$$

Practice Problem 12 Simplify.

(a) $\dfrac{7x^2y^{-4}z^{-3}}{8x^{-5}y^{-6}z^2}$

(b) $\left(\dfrac{4x^2y^{-2}}{x^{-4}y^{-3}}\right)^{-3}$

EXAMPLE 13 Simplify $(-3x^2)^{-2}(2x^3y^{-2})^3$. Express your answer with positive exponents only.

Solution

$$(-3x^2)^{-2}(2x^3y^{-2})^3 = (-3)^{-2}x^{-4} \cdot 2^3x^9y^{-6}$$

$$= \frac{2^3x^9}{(-3)^2x^4y^6} = \frac{8x^9}{9x^4y^6} = \frac{8x^5}{9y^6}$$

Practice Problem 13 Simplify $(2x^{-3})^2(-3xy^{-2})^{-3}$. Express your answer with positive exponents only.

⑤ Scientific Notation

Scientific notation is a convenient way to write very large or very small numbers. For example, we can write 50,000,000 as 5×10^7 since $10^7 = 10,000,000$, and we can write $0.0000000005 = 5 \times 10^{-10}$ since $10^{-10} = \frac{1}{10^{10}} = \frac{1}{10,000,000,000} = 0.0000000001$. In scientific notation, the first factor is a number that is greater than or equal to 1, but less than 10. The second factor is a power of 10.

SCIENTIFIC NOTATION

A positive number written in **scientific notation** has the form $a \times 10^n$, where $1 \le a < 10$ and n is an integer.

Decimal form and scientific notation are just equivalent forms of the same number. To change a number from decimal notation to scientific notation, follow the steps below. Remember that the first factor must be a number between 1 and 10. This determines where to place the decimal point.

RULE 1.11 CONVERTING FROM DECIMAL NOTATION TO SCIENTIFIC NOTATION

1. Move the decimal point from its original position to the right of the first nonzero digit.
2. Count the number of places that you moved the decimal point. This number is the absolute value of the power of 10 (that is, the exponent).
3. If you moved the decimal point to the right, the exponent is negative; if you moved it to the left, the exponent is positive.

EXAMPLE 14 Write in scientific notation.

(a) 7816   (b) 15,200,000   (c) 0.0123   (d) 0.00046

Solution

(a) $7816 = 7.816 \times 10^3$ — We moved the decimal point three places to the left, so the power of 10 is 3.

(b) $15,200,000 = 1.52 \times 10^7$

(c) $0.0123 = 1.23 \times 10^{-2}$ — We moved the decimal point two places to the right, so the power of 10 is −2.

(d) $0.00046 = 4.6 \times 10^{-4}$

NOTE TO STUDENT: Fully worked-out solutions to all of the Practice Problems can be found at the back of the text starting at page SP-1

Practice Problem 14 Write in scientific notation.

(a) 128,320   (b) 476   (c) 0.0786   (d) 0.007

We can also change a number from scientific notation to decimal notation. We simply move the decimal point to the right or to the left the number of places indicated by the power of 10.

EXAMPLE 15 Write in decimal form.

(a) 8.8632×10^4 **(b)** 6.032×10^{-2} **(c)** 4.4861×10^{-5}

Solution

(a) $8.8632 \times 10^4 = 88,632$

Move the decimal point two places to the left.

(b) $6.032 \times 10^{-2} = 0.06032$
(c) $4.4861 \times 10^{-5} = 0.000044861$

Practice Problem 15 Write in decimal form.

(a) 4.62×10^6 **(b)** 1.973×10^{-3} **(c)** 4.931×10^{-1}

Using scientific notation and the laws of exponents can greatly simplify calculations.

EXAMPLE 16 Evaluate using scientific notation. $\dfrac{(0.000000036)(0.002)}{0.000012}$

Solution Rewrite the expression in scientific notation.

$$\frac{(3.6 \times 10^{-8})(2 \times 10^{-3})}{1.2 \times 10^{-5}}$$

Now rewrite using the commutative property.

$$\frac{\overset{3}{\cancel{(3.6)}}(2)(10^{-8})(10^{-3})}{\underset{1}{\cancel{(1.2)}}(10^{-5})} = \frac{6.0}{1} \times \frac{10^{-11}}{10^{-5}} \qquad \text{Simplify and use the laws of exponents.}$$

$$= 6.0 \times 10^{-11-(-5)}$$
$$= 6.0 \times 10^{-6}$$

Practice Problem 16 Evaluate using scientific notation.

$$\frac{(55,000)(3,000,000)}{5,500,000}$$

EXAMPLE 17 In a scientific experiment, a scientist stated that a proton is theoretically traveling at 3.36×10^5 meters per second. If that is the correct speed, how far would the proton travel in 2×10^4 seconds?

Solution Here we use the idea that the rate times the time equals the distance. So we multiply 3.36×10^5 by 2×10^4. Using the commutative and associative properties, we can write this as

$$3.36 \times 2 \times 10^5 \times 10^4 = 6.72 \times 10^9.$$

The proton would travel 6.72×10^9 meters.

Practice Problem 17 The mass of Earth is considered to be 6.0×10^{24} kilograms. A scientist is studying a star whose mass is 3.4×10^5 times larger than the mass of Earth. If the scientist is correct, what is the mass of this star?

Calculator

Scientific Notation

Most scientific calculators can display only eight digits at one time. Numbers with more than eight digits are shown in scientific notation. 1.12 E 08 means 1.12×10^8. Note that the display on your calculator may be slightly different. You can use the calculator to compute large numbers by entering the numbers using scientific notation. For example, to compute

$$(7.48 \times 10^{24}) \times (3.5 \times 10^8)$$

on a scientific calculator, press these keys:

7.48 [EE] 24 [×]

3.5 [EE] 8 [=]

The display should read:

2.618 E 33

Simplify. Rewrite all expressions with positive exponents only.

1. 3^{-2}
   2. 4^{-3}
   3. x^{-5}
   4. y^{-4}

5. $(-7)^{-2}$
   6. $(-2)^{-5}$
   7. $\left(-\dfrac{1}{9}\right)^{-1}$
   8. $\left(-\dfrac{1}{2}\right)^{-4}$

Multiply. Leave your answer in exponential form.

9. $x^4 \cdot x^8$
   10. $y^{10} \cdot y$
   11. $17^4 \cdot 17$

12. $12^5 \cdot 12^9$
   13. $(3x)(-2x^5)$
   14. $(5y^2)(3y)$

$(3(-2)) \quad -6x^4$

15. $(-11x^2y^2)(-x^4y^7)$
   16. $(-15x^4y)(-6xy^5)$
   17. $4x^0y$

18. $-6a^2b^0$
   19. $(3xy)^0(7xy)$
   20. $-8a^2b^3(-6a)^0$

21. $(-6x^2yz^0)(-4x^0y^2z)$
   22. $(5^0a^3b^4)(-2a^3b^0)$

23. $\left(-\dfrac{3}{5}m^{-2}n^4\right)(5m^2n^{-5})$
   24. $\left(\dfrac{2}{3}m^2n^{-3}\right)(6m^{-5}n)$

Divide. Simplify your answers.

25. $\dfrac{x^{16}}{x^5}$
   26. $\dfrac{x^{17}}{x^3}$
   27. $\dfrac{a^{20}}{a^{25}}$
   28. $\dfrac{x^4}{x^7}$

29. $\dfrac{2^8}{2^5}$
   30. $\dfrac{3^{16}}{3^{18}}$
   31. $\dfrac{2x^3}{x^8}$
   32. $\dfrac{24y^3}{8y}$

33. $\dfrac{-15x^4yz}{3xy}$
   34. $\dfrac{40a^3b}{-5a^3}$
   35. $\dfrac{-20a^{-3}b^{-8}}{14a^{-5}b^{-12}}$
   36. $\dfrac{-27x^7y^{10}}{-6x^{-2}y^{-3}}$

Use the power rules to simplify each expression.

37. $(x^2)^8$
   38. $(a^5)^7$
   39. $(3a^5b)^4$

40. $(2xy^6)^5$
   41. $\left(\dfrac{x^2y^3}{z}\right)^6$
   42. $\left(\dfrac{x^3}{y^5z^8}\right)^4$

43. $\left(\dfrac{3ab^{-2}}{4a^0b^4}\right)^2$
   44. $\left(\dfrac{5a^3b}{-3a^{-2}b^0}\right)^3$

45. $\left(\dfrac{2xy^2}{x^{-3}y^{-4}}\right)^{-3}$
   46. $\left(\dfrac{3x^{-4}y}{x^{-3}y^2}\right)^{-2}$

47. $(x^{-1}y^3)^{-2}(2x)^2$
   48. $(x^3y^{-2})^{-2}(5x^{-5})^2$

49. $\dfrac{(-3m^5n^{-1})^3}{(mn)^2}$
   50. $\dfrac{(m^4n^3)^{-1}}{(-5m^{-3}n^4)^2}$

Mixed Practice   *Simplify. Express your answers with positive exponents only.*

51. $\dfrac{2^{-3}a^2}{2^{-4}a^{-2}}$

52. $\dfrac{3^4a^{-3}}{3^3a^4}$

53. $\left(\dfrac{1}{3}y\right)^{-3}$

54. $\left(\dfrac{2}{5}x^3\right)^{-2}$

55. $\left(\dfrac{y^{-3}}{x}\right)^{-3}$

56. $\left(\dfrac{z}{y^{-5}}\right)^{-2}$

57. $\dfrac{a^0b^{-4}}{a^{-3}b}$

58. $\dfrac{c^{-3}d^{-2}}{c^{-4}d^{-5}}$

59. $\left(\dfrac{14x^{-3}y^{-3}}{7x^{-4}y^{-3}}\right)^{-2}$

60. $\left(\dfrac{25x^{-1}y^{-6}}{5x^{-4}y^{-6}}\right)^{-2}$

61. $\dfrac{7^{-8}\cdot 5^{-6}}{7^{-9}\cdot 5^{-5}}$

62. $\dfrac{9^{-2}\cdot 8^{-10}}{9^{-1}\cdot 8^{-9}}$

63. $(9x^{-2}y)\left(-\dfrac{2}{3}x^3y^{-2}\right)$

64. $(-12x^5y^{-2})\left(\dfrac{3}{4}x^{-6}y^3\right)$

65. $(-3.6982x^3y^4)^7$

66. $\dfrac{1.98364\times 10^{-14}}{4.32571\times 10^{-16}}$

Write in scientific notation.

67. 38   **68.** 759   **69.** 1,730,000   **70.** 405,300,000

71. 0.83   **72.** 0.0654   **73.** 0.0008125   **74.** 0.0000048

Write in decimal notation.

75. 7.13×10^5 **76.** 4.006×10^6 **77.** 3.07×10^{-1}

78. 7.07×10^{-3} **79.** 9.01×10^{-7} **80.** 6.668×10^{-9}

Perform the calculations indicated. Express your answers in scientific notation.

81. $(3.1\times 10^{-4})(1.5\times 10^{-2})$

82. $(2.3\times 10^{-4})(3.0\times 10^9)$

83. $\dfrac{3.6\times 10^{-5}}{1.2\times 10^{-6}}$

84. $\dfrac{9.3\times 10^{-8}}{3.1\times 10^6}$

To Think About

85. *Amazon River* The Amazon River is famous for sending forth one-fifth of all the moving freshwater on Earth, amounting to 7,200,000 cubic *feet* per second. How many cubic *meters* per second pour out of the mouth of the Amazon? (Use 1 foot ≈ 0.305 meter.)

86. *Sensory Perception* Certain moths and butterflies can detect sweetness in a solution when the ratio of sugar to water is 1:300,000. Humans, on the other hand, are considered very sensitive if they are able to detect sweetness in a solution of one part sugar to two hundred parts water. How much more sensitive are moths and butterflies than the most sensitive humans?

Applications

87. *Sound from Bats* A bat emits a sound at a very high frequency that humans cannot hear. The frequency is approximately 5.1×10^4 cycles per second. How many cycles would occur in 1.5×10^2 seconds?

88. *Speed of Light* In one year, light travels 5.87×10^{12} miles. How far would light travel in 5×10^3 years?

89. *Oxygen Molecules* The weight of one oxygen molecule is 5.3×10^{-23} gram. How much would 2×10^4 molecule of oxygen weigh?

90. *Solar Probe* The average distance from Earth to the sun is 4.90×10^{11} feet. If a solar probe is launched from Earth and travels at 2×10^4 feet per second, how long would it take to reach the sun?

Cumulative Review *Evaluate.*

91. **[1.4.5]** *Planet Mass* The mass of Mercury is about 3.64×10^{20} tons. The mass of Jupiter, the largest planet, is about 2.09×10^{24} tons. How many times greater than the mass of Mercury is the mass of Jupiter? Round to the nearest whole number.

92. **[1.4.5]** *Calories* One calorie is equal to 2.78×10^{-7} kilowatt-hours. How many calories are in 5.56×10^3 kilowatt-hours?

93. **[1.3.3]** $-9 + 14 \div (-2) + 5^2$

94. **[1.3.3]** $-6^2 + 16 \div 2$

Quick Quiz 1.4 Simplify. Do not leave negative exponents in your answers.

1. $\dfrac{20x^4y^3}{25x^{-2}y^6}$

2. $\left(\dfrac{3a^{-4}b^2}{a^3}\right)^2$

3. Write in scientific notation. 0.000578

4. **Concept Check** Explain how you would simplify the following.

$$(3x^2y^{-3})(2x^4y^2)$$

① Combining Like Terms in an Algebraic Expression

A collection of numerical values, variables, and operation signs is called an **algebraic expression.** An algebraic expression sometimes contains the sum or difference of several *terms.* A **term** is a real number, a variable, or a product or quotient of numbers and variables.

> Terms can *always* be separated by + signs.

EXAMPLE 1 List the terms in each algebraic expression.

(a) $5x + 3y^2$ **(b)** $5x^2 - 3xy - 7$

Solution

(a) $5x$ is a product of a real number (5) and a variable (x), so $5x$ is a term. $3y^2$ is also a product of a real number and variables, so $3y^2$ is a term.

(b) Note that we can write $5x^2 - 3xy - 7$ in an equivalent form with plus signs: $5x^2 + (-3xy) + (-7)$. This second form helps us to identify more readily the three terms. The terms are $5x^2$, $-3xy$, and -7.

Practice Problem 1 List the terms in each algebraic expression.

(a) $7x - 2w^3$ **(b)** $5 + 6x + 2y$

Any factor in a term is the **coefficient** of the product of the remaining factors. For example, $4x^2$ is the coefficient of y in the term $4x^2y$, and $4y$ is the coefficient of x^2 in $4x^2y$. The numerical coefficient of a term is the numerical value multiplied by the variables. The numerical coefficient of $4x^2y$ is 4. Mathematicians often use "coefficient" to mean the "numerical coefficient." We will use it this way for the remainder of the book. Thus, in the expression $-5x^2y$, the coefficient will be considered to be -5. If no numerical coefficient appears before a variable in a term, the coefficient is understood to be 1. For example, the coefficient of xy is 1. The coefficient of $-x$ is -1.

EXAMPLE 2 Identify the numerical coefficient of each term.

(a) $5x^2 - 2x + 3xy$ **(b)** $8x^3 - 12xy^2 + y$

(c) $\frac{1}{2}x^2 + \frac{1}{4}x$ **(d)** $3.4ab - 0.5b$

Solution

(a) The coefficient of the x^2 term is 5. The coefficient of the x term is -2. The coefficient of the xy term is 3.

(b) The coefficient of the x^3 term is 8. The coefficient of the xy^2 term is -12. The coefficient of the y term is 1.

(c) The coefficient of the x^2 term is $\frac{1}{2}$. The coefficient of the x term is $\frac{1}{4}$.

(d) The coefficient of the ab term is 3.4. The coefficient of the b term is -0.5.

Practice Problem 2 Identify the numerical coefficient of each term.

(a) $5x^2y - 3.5w$ **(b)** $\frac{3}{4}x^3 - \frac{5}{7}x^2y$ **(c)** $-5.6abc - 0.34ab + 8.56bc$

We can add or subtract terms if they are **like terms;** that is, if the terms have the same variables and the same exponents. When we combine like terms, we are using the following form of the distributive property:

$$ba + ca = (b + c)a.$$

EXAMPLE 3 Combine like terms by using the distributive property.

(a) $8x + 2x$ **(b)** $5x^2y + 12x^2y$

Solution

(a) $8x + 2x = (8 + 2)x = 10x$
(b) $5x^2y + 12x^2y = (5 + 12)x^2y = 17x^2y$

Practice Problem 3 Combine like terms by using the distributive property.

(a) $9x - 12x$ **(b)** $4ab^2c + 15ab^2c$

EXAMPLE 4 Combine like terms.

(a) $7x^2 - 2x - 8 + x^2 + 5x - 12$ **(b)** $\frac{1}{3}x^2 + \frac{1}{4}x - \frac{1}{6}x^2$
(c) $2.3x^3 - 5.6x + 5.8x^3 - 7.9x$

Solution

(a) $7x^2 - 2x - 8 + x^2 + 5x - 12 = 8x^2 + 3x - 20$
Remember that the coefficient of x^2 is 1, so you are adding $7x^2 + 1x^2$.
(b) $\frac{1}{3}x^2 + \frac{1}{4}x - \frac{1}{6}x^2 = \frac{2}{6}x^2 - \frac{1}{6}x^2 + \frac{1}{4}x = \frac{1}{6}x^2 + \frac{1}{4}x$
(c) $2.3x^3 - 5.6x + 5.8x^3 - 7.9x = 8.1x^3 - 13.5x$

Practice Problem 4 Combine like terms.

(a) $12x^3 - 5x^2 + 7x - 3x^3 - 8x^2 + x$
(b) $\frac{1}{3}a^2 - \frac{1}{5}a - \frac{4}{15}a^2 + \frac{1}{2}a + 5$
(c) $4.5x^3 - 0.6x - 9.3x^3 + 0.8x$

② Multiplying Algebraic Expressions Using the Distributive Property

We can use the distributive property $a(b + c) = ab + ac$ to multiply algebraic expressions. The kinds of expressions usually encountered are called *polynomials*. **Polynomials** are variable expressions that contain terms with *nonnegative* integer exponents. Some examples of polynomials are $6x^2 + 2x - 8$, $5a + b$, $16x^3$, and $5x + 8$.

EXAMPLE 5 Use the distributive property to multiply. $-2x(x^2 + 5x)$

Solution The distributive property tells us to multiply each term in the parentheses by the term outside the parentheses.

$$-2x(x^2 + 5x) = (-2x)(x^2) + (-2x)(5x) = -2x^3 - 10x^2$$

Practice Problem 5 Use the distributive property to multiply.

$$-3x^2(2x - 5)$$

There is no limit to the number of terms we can multiply. For example, $a(b + c + d + \cdots) = ab + ac + ad + \cdots$.

EXAMPLE 6 Multiply.

(a) $7x(x^2 - 3x - 5)$ **(b)** $5ab(a^2 - ab + 8b^2 + 2)$

Solution

(a) $7x(x^2 - 3x - 5) = 7x^3 - 21x^2 - 35x$

(b) $5ab(a^2 - ab + 8b^2 + 2) = 5a^3b - 5a^2b^2 + 40ab^3 + 10ab$

Practice Problem 6 Multiply.

(a) $-5x(2x^2 - 3x - 1)$ **(b)** $3ab(4a^3 + 2b^2 - 6)$

A parenthesis preceded by no sign or a positive sign $(+)$ can be considered to have a numerical coefficient of 1. A parenthesis preceded by a negative sign $(-)$ can be considered to have a numerical coefficient of -1.

EXAMPLE 7 Simplify.

(a) $(3x^2 + 2)$ **(b)** $-(2x + 3)$

(c) $-4x(x - 2y)$ **(d)** $\dfrac{2}{3}(6x^2 - 2x + 3)$

Solution

(a) $(3x^2 + 2) = 1(3x^2 + 2) = 3x^2 + 2$

(b) $-(2x + 3) = -1(2x + 3) = -2x - 3$

(c) $-4x(x - 2y) = -4x^2 + 8xy$

(d) $\dfrac{2}{3}(6x^2 - 2x + 3) = \dfrac{2}{3}(6x^2) - \dfrac{2}{3}(2x) + \dfrac{2}{3}(3)$

$$= 4x^2 - \dfrac{4}{3}x + 2$$

Practice Problem 7 Simplify.

(a) $(7x^2 - 8)$ **(b)** $-(3x + 2y - 6)$

(c) $-5x^2(x + 2xy)$ **(d)** $\dfrac{3}{4}(8x^2 + 12x - 3)$

③ Removing Grouping Symbols to Simplify Algebraic Expressions

To simplify an expression that contains parentheses that are not placed within other parentheses, multiply and then collect like terms.

EXAMPLE 8   Simplify. $5(x - 2y) - (y + 3x) + (5x - 8y)$

Solution

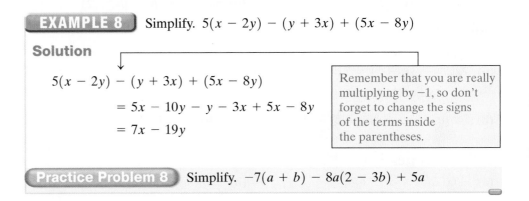

$$5(x - 2y) - (y + 3x) + (5x - 8y)$$
$$= 5x - 10y - y - 3x + 5x - 8y$$
$$= 7x - 19y$$

> Remember that you are really multiplying by -1, so don't forget to change the signs of the terms inside the parentheses.

Practice Problem 8   Simplify. $-7(a + b) - 8a(2 - 3b) + 5a$

NOTE TO STUDENT: Fully worked-out solutions to all of the Practice Problems can be found at the back of the text starting at page SP-1

To simplify an expression that contains grouping symbols within grouping symbols, work from the inside out. The grouping symbols [ ] and { } are used like parentheses.

EXAMPLE 9   Simplify. $-2\{3 + 2[x - 4(x + y)]\}$

Solution

$-2\{3 + 2[x - 4(x + y)]\}$

$= -2\{3 + 2[x - 4x - 4y]\}$ Remove the parentheses by multiplying each term of $x + y$ by -4.

$= -2\{3 + 2[-3x - 4y]\}$ Collect like terms inside the brackets.

$= -2\{3 - 6x - 8y\}$ Remove the brackets by multiplying each term of $-3x - 4y$ by 2.

$= -6 + 12x + 16y$ Remove the braces by multiplying each term by -2.

Practice Problem 9   Simplify. $-2\{4x - 3[x - 2x(1 + x)]\}$

Developing Your Study Skills

How to Review for an Exam

Reviewing adequately for an exam enables you to bring together the concepts you have learned over several sections. For your review, you will need to:

1. Reread your textbook. Make a list of any terms, rules, or formulas you need to know for the exam. Be sure you understand them all.

2. Reread your notes. Go over returned homework and quizzes. Redo the problems you missed.

3. Practice some of each type of problem covered in the chapter(s) you are to be tested on.

4. Use the end-of-chapter materials provided in your textbook. Read through the Chapter Organizer carefully. Do the Chapter Review problems. Take the Chapter Test. When you are finished, check your answers. Redo any problems you missed.

5. Get help if any concepts give you difficulty.

PRACTICE   WATCH   DOWNLOAD   READ   REVIEW

Verbal and Writing Skills

1. Explain what the coefficient of y is in $-5x^2y$.

2. Explain what the coefficient of x is in $3xy^3z$.

3. What are the terms in the expression $5x^3 - 6x^2 + 4x + 8$?

4. What are the terms in the expression $5x^3 + 3x^2 - 2y - 8$?

In exercises 5–10, list the numerical coefficient of each term.

5. $x^5 - 3x - 8y$

6. $-4xy + 9x^2 + y$

7. $5x^3 - 3x^2 + x$

8. $6x^2 - x - 6y$

9. $6.5x^3y^3 - 0.02x^2y + 3.05y$

10. $-\frac{1}{2}a^2b^2 - \frac{10}{3}a^2b - \frac{4}{5}ab$

Combine like terms.

11. $3ab + 8ab$

12. $7ab - 5ab$

13. $4y - 7x + 2x - 6y$

14. $10a + 6b - 7a + 2b$

15. $-4x^2 + 3 + x^2 - 7$

16. $-6y^2 - y + 3y + 2y^2$

17. $4ab - 3b^3 - 5ab + 3b^3$

18. $2a - ab - 2a - ab$

19. $0.1x^2 + 3x - 0.5x^2$

20. $0.7x^2 - 6x + 0.4x^2$

21. $\frac{2}{3}m + \frac{5}{6}n - \frac{1}{3}m + \frac{1}{3}n$

22. $\frac{3}{4}m + \frac{4}{5}n - \frac{1}{2}m + \frac{2}{5}n$

23. $\frac{2}{3}a^2 + 2b + \frac{1}{3}a^2 - 8b$

24. $\frac{3}{2}x^2 + 7x + \frac{1}{2}x^2 - 10y$

25. $1.2x^2 - 5.6x - 8.9x^2 + 2x$

26. $4y^2 - 2.1y - 8.6y - 2.2y^2$

Multiply. Simplify your answers wherever possible.

27. $6x(3x + y)$

28. $7y(4x - 3)$

29. $-y(y^2 - 3y + 5)$

30. $-3x(x^3 + 2x^2 - x)$

31. $-2a^2(a - 3a^2 + 2ab)$

32. $-4m^3(2m + 6 - 5mn)$

33. $2xy(x^2 - 3xy + 4y^2)$

34. $4ab(a^2 - 6ab - 2b^2)$

35. $\frac{3}{4}(8x^2 - 4x + 2)$

36. $\frac{2}{3}(4x - 3y + 6)$

37. $\frac{x}{5}(5x^2 - 2x + 1)$

38. $\frac{x}{3}(x^2 + 4x - 9)$

39. $3ab(a^4b - 3a^2 + a - b)$

40. $5xy^2(y^3 - y^2 + 3x + 1)$

41. $0.5x^2(2x - 4y + 3y^2)$

42. $1.2x^2(4x - 2xy + 5)$

Remove grouping symbols and simplify.

43. $2(x - 1) - x(x + 1) + 3(x^2 + 2)$

44. $5(x - 2) + x(3x - 8) - (x - 2)$

45. $2\{3x - 2[x - 4(x + 1)]\}$

46. $-3\{3y + 2[y + 2(y - 4)]\}$

Mixed Practice *Remove grouping symbols if necessary and simplify.*

47. $a(a - 4b) - 5a(a + b)$

48. $4(xy - 1) - (x - 3)$

49. $6x^3 - 2x^2 - 9x^3 - 12x^2$

50. $4x^4 - 7x^3 - 12x^4 + 6x^3$

51. $2[-3(2x + 4) + 8(2x - 4)]$

52. $3[3(3x + 6) - 5(5x - 9)]$

53. $3y[y - (x - 5)]$

54. $2x[4 - (3x + 2y)]$

Cumulative Review *Evaluate.*

55. [1.3.3] $3(-2)^3 - 5(-6)$

56. [1.3.3] $\sqrt{81} - 5(3 - 5 + 2)$

57. [1.2.3] $\dfrac{5(-2) - 8}{3 + 4 - (-3)}$

58. [1.3.3] $(-3)^5 + 2(-3)$

59. [1.4.5] *Bacteria Size* The smallest organism known to contain all the chemicals needed to sustain independent life is a bacterium called the *pleuropneumonia* organism. It would take 1,893,500 of them, touching side by side, to span an inch. How many *pleuropneumonia* organisms would be found if you put them in a line 1 kilometer long? Express your answer in scientific notation. (Use 1 inch = 0.0254 meter.)

60. [1.2.2] *Efficiency of Internal Combustion Engines* Efficiency of internal combustion engines is lost at the rate of 2% for every 1,000 feet of altitude above sea level. What percentage of efficiency would be lost by powerboats and cars at Lake Titicaca, 4167 *meters* above sea level? (Use 1 foot = 0.305 meter.)

Quick Quiz 1.5 Simplify.

1. $2xy(-3x^2 + 4xy - 5x)$

2. $3x^2(x + 4y) - 2(5x^3 - 2x^2y)$

3. $3[-4(x + 2) + 3(5x - 1)]$

4. **Concept Check** Explain how to simplify the following. $2x^2 - 3x + 4y - 2x^2y - 8x - 5y$

1.6 EVALUATING VARIABLE EXPRESSIONS AND FORMULAS

1 Evaluating a Variable Expression

We need to know how to **evaluate**—compute a numerical value for—variable expressions when we know the values of the variables.

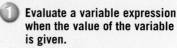
> **EVALUATING A VARIABLE EXPRESSION**
>
> **1.** Replace each variable (letter) by its numerical value. Put parentheses around the value (watch out for negative values).
>
> **2.** Carry out each step, using the correct order of operations.

EXAMPLE 1 Evaluate $x^2 - 5x - 6$ when $x = -4$.

Solution $(-4)^2 - 5(-4) - 6$ Replace x by -4 and put parentheses around it.

$= 16 - 5(-4) - 6$ Square -4.

$= 16 - (-20) - 6$ Multiply $5(-4)$.

$= 16 + 20 - 6$ Multiply $-1(-20)$.

$= 30$ Combine $16 + 20 - 6$.

Practice Problem 1   Evaluate $2x^2 + 3x - 8$ when $x = -3$.

EXAMPLE 2 Evaluate $(5 - x)^2 + 3xy$ when $x = -2$ and $y = 3$.

Solution $[5 - (-2)]^2 + 3(-2)(3)$

$= [5 + 2]^2 + 3(-2)(3) = [7]^2 + 3(-2)(3)$

$= 49 + 3(-2)(3) = 49 - 18 = 31$

Practice Problem 2   Evaluate $(x - 3)^2 - 2xy$ when $x = -3$ and $y = 4$.

EXAMPLE 3 Evaluate when $x = -3$.

(a) $(-2x)^2$ **(b)** $-2x^2$

Solution

(a) $[-2(-3)]^2 = [6]^2 = 36$ Multiply $(-2)(-3)$ and then square the result.

(b) $-2(-3)^2 = -2(9)$ Square -3.

$= -18$ Multiply $-2(9)$.

Practice Problem 3   Evaluate when $x = -4$.

(a) $(-3x)^2$ **(b)** $-3x^2$

TO THINK ABOUT: Importance of Grouping Symbols   Why are the answers to parts **(a)** and **(b)** different in Example 3? What does $(-2x)^2$ mean? What does $-2x^2$ mean? Why are the parentheses so important in this situation?

2 Evaluating Formulas

A **formula** is a rule for finding the value of a variable when the values of other variables in the expression are known. The word *formula* is usually applied to some physical situation, much like a recipe. For example, we can determine the Fahrenheit temperature F for any Celsius temperature C from the formula

$$F = \frac{9}{5}C + 32.$$

EXAMPLE 4   Find the Fahrenheit temperature when the Celsius temperature is $-30°$C.

Solution $F = \dfrac{9}{5}(-30) + 32$ Substitute the known value -30 for the variable C. Then evaluate.

$$= \frac{9}{\overset{}{\underset{1}{5}}}(-\overset{6}{30}) + 32 = 9(-6) + 32$$

$$= -54 + 32 = -22$$

Thus, when the temperature is $-30°$C, the equivalent Fahrenheit temperature is $-22°$F.

Practice Problem 4   Find the Fahrenheit temperature when the Celsius temperature is $70°$C.

EXAMPLE 5   The period T of a pendulum (the time in seconds for the pendulum to swing back and forth one time) is $T = 2\pi\sqrt{\dfrac{L}{g}}$, where L is the length of the pendulum in feet and g is the acceleration due to gravity in feet/second². Find the period when $L = 288$ ft and $g = 32$ ft/sec². Approximate the value of π as 3.14. Round your answer to the nearest tenth of a second.

Solution $T = 2(3.14)\sqrt{\dfrac{288}{32}} = 6.28\sqrt{9} = 6.28(3)$

$$= 18.84$$

The time for a 288-foot-long pendulum to swing back and forth once is approximately 18.84 seconds.

Practice Problem 5   Tarzan is swinging back and forth on a 128-foot-long rope. Use the above formula to find out how long it takes him to swing back and forth one time. Round your answer to the nearest tenth of a second.

EXAMPLE 6   An amount of money invested or borrowed (not including interest) is called the *principal*. Find the amount A to be repaid on a principal p of $1000 borrowed at a simple interest rate r of 8% for a time t of 2 years. The formula is $A = p(1 + rt)$.

Solution $A = 1000[1 + (0.08)(2)]$ Change 8% to 0.08.

$$= 1000[1 + 0.16]$$

$$= 1000(1.16) = 1160$$

The amount to be repaid (principal plus interest) is $1160.

Practice Problem 6   Find the amount to be repaid on a loan of $600 at a simple interest rate of 9% for a time of 3 years.

One of the areas of mathematics where formulas are very helpful is geometry. We use formulas to find the perimeters, areas, and volumes of common geometric figures.

▲ **EXAMPLE 7** Find the perimeter of a rectangular school playground with length 28 meters and width 16.5 meters. Use the formula $P = 2l + 2w$.

Solution We draw a picture to get a better idea of the situation.

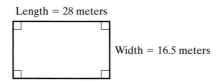

Length = 28 meters

Width = 16.5 meters

$$P = 2l + 2w = 2(28) + 2(16.5) = 56 + 33 = 89$$

The perimeter of the playground is 89 meters.

▲ **Practice Problem 7** Find the perimeter of a rectangular computer chip. The length of the chip is 0.76 centimeter and the width is 0.38 centimeter.

The most commonly used formulas for geometric figures are listed in the following box.

AREA AND PERIMETER FORMULAS

In the following formulas, A = area, P = perimeter, and C = circumference. The "squares" in the figures like ⌐ mean that the angle formed by the two lines is 90°.

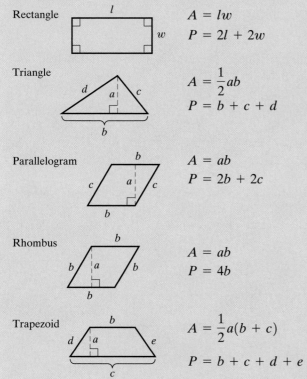

Rectangle

$A = lw$

$P = 2l + 2w$

Triangle

$A = \dfrac{1}{2}ab$

$P = b + c + d$

Parallelogram

$A = ab$

$P = 2b + 2c$

Rhombus

$A = ab$

$P = 4b$

Trapezoid

$A = \dfrac{1}{2}a(b + c)$

$P = b + c + d + e$

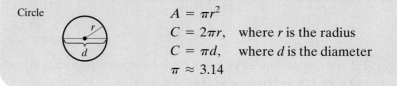

AREA AND PERIMETER FORMULAS (*continued*)

Circle

$A = \pi r^2$

$C = 2\pi r$,   where r is the radius

$C = \pi d$,   where d is the diameter

$\pi \approx 3.14$

In circle formulas we need to use π, which is an irrational number. Its value can be approximated to as many decimal places as we like, but we will use 3.14 because that is accurate enough for most calculations.

▲ **EXAMPLE 8** Find the area of a trapezoid that has a height of 6 meters and bases of 7 and 11 meters.

Solution   The formula is $A = \frac{1}{2}a(b + c)$. We are told that $a = 6, b = 7$, and $c = 11$, so we put those values into the formula.

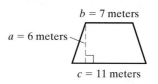

$b = 7$ meters

$a = 6$ meters

$c = 11$ meters

$A = \frac{1}{2}(6)(7 + 11) = \frac{1}{2}(6)(18) = 54$ The area is 54 square meters or 54 m².

NOTE TO STUDENT: Fully worked-out solutions to all of the Practice Problems can be found at the back of the text starting at page SP-1

▲ **Practice Problem 8** Find the area of a triangle with an altitude of 12 meters and a base of 14 meters.

VOLUME AND SURFACE AREA FORMULAS

In the following formulas, V = volume and S = surface area.

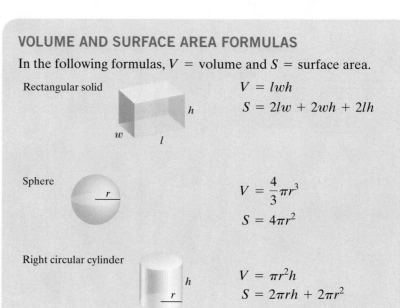

Rectangular solid

$V = lwh$

$S = 2lw + 2wh + 2lh$

Sphere

$V = \frac{4}{3}\pi r^3$

$S = 4\pi r^2$

Right circular cylinder

$V = \pi r^2 h$

$S = 2\pi rh + 2\pi r^2$

▲ **EXAMPLE 9** Find the volume of a sphere with a radius of 3 centimeters.

Solution The formula is $V = \frac{4}{3}\pi r^3$.

$r = 3$

Therefore, we have $V \approx \dfrac{4}{3}(3.14)(3)^3 = \dfrac{4}{3}(3.14)(27)$

$$= \frac{4}{\overset{}{\underset{1}{\cancel{3}}}}(3.14)(\overset{9}{\cancel{27}}) = 113.04$$

The volume is approximately 113.04 cubic centimeters or 113.04 cm³.

▲ **Practice Problem 9** Find the volume of a right circular cylinder of height 10 meters and radius 6 meters.

Developing Your Study Skills

Class Attendance in a Traditional Class

A student of mathematics needs to get started in the right direction by choosing to attend class every day, beginning with the first day of class. Statistics show that class attendance and good grades go together. Classroom activities are designed to enhance learning, and you must be in class to benefit from them. Vital information and explanations that can help you in understanding concepts are given each day. Do not be deceived into thinking that you can find out from a friend what went on in class. There is no good substitute for firsthand experience. Give yourself a push in the right direction by developing the habit of going to class every day.

Class Involvement in an Online Class

If you are enrolled in an online class, follow the schedule carefully. Even when you do not feel like doing the assignment push on and do some work anyway. Do not procrastinate. Make sure you do things on time. E-mail the professor. Do each requested assignment. Stay mentally involved with the class. This is the key to success in an online class.

In this exercise set, round all answers to the nearest hundredth unless otherwise stated. Evaluate each expression for the values given.

1. $14 + 6x$; $x = -2$

2. $11x - 7$; $x = 3$

3. $x^2 + 2x - 9$; $x = -4$

4. $x^2 + 3x - 12$; $x = -5$

5. $3 + 7x - x^2$; $x = 1$

6. $-7x - x^2 + 5$; $x = 2$

7. $-2x^2 + 5x - 3$; $x = -4$

8. $6x^2 - 3x + 5$; $x = 5$

9. $(-3y)^4$; $y = -1$

10. $(-5a)^2$; $a = -2$

11. $-3y^4$; $y = -1$

12. $-5a^2$; $a = -2$

13. $-ay + 4bx - b$; $a = -1$, $b = 3$, $x = 2$, $y = -4$

14. $-3ay + 5ab - y$; $a = -2$, $b = -1$, $y = 6$

15. $\sqrt{b^2 - 4ac}$; $b = 5$, $a = 1$, $c = -14$

16. $\sqrt{b^2 - 4ac}$; $b = 3$, $a = -1$, $c = -2$

17. Evaluate $2x^2 - 5x + 6$ when $x = -3.52176$. (Round to five decimal places.)

18. Evaluate $3x^2 - 7x - 2$ when $x = -0.56736$. (Round answer to five decimal places.)

Applications

Temperature Conversion *For exercises 19 and 20, use the formula* $F = \frac{9}{5}C + 32$.

19. Find the Fahrenheit temperature when the Celsius temperature is $-60°C$.

20. Find the Fahrenheit temperature when the Celsius temperature is $85°C$.

For exercises 21 and 22, use the formula $C = \dfrac{5F - 160}{9}$.

21. Find the Celsius temperature if the Fahrenheit temperature is $122°F$.

22. Find the Celsius temperature if the Fahrenheit temperature is $-40°F$.

Swinging of a Pendulum *For exercises 23 and 24, use* $T = 2\pi\sqrt{\dfrac{L}{g}}$. *Let* $\pi \approx 3.14$ *and* $g = 32$ *feet per second*^{2}.

23. A child is swinging on a rope 32 feet long over a river swimming hole. How long does it take (in seconds) to complete one swing back and forth?

24. A cable swinging from a skyscraper under construction acts as a pendulum. The cable is 512 feet long. How many seconds will it take for the cable to swing back and forth one time?

Simple Interest *In exercises 25–28, use the simple interest formula* $A = p(1 + rt)$.

25. Find A if $p = \$4800$, $r = 0.12$, and $t = 1.5$ years.

26. Find A if $p = \$3200$, $r = 0.07$, and $t = 2$ years.

27. Find the amount to be repaid on a loan of $2200 at a simple interest rate of 4% for 4 years.

28. Find the amount to be repaid on a loan of $1700 at a simple interest rate of 10% for 5 years.

Falling Time in Gravity *In exercises 29–32, use the fact that the distance S in feet an object falls in t seconds is given by* $S = \frac{1}{2}gt^2$, *where* $g = 32$ *feet per second*^{2}.

29. Find S if $t = 3$ seconds.

30. Find S if $t = 6$ seconds.

31. A piece of a window ledge fell from the sixty-eighth floor of the Texas Commerce Tower in Houston. It took 7 seconds to hit the ground. How far did the piece of window ledge fall?

32. A bolt fell out of a window frame on the twenty-third floor of the John Hancock Tower in Boston. The bolt took 4 seconds to hit the ground. How far did the bolt fall?

33. Find $z = \dfrac{Rr}{R + r}$ if $R = 36$ and $r = 4$.

34. Find $z = \dfrac{Rr}{R + r}$ if $R = 35$ and $r = 15$.

Dosage of Medicine *The approximate number of milligrams m of a medicine that should be given a child of age c when the usual adult dosage is x milligrams can be obtained from the following equation. Use the equation to solve exercises 35 and 36. Round your answers to the nearest whole number.*

$$m = \frac{cx}{c + 12}$$

35. Find the amount of ibuprofen to give a child of age 8 if the usual adult dosage is 400 milligrams.

36. Find the amount of aspirin to give a child of age 10 if the usual adult dosage is 325 milligrams.

Geometry *In exercises 37–49, use the geometry formulas on pages 45 and 46 to find the quantities specified. Use* $\pi \approx 3.14$.

▲ **37.** Find the area of a circle with a radius of 0.5 inches.

▲ **38.** Find the circumference of a circle with a diameter of 0.2 meters.

▲ **39.** Find the area of a triangle with a base of 16 centimeters and a height of 7 centimeters.

▲ **40.** Find the area of a triangle with a base of 12 meters and an altitude of 14 meters.

▲ **41.** Find the area of a parallelogram with an altitude of 4 yards and a base of $\frac{7}{8}$ yards.

▲ **42.** Find the area of a rhombus with an altitude of 8 centimeters and a base of $10\frac{3}{4}$ centimeters.

▲ **43.** ***Graphing Calculator Display Window*** A graphing calculator has a display window 6.1 centimeters wide and 4.05 centimeters high. What is the area of the display window?

▲ **44.** Find the surface area of an ice cube that is 1.8 inches long, 1 inch wide, and 1 inch high.

▲ **45.** A trapezoid has a height of 6 centimeters and parallel bases measuring 5 centimeters and 7 centimeters. Find its area.

▲ **46.** A trapezoid has sides 5.2, 6.1, 3.5, and 2.2 meters long. Find its perimeter.

▲ **47.** *Soup Can* A can of soup has a height of 11 centimeters and a radius of 4 centimeters.

(a) Find its volume. (Use $\pi \approx 3.14$.)

(b) Find its total surface area.

▲ **48.** *Tennis Ball* A tennis ball has a radius of 3.2 centimeters.

(a) Find its volume. (Use $\pi \approx 3.14$.)

(b) Find its surface area.

▲ **49.** *Telephone Cable* A cross section of telephone cable is a circle that has a radius of 8 centimeters. What is the area of the cross section? What is the circumference of this circular cross section? (Use $\pi \approx 3.14$.)

▲ **50.** *Stainless Steel Circle* An engineer must make a stainless steel circle with a radius of 6 centimeters. What is the area of the circle? What is the circumference of the circle? (Use $\pi \approx 3.14$.)

Cumulative Review *Simplify.*

51. [1.4.4] $(6x^{-4}y^3z^0)^2$

52. [1.4.4] $\left(\dfrac{2x^3}{3y}\right)^3$

53. [1.5.3] $2\{5 - 2[x - 3(2x + 1)]\}$

54. [1.3.3] $2^3 - 4^2 + \sqrt{9 \cdot 2 - 2}$

55. [1.2.2] *People on the Internet* Today there were twelve million Americans on the Internet at 9 P.M., eastern standard time. At 2 P.M., eastern standard time, there were nineteen million Americans on the Internet. Thirty percent of the people on the Internet at 2 P.M. were also on the Internet at 9 P.M. How many people were on the Internet at 9 P.M. but were not on the Internet at 2 P.M.?

56. [1.2.2] *High School Reunion* An invitation was sent out to Beverly High School alumni from the class of 1961. The graduating class that year had 180 students. A total of 69% of the living students went to the reunion. A total of 5% of the graduates of the class of 1961 are no longer living. Seventy-seven percent of all the students attending the reunion brought a spouse to the gathering. In addition, twenty-two faculty members returned for this big event. What was the total number of people at the reunion? Round your answer to the nearest whole number.

Quick Quiz 1.6

1. Evaluate $3x^2 - 5x - 2$ when $x = -3$.

2. Evaluate $-2x^2 + 4xy - y^2$ when $x = -2$ and $y = 4$.

3. Find the area of a circle with a radius of 5 centimeters. Use $\pi \approx 3.14$.

4. **Concept Check** Explain how you would find the amount A to be repaid on a principal of $5000 borrowed at a simple interest rate r of 8% for a time t of 2 years. The formula is $A = p(1 + rt)$.

Putting Your Skills to Work: Use Math to Save Money

GETTING MORE FOR YOUR MONEY

Expenses for Lunch

Do you buy your lunch each day? Are you getting the most for your money? At the grocery store or the restaurant it is important to find ways to get the most for your money. Consider the story of Davie.

Comparing Pizza Costs

On his way to work, Davie notices two competing pizza shops offering "lunch specials." Poppolo's advertises a 16″ diameter pizza, cut into 6 pieces, for $1.75 a slice. Across the street, Joe's has a 20″ pizza, cut into 8 pieces, at $2.10 per slice. Davie is pretty hungry, but he wants to get the most for his money. So, using his math skills, he does some quick calculations.

Helpful Calculations and Decisions

1. How many square inches of pizza is in a slice of Poppolo's pizza?

2. How many square inches of pizza is in a slice of Joe's pizza?

3. Since Davie is counting his pennies, he decides to find which deal is better in terms of money. Find the cost per square inch of each slice of pizza.

4. Which is the better lunch deal? What are some other considerations that Davie should think about?

5. Davie decides to start packing a lunch each day to save money. He figures that he will save $20 a week by bringing his lunch each day. How much would Davie save in a year?

Making a Personal Application

6. How much could you save in a week by packing and bringing your lunch, rather than buying cafeteria lunch? What are some things to consider when calculating your savings?

7. Do you comparison shop for food, or do you just buy what you want when you want it? What things should you think about before you make food buying decisions?

Topic	Procedure	Examples
Commutative property of addition, p. 4.	$a + b = b + a$	$12 + 13 = 13 + 12$
Commutative property of multiplication, p. 4.	$a \cdot b = b \cdot a$	$11 \cdot 19 = 19 \cdot 11$
Associative property of addition, p. 4.	$a + (b + c) = (a + b) + c$	$4 + (3 + 6) = (4 + 3) + 6$
Associative property of multiplication, p. 4.	$a \cdot (b \cdot c) = (a \cdot b) \cdot c$	$7 \cdot (3 \cdot 2) = (7 \cdot 3) \cdot 2$
Identity property of addition, p. 4.	$a + 0 = a$	$9 + 0 = 9$
Identity property of multiplication, p. 4.	$a \cdot 1 = a$	$7 \cdot 1 = 7$
Inverse property of addition, p. 4.	$a + (-a) = 0$	$8 + (-8) = 0$
Inverse property of multiplication, p. 4.	If $a \neq 0, a\left(\dfrac{1}{a}\right) = 1$	$15\left(\dfrac{1}{15}\right) = 1$
Distributive property of multiplication over addition, p. 5.	$a(b + c) = a \cdot b + a \cdot c$	$7(9 + 4) = 7 \cdot 9 + 7 \cdot 4$
Addition of real numbers, p. 10.	*Addition:* To add two real numbers with the *same sign,* add their absolute values and use the common sign. To add two real numbers with *different signs,* find the difference between their absolute values, and the answer takes the sign of the number with the larger absolute value.	$9 + 5 = 14 \qquad -7 + (-3) = -10$ $-\dfrac{1}{5} + \dfrac{3}{5} = \dfrac{2}{5} \qquad -42 + 19 = -23$
Subtraction of real numbers, p. 11.	*Subtraction:* To subtract b from a, *add the opposite* of b to a. $$a - b = a + (-b)$$	$12 - (-3) = 12 + (+3) = 15$ $-7.2 - (+1.6) = -7.2 + (-1.6) = -8.8$
Multiplication and division of real numbers, p. 12.	*Multiplication and Division:* When you multiply or divide two real numbers with like signs, the answer is a *positive* number. When you multiply or divide two real numbers whose signs are *different,* the answer is a *negative* number.	$(-6)(-3) = 18 \qquad -20 \div (-4) = 5$ $(-8)(5) = -40 \qquad 16 \div (-2) = -8$
Order of operations of real numbers, p. 20.	To simplify numerical expressions, use this order of operations. 1. Combine numbers inside grouping symbols. 2. Raise numbers to their indicated powers and take any indicated roots. 3. Multiply and divide numbers from left to right. 4. Add and subtract numbers from left to right.	Simplify. $5 + 2(5 - 8)^3 - 12 \div (-4)$ $5 + 2(-3)^3 - 12 \div (-4)$ $= 5 + 2(-27) - 12 \div (-4)$ $= 5 + (-54) - (-3)$ $= -49 + 3$ $= -46$
Absolute value of a number, p. 9.	$$\lvert x \rvert = \begin{cases} x, & \text{if } x \geq 0 \\ -x, & \text{if } x < 0 \end{cases}$$	$\lvert 6 \rvert = 6 \qquad \lvert 0 \rvert = 0$ $\lvert -2 \rvert = 2 \qquad \lvert -3.6 \rvert = 3.6$ $\left\lvert -\dfrac{4}{7} \right\rvert = \dfrac{4}{7}$
Rules of Exponents for multiplication and division, pp. 27 and 28.	If x and y are any nonzero real numbers and m and n are integers, $$x^m x^n = x^{m+n} \quad \text{and} \quad \frac{x^m}{x^n} = x^{m-n}.$$	$(2x^5)(3x^6) = 6x^{11}$ $\dfrac{15x^8}{5x^3} = 3x^5$

Topic	Procedure	Examples
Negative exponents, p. 26.	$x^{-n} = \dfrac{1}{x^n}$ $\dfrac{x^{-n}}{y^{-m}} = \dfrac{y^m}{x^n}$	$x^{-6} = \dfrac{1}{x^6}$ $2^{-8} = \dfrac{1}{2^8}$ $\dfrac{x^{-4}}{y^{-5}} = \dfrac{y^5}{x^4}$
Zero exponent, p. 29.	$x^0 = 1$ when $x \neq 0$.	$x^0 = 1$ $5^0 = 1$ $(3ab)^0 = 1$
Power rules, p. 30.	$(x^m)^n = x^{mn}$ $(xy)^n = x^n y^n$ $\left(\dfrac{x}{y}\right)^n = \dfrac{x^n}{y^n},$ if $y \neq 0$	$(7^3)^4 = 7^{12}$ $(3x^{-2})^4 = 3^4 x^{-8} = \dfrac{3^4}{x^8}$ $\left(\dfrac{2a^2}{b^3}\right)^{-4} = \dfrac{2^{-4}a^{-8}}{b^{-12}} = \dfrac{b^{12}}{2^4 a^8}$
Scientific notation, p. 32.	A positive number is written in scientific notation if it is in the form $a \times 10^n$, where $1 \leq a < 10$ and n is an integer.	$128 = 1.28 \times 10^2$ $2{,}568{,}000 = 2.568 \times 10^6$ $13{,}200{,}000{,}000 = 1.32 \times 10^{10}$ $0.16 = 1.6 \times 10^{-1}$ $0.00079 = 7.9 \times 10^{-4}$ $0.0000034 = 3.4 \times 10^{-6}$
Combining like terms, p. 37.	Combine terms that have identical variables and exponents.	$7x^2 - 3x + 4y + 2x^2 - 8x - 9y = 9x^2 - 11x - 5y$
Using the distributive property, p. 38.	Use the distributive property to remove parentheses. $a(b + c) = ab + ac$	$3(2x^2 - 6x - 8) = 6x^2 - 18x - 24$
Removing grouping symbols, p. 40.	1. Remove grouping symbols from the inside out. 2. Combine like terms inside grouping symbols whenever possible. 3. Continue until all grouping symbols are removed. 4. Combine like terms.	$5\{3x - 2[4 + 3(x - 1)]\} = 5\{3x - 2[4 + 3x - 3]\}$ $= 5\{3x - 2[1 + 3x]\}$ $= 5\{3x - 2 - 6x\}$ $= 5\{-3x - 2\}$ $= -15x - 10$
Evaluating variable expressions, p. 43.	1. Replace each variable by the numerical value given. Put parentheses around the value. 2. Follow the order of operations in evaluating the expression.	Evaluate $2x^3 + 3xy + 4y^2$ for $x = -3$ and $y = 2$. $2(-3)^3 + 3(-3)(2) + 4(2)^2$ $= 2(-27) + 3(-3)(2) + 4(4)$ $= -54 - 18 + 16$ $= -56$
Using formulas, p. 44.	1. Replace the variables in the formula by the given values. 2. Evaluate the expression. 3. Label units carefully.	Find the area of a circle with a radius of 4 feet. Use $A = \pi r^2$ and $\pi \approx 3.14$. $A \approx (3.14)(4 \text{ feet})^2$ $= (3.14)(16 \text{ feet}^2)$ $= 50.24 \text{ feet}^2$ The area of the circle is approximately 50.24 square feet.

Chapter 1 Review Problems

Check the set to which the number belongs.

		Natural Numbers	Whole Numbers	Integers	Rational Numbers	Irrational Numbers	Real Numbers
1.	-5						
2.	$\dfrac{7}{8}$						
3.	3						
4.	$0.\overline{3}$						
5.	$2.1652384\ldots$ (no discernible pattern)						

In exercises 6 and 7, name the property of real numbers that justifies each statement.

6. $4 + a = a + 4$

7. $5(2 \cdot x) = (5 \cdot 2) \cdot x$

8. Are all rational numbers also real numbers?

Compute, if possible.

9. $-15 - (-20)$

10. $-7.3 + (-16.2)$

11. $-8(-6)$

12. $-12 \div 3$

13. $-\dfrac{4}{5} \div \left(-\dfrac{12}{5}\right)$

14. $-\dfrac{5}{6}\left(\dfrac{7}{10}\right)$

15. $4(-3)(-10)$

16. $5 + 6 - 2 - 5$

17. $-3.6(-1.5)$

18. $0 \div (-14)$

19. $7 \div 0$

20. $-17 + (+17)$

21. $17 - 3(6)$

22. $\dfrac{5 - 8}{2 - 7 - (-2)}$

23. $2\sqrt{49} - 3^2 + 5$

24. $4(6) - |-8| + (-1)^3$

25. $4 - 2 + 6\left(-\dfrac{1}{3}\right)$

26. $\sqrt{(-1)^2 + 6(4)} + 8 \div (-2)$

27. $\sqrt{\dfrac{25}{36}} - 2\left(\dfrac{1}{12}\right)$

28. $6|-3 - 1| + 5(-3)(0) - 4^2$

29. $(-0.4)^3$

Simplify. Do not leave negative exponents in your answers.

30. $(3xy^2)(-2x^0y)(4x^3y^3)$

31. $(5a^4bc^2)(-6ab^2)$

32. $\dfrac{36x^5y^4}{60x^2y^5}$

33. $\dfrac{16abc^0}{48ab^4c^2}$

34. $\left(\dfrac{-3x^3y}{2x^4z^2}\right)^4$

35. $(-2xy^6z^0)^3$

36. $(2x^2y^{-4})(-5x^{-1}y)$

37. $\dfrac{3x^5y^{-6}}{12x^{-2}y}$

38. $\dfrac{(5^{-1}x^{-2})^{-1}}{(2^{-2}y)^{-3}}$

39. $\dfrac{(3a)^{-2}}{(4b^{-3})^{-2}}$

40. $\left(\dfrac{a^5b^2}{3^{-1}a^{-5}b^{-4}}\right)^3$

41. $\left(\dfrac{x^3y^4}{5x^6y^8}\right)^3$

42. Write in scientific notation. 0.00721

43. Change to scientific notation and multiply. Express your answer in scientific notation. (5,300,000) (2,000,000,000)

44. Write in decimal form. 3.48×10^{-7}

45. Write in decimal form. 5.82×10^{13}

46. Combine like terms. $-x + 8 + 6x^2 + 7x - 4$

47. Multiply. $-5ab^2(a^3 + 2a^2b - 3b - 4)$

In exercises 48 and 49, remove grouping symbols and simplify.

48. $3x(x - 7) - (x^2 + 1)$

49. $2x^2 - \{2 + x[3 - 2(x - 1)]\}$

50. Evaluate $5x^2 - 3xy - 2y^3$ when $x = 2$ and $y = -1$.

▲ **51.** A coffee can has a height of 8 inches and a radius of 3 inches. Find its volume. (Use $V = \pi r^2 h$ and $\pi \approx 3.14$.)

▲ **52.** *Park Measurements* A triangular park in a Minneapolis neighborhood has an altitude of 88 yards and a base of 52 yards. Find the area of this triangular region.

▲ **53.** *Aircraft Windshield* The front view of the windshield of a stealth bomber appears to the observer to be shaped like a trapezoid. The altitude of the trapezoid is 14 inches. The bases of the trapezoid are 26 inches and 34 inches. What is the area of this apparent trapezoid?

Mixed Practice

Evaluate.

54. $9(-2) + (-28 \div 7)^3 - 5$

55. $\sqrt{20 + 5} - 12(2) \div 8$

Simplify each expression.

56. $(-7a^2b)(-2a^0b^3c^2)$

57. $\dfrac{(3x^{-1}y^2)^3}{(4x^2y^{-2})^2}$

58. $\dfrac{4x^3y^2}{-16x^2y^{-3}}$

59. $\left(\dfrac{3a^{-5}b^0}{2a^{-2}b^3}\right)^2$

60. Write 0.000058 in scientific notation.

61. Write 8.95×10^7 in decimal form.

Simplify.

62. $4x^2 - x^3 + 7x - 5x^2 + 6x^3 - 2x$

63. $2a^3b(5a - ab - 3)$

64. $-2\{x + 3[y - 5(x + y)]\}$

65. Evaluate $5a^2 - 3ab + 4b$ when $a = -3$ and $b = -2$.

▲ **66.** *Geometry* Use the formula $A = \pi r^2$ to find the area of a circle with a radius of 4 meters. (Use $\pi \approx 3.14$.)

67. *Swinging Pendulum* Use the formula $T = 2\pi\sqrt{L/g}$ to find the period of a pendulum T in seconds where the length of the pendulum $L = 512$ feet and gravity $g = 32$ ft/sec². (Use $\pi \approx 3.14$.)

68. Simplify. $\dfrac{2}{3}(6x - 9y) - (x - 2y)$

69. Find the amount to be repaid on a loan of $3200 at a simple interest rate of 9% for 2 years.

Remember to use your Chapter Test Prep Video CD to see the worked-out solutions to the test problems you want to review.

Exercises 1 and 2 refer to the set $\left\{-2, 12, \frac{9}{3}, \frac{25}{25}, 0, 2.585858, \ldots, \pi, \sqrt{4}, 2\sqrt{5}\right\}$.

1. List the real numbers that are not rational.

2. List the integers.

3. Name the property that justifies the statement $(8 \cdot x)3 = 3(8 \cdot x)$.

In exercises 4–11, do not leave negative exponents in your answers.

4. Evaluate. $(7 - 5)^3 - 18 \div (-3) + \sqrt{10 + 6}$

5. Evaluate. $(4 - 5)^2 - 3(-2) \div 3$

6. Simplify. $\dfrac{16x^3y}{20x^{-1}y^5}$

7. Simplify. $(5x^{-3}y^{-5})(-2x^3y^0)$

8. Simplify. $\left(\dfrac{5a^{-2}b}{a}\right)^2$

9. Simplify. $7x - 9x^2 - 12x - 8x^2 + 5x$

10. Simplify. $5a + 4b - 6a^2 + b - 7a - 2a^2$

11. Simplify. $3xy^2(4x - 3y + 2x^2)$

12. Write in scientific notation. 0.000002186

13. Write in decimal form. 2.158×10^9

14. Perform the calculation indicated. Leave your answer in scientific notation. $(3.8 \times 10^{-5})(4 \times 10^{-2})$

15. Simplify. $2x^2(x - 3y) - x(4 - 8x^2)$

16. Simplify. $2[-3(2x + 4) + 8(3x - 2)]$.

17. Evaluate $2x^2 - 3x - 6$ when $x = -4$.

18. Evaluate $5x^2 + 3xy - y^2$ when $x = 3$ and $y = -3$.

▲**19.** Find the area of a trapezoid with an altitude of 12 meters and bases of 6 and 7 meters.

▲**20.** Find the area of a circle with a radius of 6 meters. Use $\pi \approx 3.14$.

21. Find the amount A to be repaid on a principal of $8000 borrowed at a simple interest rate r of 5% for a time t of 3 years. The formula is $A = p(1 + rt)$.

1. _____

2. _____

3. _____

4. _____

5. _____

6. _____

7. _____

8. _____

9. _____

10. _____

11. _____

12. _____

13. _____

14. _____

15. _____

16. _____

17. _____

18. _____

19. _____

20. _____

21. _____

How do you count the number of walruses? Are these ice-dependent animals being endangered by global warming? Recently scientists from Russia and Alaska were involved in a joint venture to count the number of these animals located on ice floes between these two regions. The count used infrared technology and the kind of mathematical equations you will study in this chapter.

Linear Equations and Inequalities

2.1   FIRST-DEGREE EQUATIONS WITH ONE UNKNOWN   58

2.2   LITERAL EQUATIONS AND FORMULAS   66

2.3   ABSOLUTE VALUE EQUATIONS   72

2.4   USING EQUATIONS TO SOLVE WORD PROBLEMS   78

       HOW AM I DOING? SECTIONS 2.1–2.4   84

2.5   SOLVING MORE-INVOLVED WORD PROBLEMS   85

2.6   LINEAR INEQUALITIES   94

2.7   COMPOUND INEQUALITIES   103

2.8   ABSOLUTE VALUE INEQUALITIES   109

       CHAPTER 2 ORGANIZER   117

       CHAPTER 2 REVIEW PROBLEMS   119

       HOW AM I DOING? CHAPTER 2 TEST   123

       CUMULATIVE TEST FOR CHAPTERS 1–2   124

Student Learning Objective

After studying this section, you will be able to:

1 Solve first-degree equations with one unknown.

1 ## Solving First-Degree Equations with One Unknown

An **equation** is a mathematical statement that two quantities are equal. A **first-degree equation with one unknown** is an equation in which only one kind of variable appears, and that variable has an exponent of 1. This is also called a **linear equation with one unknown.** The variable itself may appear more than once. The equation $-6x + 7x + 8 = 2x - 4$ is a first-degree equation with one unknown because there is one variable, x, and that variable has an exponent of 1. Other examples are $5(6y + 1) = 2y$ and $5z + 3 = 10z$.

To **solve** a first-degree equation with one unknown, we need to find the value of the variable that makes the equation a true mathematical statement. This value is the **solution** or **root** of the equation.

EXAMPLE 1 Is $\frac{1}{3}$ a solution of the equation $2a + 5 = a + 6$?

Solution We replace a by $\frac{1}{3}$ in the equation $2a + 5 = a + 6$.

$$2\left(\frac{1}{3}\right) + 5 \stackrel{?}{=} \frac{1}{3} + 6$$

$$\frac{2}{3} + 5 \stackrel{?}{=} \frac{1}{3} + 6$$

$$\frac{17}{3} \neq \frac{19}{3}$$

This last statement is not true. Thus, $\frac{1}{3}$ is not a solution of $2a + 5 = a + 6$.

Practice Problem 1 Is $-\frac{3}{2}$ a solution of the equation $6a - 5 = -4a + 10$?

NOTE TO STUDENT: Fully worked-out solutions to all of the Practice Problems can be found at the back of the text starting at page SP-1

This procedure of verifying that a value is indeed a root of an equation is very valuable. We will use it throughout Chapter 2 to check our answers when solving equations.

Equations that have the same solution are said to be **equivalent.** The equations

$$7x - 2 = 12, \qquad 7x = 14, \quad \text{and} \quad x = 2$$

are equivalent because the solution of each equation is $x = 2$.

To solve an equation, we perform algebraic operations on it to obtain a simpler, equivalent equation of the form variable = constant or constant = variable.

PROPERTIES OF EQUIVALENT EQUATIONS

1. If $a = b$, then $a + c = b + c$ and $a - c = b - c$.

If the same number is added to or subtracted from both sides of an equation, the result is an equivalent equation.

2. If $a = b$ and $c \neq 0$, then $ac = bc$ and $\dfrac{a}{c} = \dfrac{b}{c}$.

If both sides of an equation are multiplied or divided by the same nonzero number, the result is an equivalent equation.

To solve a first-degree equation, we isolate the variable on one side of the equation and the constants on another, using the properties of equivalent equations.

EXAMPLE 2   Solve. $x - 8.2 = 5.0$

Solution $x - 8.2 + 8.2 = 5.0 + 8.2$ Add 8.2 to each side.

$$x = 13.2$$

Check: $x - 8.2 = 5.0$

$$13.2 - 8.2 \overset{?}{=} 5.0$$

$$5.0 = 5.0 \ \checkmark$$

The statement is valid, so our answer is correct. 13.2 is the root of the equation.

Practice Problem 2   Solve. $x + 5.2 = -2.8$

EXAMPLE 3   Solve. $\dfrac{1}{3}y = -6$

Solution $3\left(\dfrac{1}{3}\right)y = 3(-6)$ Multiply each side by 3 to eliminate the fraction.

$$y = -18$$

The solution is -18. Be sure to check this value.

Practice Problem 3   Solve.

$$\frac{1}{5}w = -6$$

EXAMPLE 4   Solve. $6x - 2 - 4x = 8x + 3$

Solution $2x - 2 = 8x + 3$ Combine like terms.

$2x - 8x - 2 = 8x - 8x + 3$ Subtract 8x from (or add $-8x$ to) each side.

$$-6x - 2 = 3$$

$-6x - 2 + 2 = 3 + 2$ Add 2 to each side.

$$-6x = 5$$

$$\frac{-6x}{-6} = \frac{5}{-6}$$ Divide each side by -6.

$$x = -\frac{5}{6}$$

When checking fractional values like $x = -\frac{5}{6}$, take extra care in order to perform the operations correctly.

Check: $6\left(-\dfrac{5}{6}\right) - 2 - 4\left(-\dfrac{5}{6}\right) \overset{?}{=} 8\left(-\dfrac{5}{6}\right) + 3$ Replace x by $-\frac{5}{6}$ in the *original* equation.

$$-5 - 2 + \frac{10}{3} \overset{?}{=} \frac{-20}{3} + 3$$

$$\frac{-21}{3} + \frac{10}{3} \overset{?}{=} \frac{-20}{3} + \frac{9}{3}$$

$$-\frac{11}{3} = -\frac{11}{3} \ \checkmark$$ Thus, $-\frac{5}{6}$ is the solution.

Practice Problem 4   Solve.

$$8w - 3 = 2w - 7w + 4$$

When an equation contains grouping symbols and fractions, use the following procedure to solve it.

PROCEDURE FOR SOLVING FIRST-DEGREE EQUATIONS WITH ONE UNKNOWN

1. Remove grouping symbols in the proper order.
2. If fractions exist, multiply both sides of the equation by the least common denominator (LCD) of all the fractions.
3. Combine like terms if possible.
4. Add or subtract a variable term on both sides of the equation to obtain all variable terms on one side of the equation.
5. Add or subtract a numerical value on both sides of the equation to obtain all numerical values on the other side of the equation.
6. Divide each side of the equation by the coefficient of the variable.
7. Simplify the solution (if possible).
8. Check your solution.

EXAMPLE 5 Solve and check $3(3x + 2) - 4x = -2(x - 3)$.

Solution

$$9x + 6 - 4x = -2x + 6 \qquad \text{Remove the parentheses.}$$

$$5x + 6 = -2x + 6 \qquad \text{Combine like terms.}$$

$$5x + 2x + 6 = -2x + 2x + 6 \qquad \text{Add } 2x \text{ to each side.}$$

$$7x + 6 = 6$$

$$7x + 6 - 6 = 6 - 6 \qquad \text{Subtract 6 from each side.}$$

$$7x = 0$$

$$\frac{7x}{7} = \frac{0}{7} \qquad \text{Divide each side by 7.}$$

$$x = 0$$

Check: $3[3(0) + 2] - 4(0) \overset{?}{=} -2(0 - 3)$

$$3(0 + 2) - 0 \overset{?}{=} -2(-3)$$

$$6 - 0 \overset{?}{=} 6$$

$$6 = 6 \quad ✓$$

Thus, 0 is the solution.

Practice Problem 5 Solve and check. $a - 4(2a - 7) = 3(a + 6)$

NOTE TO STUDENT: Fully worked-out solutions to all of the Practice Problems can be found at the back of the text starting at page SP-1

It is not necessary to memorize this eight-step procedure. However, you should refer to it often as you work the homework exercises in Section 2.1 until the sequence of steps is second nature to you.

EXAMPLE 6 Solve. $\dfrac{x}{5} + \dfrac{1}{2} = \dfrac{4}{5} + \dfrac{x}{2}$

Solution

$$10\left(\dfrac{x}{5} + \dfrac{1}{2}\right) = 10\left(\dfrac{4}{5} + \dfrac{x}{2}\right)$$ Multiply each term by the LCD 10.

$$10\left(\dfrac{x}{5}\right) + 10\left(\dfrac{1}{2}\right) = 10\left(\dfrac{4}{5}\right) + 10\left(\dfrac{x}{2}\right)$$ Use the distributive property.

$$2x + 5 = 8 + 5x$$ Simplify.

$$2x - 2x + 5 = 8 + 5x - 2x$$ Subtract $2x$ from each side.

$$5 = 8 + 3x$$

$$5 - 8 = 8 - 8 + 3x$$ Subtract 8 from each side.

$$-3 = 3x$$

$$\dfrac{-3}{3} = \dfrac{3x}{3}$$ Divide each side by 3.

$$-1 = x$$

Check: See if you can verify this solution.

Practice Problem 6 Solve and check.

$\dfrac{y}{3} + \dfrac{1}{2} = 5 + \dfrac{y-9}{4}$ $\left(\textit{Hint:} \text{ You can write } \dfrac{y-9}{4} \text{ as } \dfrac{y}{4} - \dfrac{9}{4}.\right)$

An equation that contains many decimals can be multiplied by an appropriate power of 10 to clear it of decimals.

EXAMPLE 7 Solve and check. $0.9 + 0.2(x + 4) = -3(0.1x - 0.4)$

Solution

$$0.9 + 0.2x + 0.8 = -0.3x + 1.2$$ Remove the parentheses.

$$10(0.9) + 10(0.2x) + 10(0.8) = 10(-0.3x) + 10(1.2)$$ Multiply each side of the equation by 10 and use the distributive property.

$$9 + 2x + 8 = -3x + 12$$ Simplify.

$$2x + 17 = -3x + 12$$ Combine like terms.

$$2x + 3x + 17 = -3x + 3x + 12$$ Add $3x$ to each side.

$$5x + 17 = 12$$

$$5x + 17 - 17 = 12 - 17$$ Add -17 to each side.

$$5x = -5$$

$$x = -1$$ Divide each side by 5.

Check: $0.9 + 0.2(-1 + 4) \overset{?}{=} -3[0.1(-1) - 0.4]$

$$0.9 + 0.2(3) \overset{?}{=} -3(-0.1 - 0.4)$$

$$0.9 + 0.6 \overset{?}{=} -3(-0.5)$$

$$1.5 = 1.5 \ \checkmark$$

Thus, -1 is the solution.

Practice Problem 7 Solve and check.

$$4(0.01x + 0.09) - 0.07(x - 8) = 0.83$$

Not every equation has a solution. Some equations have no solution at all.

EXAMPLE 8 Solve. $7x + 3 - 9x = 14 - 2x + 5$

Solution

$-2x + 3 = -2x + 19$	Combine like terms.
$-2x + 3 + 2x = -2x + 19 + 2x$	Add $2x$ to each side.
$3 = 19$	Simplify. We obtain a false equation.

No matter what value we use for x in this equation, we get a false sentence. This equation has **no solution.**

NOTE TO STUDENT: *Fully worked-out solutions to all of the Practice Problems can be found at the back of the text starting at page SP-1*

Practice Problem 8 Solve. $7 + 14x - 3 = 2(x - 4) + 12x$

Now we examine a totally different situation. There are some equations for which any real number is a solution.

EXAMPLE 9 Solve. $5(x - 2) + 3x = 10x - 2(x + 5)$

Solution

$5x - 10 + 3x = 10x - 2x - 10$	Remove parentheses.
$8x - 10 = 8x - 10$	Combine like terms.
$8x - 10 - 8x = 8x - 10 - 8x$	Subtract $8x$ from each side of the equation.
$-10 = -10$	Simplify. This is an equation that is always true.

Replacing x in this equation by any real number will always result in a true sentence. This is an equation for which **any real number is a solution.**

Practice Problem 9 Solve. $13x - 7(x + 5) = 4x - 35 + 2x$

As you work through the exercises in this section you will notice that solutions to equations can be integers, fractions, or decimals. Recall from page 2 that a **terminating decimal** is one that has a definite number of digits. Throughout this book, the decimal form of a solution will be given only if is a terminating decimal.

Developing Your Study Skills

Class Participation

People learn mathematics through active participation, not through observation from the sidelines. If you want to do well in this course, be involved in classroom activities. Sit near the front where you can see and hear well, and where your focus is on the instruction process and not on the students around you. Ask questions, be ready to contribute toward solutions, and take part in all classroom activities. Your contributions are valuable to the class and to yourself. Class participation requires an investment of yourself in the learning process, which you will find pays huge dividends.

Verbal and Writing Skills

1. Is -20 a root of the equation $3x - 15 = 45$? Why or why not?

2. Is 21 a root of the equation $2x + 12 = -30$? Why or why not?

3. Is $\frac{2}{7}$ a solution to the equation $7x - 8 = -6$? Why or why not?

4. Is $\frac{3}{5}$ a solution to the equation $5y + 9 = 12$? Why or why not?

5. What is the first step in solving the equation $\frac{x}{3} + \frac{3}{4} = 2 - \frac{x}{2}$? Why?

6. What is the first step in solving the equation $0.7 + 0.03x = 4$? Why?

7. When solving the equation $x + 3.6 = 8$, would you clear it of decimals by multiplying each term by 10? Why or why not?

8. When solving the equation $x - \frac{1}{4} = 3$, would you clear it of fractions by multiplying each term by 4? Why or why not?

Solve exercises 9–44. Check your solutions.

9. $-11 + x = -3$

10. $26 + x = -35$

11. $-8x = 56$

12. $-16x = -64$

13. $-14x = -70$

14. $-12x = 72$

15. $8x - 1 = 11$

16. $10x + 3 = 15$

17. $8x + 5 = 2x - 13$

18. $14x + 3 = 11x - 3$

19. $16 - 2x = 5x - 5$

20. $-12x - 8 = 10 - 3x$

21. $3a - 5 - 2a = 2a - 3$

22. $5a - 2 + 4a = 2a + 12$

23. $4(y - 1) = -2(3 + y)$

24. $3(4 - y) = 5(y + 2)$

25. $3 + y = 10 - 3(y + 1)$

Solve the following exercises. You may leave answers in either decimal or fraction form.

26. $3y + 3 = 7(y + 2) - 3y$

27. $\frac{2}{3}x = 8$

28. $-\frac{5}{6}x = 5$

29. $\frac{y}{2} + 4 = \frac{1}{6}$

30. $\frac{y}{3} + 2 = \frac{4}{5}$

31. $\frac{2}{3} - \frac{x}{6} = 1$

32. $\dfrac{4x}{5} + \dfrac{3}{2} = 2x$

33. $\dfrac{1}{2}(x + 3) - 2 = 1$

34. $5 - \dfrac{2}{3}(x + 2) = 3$

35. $5 - \dfrac{2x}{7} = 1 - (x - 4)$

36. $6 + 2(x - 1) = \dfrac{3x}{5} + 4$

37. $0.3x + 0.4 = 0.5x - 0.8$

38. $0.7x - 0.2 = 0.5x + 0.8$

39. $0.3 - 0.05x = 0.2x - 0.1$

40. $0.1x - 0.12 = 0.04x + 0.03$

41. $0.2(x - 4) = 3$

42. $0.6(2x + 1) = 1$

43. $0.05x - 2 = 0.3(x - 5)$

44. $0.3(x + 2) - 2 = 0.05x$

Mixed Practice *Solve each of the following.*

45. $2x + 12 = 3 + 4x - 7$

46. $6y - 15 - 8y = 24 - 5$

47. $\dfrac{1}{2} - \dfrac{x}{3} = \dfrac{2x - 3}{3}$

48. $\dfrac{1}{6} - \dfrac{x}{2} = \dfrac{x - 5}{3}$

49. $\dfrac{1}{3} - \dfrac{x + 1}{5} = \dfrac{x}{3}$

50. $\dfrac{y + 5}{12} = \dfrac{3}{4} - \dfrac{y + 1}{8}$

51. $2 + 0.1(5 - x) = 1.3x - (0.4x - 2.5)$

52. $3(0.4 - x) + 2 = x + 0.4(x + 8)$

To Think About *Some of the equations in this section have no solution. For some of the equations, any real number is a solution. Other equations have one real solution. Work carefully and solve each of the following equations.*

53. $6x - 8 = 9 - 2x + 5 - 3x$

54. $7x - 5 = -2x - 15 + 10x + 6$

55. $2x - 4(x + 1) = -2x + 14$

56. $3x - 17 = 8x - 5(x - 2)$

57. $-(x - 2) + 1 = 2x + 3(1 - x)$

58. $8(x + 2) - 7 = 3(x + 3) + 5x$

59. $6 + 8(x - 2) = 10x - 2(x + 4)$

60. $2x + 4(x - 5) = -x + 7(x - 1) + 3$

61. $x - 2 + \dfrac{2x}{5} = -2 + \dfrac{7x}{5}$

62. $x + \dfrac{2x + 8}{3} = \dfrac{5x + 5}{3} + 1$

Cumulative Review *Simplify. Do not leave negative exponents in your answers.*

63. **[1.3.3]** $5 - (4 - 2)^2 + 3(-2)$

64. **[1.3.3]** $(-2)^4 - 12 - 6(-2)$

65. **[1.4.4]** $\left(\dfrac{3xy^2}{2x^2y}\right)^3$

66. **[1.4.2]** $(2x^{-2}y^{-3})^2(4xy^{-2})^{-2}$

Quick Quiz 2.1 Solve for x.

1. $4(7 - 3x) = 16 - 3(3x + 1)$

2. $\dfrac{2}{3}(2x - 1) + 3 = 4(2x - 4)$

3. $0.7x + 1.3 = 5x - 2.14$

4. **Concept Check** Explain how you would solve the following equation for x.

$$\dfrac{3x + 1}{2} + \dfrac{2}{3} = \dfrac{4x}{5}$$

1 Solving Literal Equations for the Desired Unknown

A first-degree **literal equation** is an equation that contains variables other than the variable that we are solving for. When you solve for an unknown in a literal equation, the final expression will contain these other variables. We use this procedure to deal with formulas in applied problems.

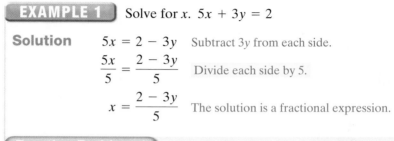

EXAMPLE 1 Solve for x. $5x + 3y = 2$

Solution

$5x = 2 - 3y$ Subtract $3y$ from each side.

$\dfrac{5x}{5} = \dfrac{2 - 3y}{5}$ Divide each side by 5.

$x = \dfrac{2 - 3y}{5}$ The solution is a fractional expression.

Practice Problem 1 Solve for W. $P = 2L + 2W$

$P - 2L = 2W$

$\dfrac{P - 2L}{2} = W$

Where possible, combine like terms as you solve the equation.

When solving more complicated first-degree literal equations, use the following procedure.

PROCEDURE FOR SOLVING FIRST-DEGREE (OR LINEAR) LITERAL EQUATIONS

1. Remove grouping symbols in the proper order.
2. If fractions exist, multiply all terms on both sides by the LCD.
3. Combine like terms if possible.
4. Add or subtract a term with the desired unknown on both sides of the equation to obtain all terms with the desired unknown on one side of the equation.
5. Add or subtract appropriate terms on both sides of the equation to obtain all other terms on the other side of the equation.
6. Divide each side of the equation by the coefficient of the desired unknown.
7. Simplify the solution (if possible).

Some equations appear difficult to solve because they contain fractions and parentheses. Immediately remove the parentheses. Then multiply each term by the LCD. The equation will then appear less threatening.

EXAMPLE 2 Solve for b. $A = \dfrac{2}{3}(a + b + 3)$

Solution

$A = \dfrac{2}{3}a + \dfrac{2}{3}b + 2$ Remove parentheses.

$3A = 3\left(\dfrac{2}{3}a\right) + 3\left(\dfrac{2}{3}b\right) + 3(2)$ Multiply all terms by the LCD 3.

$3A = 2a + 2b + 6$ Simplify.

$3A - 2a - 6 = 2b$ Subtract $2a$ from each side.
Subtract 6 from each side.

$$\frac{3A - 2a - 6}{2} = \frac{2b}{2} \qquad \text{Divide each side of the}$$
equation by the coefficient of b.

$$\frac{3A - 2a - 6}{2} = b \qquad \text{Simplify.}$$

Practice Problem 2 Solve for a.

$$H = \frac{3}{4}(a + 2b - 4)$$

NOTE TO STUDENT: *Fully worked-out solutions to all of the Practice Problems can be found at the back of the text starting at page SP-1*

Be sure to combine like terms after removing the parentheses. This will simplify the equation and make it much easier to solve.

EXAMPLE 3 Solve for x. $5(2ax + 3y) - 4ax = 2(ax - 5)$

Solution

$$10ax + 15y - 4ax = 2ax - 10 \qquad \text{Remove parentheses.}$$

$$6ax + 15y = 2ax - 10 \qquad \text{Combine like terms.}$$

$$6ax - 2ax + 15y = -10 \qquad \begin{array}{l}\text{Subtract } 2ax \text{ from each side to obtain terms} \\ \text{containing } x \text{ on one side.}\end{array}$$

$$4ax = -10 - 15y \qquad \text{Simplify and subtract } 15y \text{ from each side.}$$

$$\frac{4ax}{4a} = \frac{-10 - 15y}{4a} \qquad \text{Divide each side by the coefficient of } x.$$

$$x = \frac{-10 - 15y}{4a}$$

Practice Problem 3 Solve for b. $-2(ab - 3x) + 2(8 - ab) = 5x + 4ab$

The Boston Marathon is a grueling 26-mile endurance race over city streets from Hopkinton to Boston, Massachusetts. The race has been held annually for over one hundred years and has been watched by millions of people. The winning time on April 17, 2000, was 2 hours 9 minutes 47 seconds. If we round to the nearest minute, this time is approximately equal to 130 minutes. The bar graph to the right shows the approximate winning time in minutes for this race for selected years from 1900 to 2000.

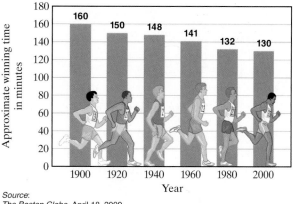

Winning Times in the Boston Marathon

Source:
The Boston Globe, April 18, 2000

EXAMPLE 4 The winning times in minutes for the Boston Marathon are approximated by the equation $t = -0.3x + 159$, where x is the number of years since 1900. Solve this equation for x. Determine approximately what year it will be when the winning time for the Boston Marathon is 126 minutes.

Solution

$$t = -0.3x + 159$$

$$0.3x = 159 - t \qquad \text{Add } 0.3x - t \text{ to each side.}$$

$$3x = 1590 - 10t \qquad \text{Multiply each side by 10 to clear decimals.}$$

$$x = \frac{1590 - 10t}{3} \qquad \text{Divide each side by 3.}$$

Now we will use this equation, which has been solved for *x* (the number of years since 1900), to find when the winning time is 126 minutes.

$$x = \frac{1590 - 10(126)}{3} \qquad \text{We substitute in 126 for } t.$$

$$x = \frac{1590 - 1260}{3}$$

$$x = \frac{330}{3} = 110$$

It will be 110 years from 1900. Thus, we estimate that this winning time will occur in the year 2010.

NOTE TO STUDENT: *Fully worked-out solutions to all of the Practice Problems can be found at the back of the text starting at page SP-1*

Practice Problem 4   The winning time in minutes for a 15-mile race to benefit breast cancer research is given by the equation $t = -0.4x + 81$, where *x* is the number of years since 1990. Solve this equation for *x*. Determine approximately what year it will be when the winning time for this 15-mile race is 73 minutes.

EXAMPLE 5

(a) Solve the formula for the area of a trapezoid, $A = \frac{1}{2}a(b + c)$, for *c*.

(b) Find *c* when $A = 20$ square inches, $a = 3$ inches, and $b = 4$ inches.

Solution

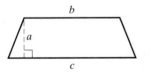

(a)

$$A = \frac{1}{2}ab + \frac{1}{2}ac \qquad \text{Remove the parentheses.}$$

$$2A = 2\left(\frac{1}{2}ab\right) + 2\left(\frac{1}{2}ac\right) \qquad \text{Multiply each term by 2.}$$

$$2A = ab + ac \qquad \text{Simplify.}$$

$$2A - ab = ac \qquad \text{Subtract } ab \text{ from each side to isolate the term containing } c.$$

$$\frac{2A - ab}{a} = c \qquad \text{Divide each side by } a.$$

(b) We use the equation we derived in part **(a)** to find *c* for the given values.

$$c = \frac{2A - ab}{a}$$

$$= \frac{2(20) - (3)(4)}{3} \qquad \text{Substitute the given values of } A, a, \text{ and } b \text{ to find } c.$$

$$= \frac{40 - 12}{3} = \frac{28}{3} \qquad \text{Simplify.}$$

Thus, side $c = \frac{28}{3}$ inches or $9\frac{1}{3}$ inches.

Practice Problem 5

(a) Solve for *h*. $A = 2\pi rh + 2\pi r^2$

(b) Find *h* when $A = 100$, $\pi \approx 3.14$, and $r = 2.0$. Round your answer to the nearest hundredth.

Solve for x.

1. $6x + 5y = 3$

2. $7x + 2y = 5$

3. $4x + y = 18 - 3x$

4. $11x - 8 = 5y + 7x$

5. $y = \dfrac{2}{3}x - 4$

6. $y = -\dfrac{1}{3}x + 2$

Solve for the variable specified.

7. $x = -\dfrac{3}{4}y + \dfrac{2}{3}$; for y

8. $x = \dfrac{5}{2}y - \dfrac{1}{5}$; for y

▲ 9. $A = lw$; for l

▲ 10. $V = lwh$; for w

▲ 11. $A = \dfrac{h}{2}(B + b)$; for B

12. $C = \dfrac{5}{9}(F - 32)$; for F

▲ 13. $A = 2\pi rh$; for r

▲ 14. $V = \pi r^2 h$; for h

15. $H = \dfrac{2}{3}(a + 2b)$; for b

16. $H = \dfrac{3}{4}(5a + b)$; for a

17. $2(2ax + y) = 3ax - 4y$; for x

18. $3(4ax + y) = 2ax - 3y$; for x

Follow the directions given.

▲ 19. (a) Solve for b. $A = \dfrac{1}{2}ab$

   (b) Evaluate when $A = 18$ and $a = \dfrac{3}{2}$.

20. (a) Solve for C. $F = \dfrac{9}{5}C + 32$

   (b) Evaluate when $F = 23°$.

21. (a) Solve for n. $A = a + d(n - 1)$

▲ 22. (a) Solve for h. $V = \dfrac{1}{3}\pi r^2 h$

   (b) Evaluate when $A = 28$, $a = 3$, and $d = 15$.

   (b) Evaluate when $V = 6.28$, $r = 3$, and $\pi \approx 3.14$.

Applications

23. ***Heart Disease Deaths*** The heart disease death rate (number of deaths per 100,000 people) in the United States can be approximated by the equation $y = -7.4x + 322$, where x is the number of years since 1990. Solve this equation for x. Use this new equation to determine in which year the approximate heart disease death rate will be 137. (*Source:* www.cdc.gov)

24. ***Cancer Deaths*** The cancer death rate (number of deaths per 100,000 people) in the United States can be approximated by the equation $y = -2.2x + 216$, where x is the number of years since 1990. Solve this equation for x. Use this new equation to determine in which year the approximate cancer death rate will be 150. (*Source:* www.cdc.gov)

25. ***Mariner's Formula*** The mariner's formula can be written in the form $\frac{m}{1.15} = k$, where m is the speed of a ship in miles per hour and k is the speed of the ship in knots (nautical miles per hour).

   (a) Solve the formula for m.

   (b) Use this result to find the number of miles per hour a ship is traveling if its speed is 29 knots.

26. ***Patient Appointments*** Some doctors use the formula $ND = 1.08T$ to relate the variables N (the number of patient appointments the doctor schedules in one day), D (the duration of each patient appointment), and T (the total number of minutes the doctor can use to see patients in one day).

   (a) Solve the formula for N.

   (b) Use this result to find the number of patient appointments N a doctor should make if she has 6 hours available for patients and each appointment is 15 minutes long. (Round your answer to the nearest whole number.)

In exercises 27 and 28, the variable C represents the consumption of products in the United States in billions of dollars, and D represents disposable income in the United States in billions of dollars.

27. ***An Economic Model*** Suppose economists use as a model of the country's economy the equation $C = 0.6547D + 5.8263$.

   (a) Solve the equation for D.

   (b) Use this result to determine the disposable income D if the consumption C is $9.56 billion. Round your answer to the nearest tenth of a billion.

28. ***An Economic Model*** Suppose economists use as a model of the country's economy the equation $C = 0.7649D + 6.1275$.

   (a) Solve the equation for D.

   (b) Use this result to determine the disposable income D if the consumption C is $12.48 billion. Round your answer to the nearest tenth of a billion.

Cumulative Review *Write with positive exponents in simplest form.*

29. **[1.4.4]** $(2x^{-3}y)^{-2}$

30. **[1.4.3]** $\left(\dfrac{5x^2y^{-3}}{x^{-4}y^2}\right)^{-3}$

31. **[1.3.3]** Simplify. $1 + 16 \div (2 - 4)^3 - 3$

32. **[1.5.3]** Simplify. $2[a - (3 - 2b)] + 5$

33. **[1.2.2]** *Education Fund* Sharon and James want to begin an education fund for their two daughters. They invest $5000 in a certificate of deposit for one year, with an annual return of 5%. $4000 is invested in a more risky venture that they hope will have an annual return of 9%. How much money will they have at the end of 1 year if their risky investment does well? (*Hint:* Use $I = prt$.)

34. **[1.2.2]** *Automobile Costs* Drew wants to go to college in Pennsylvania. He and his parents take a long week-end to drive there from Kansas City, Missouri. His odometer read 45,711.3 when he left the college in Pennsylvania and 46,622.1 when he arrived back home. He started and ended his trip on a full tank of gas. He made gas purchases of 9.9 gallons, 11.7 gallons, 10.6 gallons, 5.8 gallons, and 8 gallons during the trip. How many miles per gallon did the car get on the trip?

Quick Quiz 2.2

1. Solve for b. $A = d + g(b + 3)$

2. Solve for x. $V = \dfrac{1}{3}xy$

3. Solve for w. $G = 4x + \dfrac{1}{2}w + \dfrac{3}{4}$

4. **Concept Check** Explain how you would solve for x in the following equation. $3(3ax + y) = 2ax - 5y$

Developing Your Study Skills

Taking Notes in Class

An important part of learning mathematics is taking notes. To take meaningful notes, you must be an active listener. Keep your mind on what the instructor is saying, and be ready with questions whenever you do not understand something.

If you have previewed the lesson material, you will be prepared to take good notes. The important concepts will seem somewhat familiar. You will have a better idea of what needs to be written down. If you frantically try to write all that the instructor says or copy all the examples done in class, you may find your notes to be nearly worthless when you are home alone. You may find that you are unable to make sense of what you have written.

Write down *important* ideas and examples as the instructor lectures, making sure that you are listening and following the logic. Include any helpful hints or suggestions that your instructor gives you or refers to in your text. You will be amazed at how easily these are forgotten if they are not written down.

Successful note taking requires active listening and processing. Stay alert in class. You will realize the advantages of taking your own notes over copying those of someone else.

Student Learning Objectives

After studying this section, you will be able to:

1. Solve absolute value equations of the form $|ax + b| = c$.

2. Solve absolute value equations of the form $|ax + b| + c = d$.

3. Solve absolute value equations of the form $|ax + b| = |cx + d|$.

Solving Absolute Value Equations of the Form $|ax + b| = c$

From Section 1.2, you know that the absolute value of a number x can be pictured as the distance between 0 and x on the number line. Let's look at a simple absolute value equation, $|x| = 4$, and draw a picture.

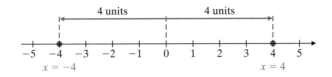

Thus, the equation $|x| = 4$ has two solutions, $x = 4$ and $x = -4$. Let's look at another example.

$$\text{If} \quad |x| = \frac{2}{3},$$

$$\text{then} \quad x = \frac{2}{3} \quad \text{or} \quad x = -\frac{2}{3},$$

$$\text{because} \quad \left|\frac{2}{3}\right| = \frac{2}{3} \quad \text{and} \quad \left|-\frac{2}{3}\right| = \frac{2}{3}.$$

We can solve these relatively simple absolute value equations by recalling the definition of absolute value.

$$|x| = \begin{cases} x, & \text{if } x \geq 0 \\ -x, & \text{if } x < 0 \end{cases}$$

Now let's take a look at a more complicated absolute value equation: $|ax + b| = c$.

> The solutions of an equation of the form $|ax + b| = c$, where $a \neq 0$ and c is a positive number, are those values that satisfy
>
> $$ax + b = c \quad \text{or} \quad ax + b = -c.$$

EXAMPLE 1 Solve $|2x + 5| = 11$ and check your solutions.

Solution Using the rule established in the box, we have the following:

$$\begin{array}{ll} 2x + 5 = 11 & \text{or} \quad 2x + 5 = -11 \\ 2x = 6 & \qquad 2x = -16 \\ x = 3 & \qquad x = -8 \end{array}$$

The two solutions are 3 and -8.

Check: **if $x = 3$**   **if $x = -8$**

$$\begin{array}{ll} |2x + 5| = 11 & \qquad |2x + 5| = 11 \\ |2(3) + 5| \stackrel{?}{=} 11 & \qquad |2(-8) + 5| \stackrel{?}{=} 11 \\ |6 + 5| \stackrel{?}{=} 11 & \qquad |-16 + 5| \stackrel{?}{=} 11 \\ |11| \stackrel{?}{=} 11 & \qquad |-11| \stackrel{?}{=} 11 \\ 11 = 11 \ \checkmark & \qquad 11 = 11 \ \checkmark \end{array}$$

Practice Problem 1 Solve $|3x - 4| = 23$ and check your solutions.

NOTE TO STUDENT: Fully worked-out solutions to all of the Practice Problems can be found at the back of the text starting at page SP-1

EXAMPLE 2 Solve $\left|\dfrac{1}{2}x - 1\right| = 5$ and check your solutions.

Solution The solutions of the given absolute value equation must satisfy

$$\frac{1}{2}x - 1 = 5 \quad \text{or} \quad \frac{1}{2}x - 1 = -5.$$

If we multiply each term of both equations by 2, we obtain the following:

$$x - 2 = 10 \quad \text{or} \quad x - 2 = -10$$
$$x = 12 \qquad\qquad x = -8$$

Check: **if $x = 12$**

$$\left|\frac{1}{2}(12) - 1\right| \overset{?}{=} 5$$
$$|6 - 1| \overset{?}{=} 5$$
$$|5| \overset{?}{=} 5$$
$$5 = 5 \ \checkmark$$

if $x = -8$

$$\left|\frac{1}{2}(-8) - 1\right| \overset{?}{=} 5$$
$$|-4 - 1| \overset{?}{=} 5$$
$$|-5| \overset{?}{=} 5$$
$$5 = 5 \ \checkmark$$

Practice Problem 2 Solve and check your solutions.

$$\left|\frac{2}{3}x + 4\right| = 2$$

Solving Absolute Value Equations of the Form $|ax + b| + c = d$

Notice that in each of the previous examples the absolute value expression is on one side of the equation and a positive real number is on the other side of the equation. What happens when we encounter an equation of the form $|ax + b| + c = d$?

EXAMPLE 3 Solve $|3x - 1| + 2 = 5$ and check your solutions.

Solution First we will rewrite the equation so that the absolute value expression is alone on one side of the equation.

$$|3x - 1| + 2 - 2 = 5 - 2$$
$$|3x - 1| = 3$$

Now we solve $|3x - 1| = 3$.

$$3x - 1 = 3 \quad \text{or} \quad 3x - 1 = -3$$
$$3x = 4 \qquad\qquad 3x = -2$$
$$x = \frac{4}{3} \qquad\qquad x = -\frac{2}{3}$$

Check: **if $x = \dfrac{4}{3}$**

$$\left|3\left(\frac{4}{3}\right) - 1\right| + 2 \overset{?}{=} 5$$
$$|4 - 1| + 2 \overset{?}{=} 5$$
$$|3| + 2 \overset{?}{=} 5$$
$$3 + 2 \overset{?}{=} 5$$
$$5 = 5 \ \checkmark$$

if $x = -\dfrac{2}{3}$

$$\left|3\left(-\frac{2}{3}\right) - 1\right| + 2 \overset{?}{=} 5$$
$$|-2 - 1| + 2 \overset{?}{=} 5$$
$$|-3| + 2 \overset{?}{=} 5$$
$$3 + 2 \overset{?}{=} 5$$
$$5 = 5 \ \checkmark$$

Practice Problem 3 Solve $|2x + 1| + 3 = 8$ and check your solutions.

③ Solving Absolute Value Equations of the Form $|ax + b| = |cx + d|$

Let us now consider the possibilities for a and b if $|a| = |b|$.

Suppose $a = 5$; then $b = 5$ or -5.
If $a = -5$, then $b = 5$ or -5.

> To generalize, if $|a| = |b|$, then $a = b$ or $a = -b$.

We now apply this property to solve more complex equations.

EXAMPLE 4 Solve and check: $|3x - 4| = |x + 6|$

Solution The solutions of the given equation must satisfy

$$3x - 4 = x + 6 \quad \text{or} \quad 3x - 4 = -(x + 6).$$

Now we solve each equation in the normal fashion.

$$
\begin{array}{ll}
3x - 4 = x + 6 \quad \text{or} & 3x - 4 = -x - 6 \\
3x - x = 4 + 6 & 3x + x = 4 - 6 \\
2x = 10 & 4x = -2 \\
x = 5 & x = -\dfrac{1}{2}
\end{array}
$$

We will check each solution by substituting it into the *original equation*.

Check: **if $x = 5$**

$$|3(5) - 4| \overset{?}{=} |5 + 6|$$
$$|15 - 4| \overset{?}{=} |11|$$
$$|11| \overset{?}{=} |11|$$
$$11 = 11 \ \checkmark$$

if $x = -\dfrac{1}{2}$

$$\left|3\left(-\frac{1}{2}\right) - 4\right| \overset{?}{=} \left|-\frac{1}{2} + 6\right|$$
$$\left|-\frac{3}{2} - 4\right| \overset{?}{=} \left|-\frac{1}{2} + 6\right|$$
$$\left|-\frac{3}{2} - \frac{8}{2}\right| \overset{?}{=} \left|-\frac{1}{2} + \frac{12}{2}\right|$$
$$\left|-\frac{11}{2}\right| \overset{?}{=} \left|\frac{11}{2}\right|$$
$$\frac{11}{2} = \frac{11}{2} \ \checkmark$$

NOTE TO STUDENT: Fully worked-out solutions to all of the Practice Problems can be found at the back of the text starting at page SP-1

Practice Problem 4 Solve and check: $|x - 6| = |5x + 8|$

TO THINK ABOUT: Two Other Absolute Value Equations Explain how you would solve an absolute value equation of the form $|ax + b| = 0$. Give an example. Does $|3x + 2| = -4$ have a solution? Why or why not?

Verbal and Writing Skills

1. The equation $|x| = b$, where b is a positive number, will always have how many solutions? Why?

2. The equation $|x| = b$ might have only one solution. How could that happen?

3. To solve an equation like $|x + 7| - 2 = 8$, what is the first step that must be done? What will be the result?

4. To solve an equation like $|3x - 1| + 5 = 14$, what is the first step that must be done? What will be the result?

Solve each absolute value equation. Check your solutions for exercises 5–24.

5. $|x| = 30$

6. $|x| = 14$

7. $|x + 4| = 10$

8. $|x + 6| = 13$

9. $|2x - 5| = 13$

10. $|7x - 3| = 11$

11. $|5 - 4x| = 11$

12. $|2 - 3x| = 13$

13. $\left|\dfrac{1}{2}x - 3\right| = 2$

14. $\left|\dfrac{1}{4}x + 5\right| = 3$

15. $|1.8 - 0.4x| = 1$

16. $|0.9 - 0.7x| = 4$

17. $|x + 2| - 1 = 7$

18. $|x + 3| - 4 = 8$

19. $\left|\dfrac{1}{2} - \dfrac{3}{4}x\right| + 1 = 3$

20. $\left|\dfrac{2}{3} - \dfrac{1}{2}x\right| - 2 = -1$

21. $\left|2 - \dfrac{2}{3}x\right| - 3 = 5$

22. $\left|5 - \dfrac{7}{2}x\right| + 1 = 11$

23. $\left|\dfrac{1 - 3x}{2}\right| = \dfrac{4}{5}$

24. $\left|\dfrac{3x - 2}{3}\right| = \dfrac{1}{2}$

Solve each absolute value equation.

25. $|x + 4| = |2x - 1|$

26. $|x - 8| = |2x - 7|$

27. $\left|\dfrac{x - 1}{2}\right| = |2x + 3|$

28. $\left|\dfrac{2x + 7}{3}\right| = |x + 2|$

29. $|1.5x - 2| = |x - 0.5|$

30. $|2.2x + 2| = |1 - 2.8x|$

31. $|3 - x| = \left|\dfrac{x}{2} + 3\right|$

32. $\left|\dfrac{2x}{5} + 1\right| = |1 - x|$

Solve for x. Round to the nearest hundredth.

33. $|1.62x + 3.14| = 2.19$

34. $|-0.74x - 8.26| = 5.36$

Mixed Practice *Solve each equation, if possible. Check your solutions.*

35. $|3(x + 4)| + 2 = 14$

36. $|4(x - 2)| + 1 = 19$

37. $\left|\dfrac{5x}{3} - 1\right| = 0$

38. $\left|\dfrac{3}{4}x + 9\right| = 0$

39. $\left|\dfrac{4}{3}x - \dfrac{1}{8}\right| = -5$

40. $\left|\dfrac{3}{4}x - \dfrac{2}{3}\right| = -8$

41. $\left|\dfrac{3x - 1}{3}\right| = \dfrac{2}{5}$

42. $\left|\dfrac{2x + 3}{3}\right| = \dfrac{1}{4}$

Cumulative Review

43. **[1.4.2]** Simplify. $(3x^{-3}yz^0)\left(\dfrac{5}{3}x^4y^2\right)$

44. **[1.3.3]** Evaluate. $\dfrac{\sqrt{3 - 2 \cdot 1^2} + 5}{4^2 - 2 \cdot 3}$

Quick Quiz 2.3 Solve for x.

1. $|3x - 4| = 49$

2. $\left|\dfrac{2}{3}x + 1\right| - 3 = 5$

3. $|2x + 5| = |x - 4|$

4. **Concept Check** Explain how you would solve for x in the following equation.

$$|2x + 4| = \dfrac{1}{2}$$

Developing Your Study Skills

How to Do Homework

Set aside time each day for your homework assignments. Do not attempt to do a whole week's worth on the weekend. Two hours spent studying outside of class for each hour in class is usual for college courses, and you may need more than that for mathematics.

Before beginning your homework exercises, read your textbook very carefully. Expect to spend much more time reading a few pages of a mathematics textbook than several pages of another text. Read for complete understanding, not just for the general idea.

As you begin your homework assignments, read the directions carefully. You need to understand what is being asked for. Concentrate on each exercise, taking time to solve it accurately. Rushing through your work will cause you to make errors. Check your answers with those given in the back of the textbook. If your answer is incorrect, check to see that you are doing the right exercise. Redo the exercise, watching

for little errors. If it is still wrong, check with a friend. Perhaps the two of you can figure out where you made an error.

Also, check the examples in the textbook or in your notes for a similar exercise. Can this one be solved in the same way? Give it some thought. You may want to leave it for a while and take a break or do a different exercise. But come back later and try again. If you are still unable to figure it out, ask your instructor for help during office hours or in class.

Work on assignments every day, and do as many exercises as it takes for you to know what you are doing. Begin by doing all the exercises that have been assigned. If there are more available in the section in your text, then do more. When you think you have done enough exercises to understand fully the kind at hand, do a few more to be sure. This may mean that you do many more exercises than the instructor assigns, but you can never practice too much. Practice improves your skills and increases your accuracy, speed, competence, and confidence.

1 Solving Applied Problems by Using Equations

The skills you have developed in solving equations will allow you to solve a variety of applied problems. The following steps may help you to organize your thoughts and provide you with a procedure to solve such problems.

1. **Understand the problem.**
   (a) Read the word problem carefully to get an overview.
   (b) Determine what information you will need to solve the problem.
   (c) Draw a sketch. Label it with the known information. Determine what needs to be found.
   (d) Choose a variable to represent one unknown quantity.
   (e) If necessary, represent other unknown quantities in terms of the same variable.

2. **Write an equation.**
   (a) Look for key words to help you translate the words into algebraic symbols.
   (b) Use a relationship given in the problem or an appropriate formula in order to write an equation.

3. **Solve the equation and state the answer.**

4. **Check.**
   (a) Check the solution in the original situation.
   (b) Be sure the solution to the equation answers the question in the word problem. You may need to do some additional calculations if it does not.

Symbolic Equivalents of English Phrases

English Phrase	Mathematical Symbol
and, added to, increased by, greater than, plus, more than, sum of	$+$
decreased by, subtracted from, less than, diminished by, minus, difference between	$-$
product of, multiplied by, of, times	$\cdot$ or $(\ )(\ )$ or $\times$
divided by, quotient of, ratio of	$\div$ or fraction bar
equals, are, is, will be, yields, gives, makes, is the same as, has a value of	$=$

Although we often use x to represent the unknown quantity when we write equations, any letter can be used. It is a good idea to use a letter that helps us remember what the variable represents. (For example, we might use s for speed or h for hours.) We now look at some translations of English sentences into algebraic equations.

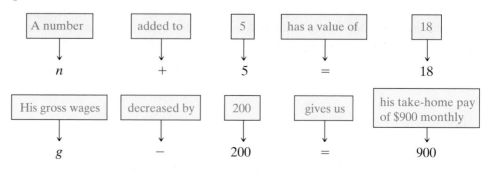

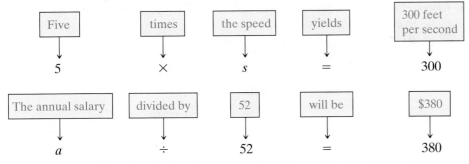

Be careful when translating the expressions "more than" and "less than." The order in which you write the symbols does not follow the order found in the English sentence. For example, "5 less than a number will be 40" is written as follows:

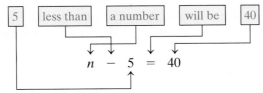

"Two more than a number is −5" is written as shown next.

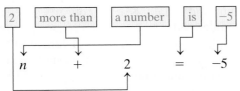

Since addition is commutative, we can also write $2 + n = -5$ for "2 more than a number is −5." Be careful, however, with subtraction. "5 less than a number will be 40" must be written as $n - 5 = 40$.

We will now illustrate the use of the four-step procedure in solving a number of word problems.

EXAMPLE 1 Nancy works for an educational services company that provides computers and software to local public schools. She went to a local truck rental company to rent a truck to deliver some computers to the Newbury Elementary School. The truck rental company has a fixed rate of $40 per day plus 20¢ per mile. Nancy rented a truck for 3 days and was billed $177. How many miles did she drive?

Solution

1. **Understand the problem.** Let $n =$ the number of miles driven.
   Since each mile driven costs 20¢, we multiply the 20¢- (or $0.20) per-mile cost by the number of miles n.
   Thus, $0.20n =$ the cost of driving n miles at 20¢ per mile.

2. **Write an equation.**

Fixed costs for 3 days	plus	mileage charge	equals	$177
$(40)(3)$	$+$	$(0.20)(n)$	$=$	177

3. **Solve the equation and state the answer.**

$$120 + 0.20n = 177 \qquad \text{Multiply } (40)(3).$$
$$0.20n = 57 \qquad \text{Subtract 120 from each side.}$$
$$\frac{0.20n}{0.20} = \frac{57}{0.20} \qquad \text{Divide each side by 0.20.}$$
$$n = 285 \qquad \text{Simplify.}$$

Nancy drove the truck for 285 miles.

4. Check. Does a truck rental for 3 days at $40 per day plus 20¢ per mile for 285 miles come to a total of $177?

We will check our values in the original equation.

$$(40)(3) + (0.20)(n) = 177$$
$$120 + (0.20)(285) \stackrel{?}{=} 177$$
$$120 + 57 \stackrel{?}{=} 177$$
$$177 = 177 \quad \checkmark$$

It checks. Our answer is correct.

NOTE TO STUDENT: *Fully worked-out solutions to all of the Practice Problems can be found at the back of the text starting at page SP-1*

Practice Problem 1 Western Laboratories rents a computer terminal for $400 per month plus $8 per hour of computer use time. The bill for 1 year's computer use was $7680. How many hours did Western Laboratories actually use the computer?

EXAMPLE 2 The Acetones are a barbershop quartet. They travel across the country in a special bus and usually give six concerts a week. This popular group always sings to a sell-out crowd. The concert halls have an average seating capacity of three thousand people each. Concert tickets average $12 per person. The one-time expenses for each concert are $15,000. The cost of meals, motels, security and sound people, bus drivers, and other expenses totals $100,000 per week. How many weeks per year will the Acetones need to be on tour if each wants to earn $71,500 per year?

Solution

1. Understand the problem. The item we are trying to find is the number of weeks the quartet needs to be on tour. So we let

$$x = \text{the number of weeks on tour.}$$

Now we need to find an expression that describes the quartet's income. Each week there are six concerts with three thousand people paying $12 per person. Thus,

$$\text{weekly income} = (6)(3000)(12) = \$216{,}000, \quad \text{and}$$
$$\text{income for } x \text{ weeks} = \$216{,}000x.$$

Now we need to find an expression for the total expenses. The expenses for each concert are $15,000, and there are six concerts per week. Thus, concert expenses will be $(6)(15{,}000) = \$90{,}000$ per week. Now the cost of meals, motels, security and sound people, bus drivers, and other expenses total $100,000 per week. Thus,

$$\text{total weekly expenses} = \$90{,}000 + \$100{,}000 = \$190{,}000, \quad \text{and}$$
$$\text{total expenses for } x \text{ weeks} = \$190{,}000x.$$

The four Acetones each want to earn $71,500 per year, so the group will need $4 \times \$71{,}500 = \$286{,}000$ for the year.

2. Now we can write an equation.

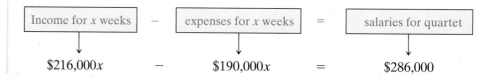

Income for x weeks	−	expenses for x weeks	=	salaries for quartet
$216,000x	−	$190,000x	=	$286,000

3. Solve the equation.

$$26{,}000x = 286{,}000 \quad \text{Combine like terms.}$$

$$\frac{26{,}000x}{26{,}000} = \frac{286{,}000}{26{,}000} \quad \text{Divide each side of the equation by 26,000.}$$

$$x = 11$$

Thus, the Acetones will need to be on tour for 11 weeks to meet their salary goal.

4. Check. The check is left to the student.

Practice Problem 2 A group of five women in a rock band plans to go on tour. They are scheduled to give five concerts a week with an average audience of six thousand people at each concert. The average ticket price for a concert is $14. The onetime expenses for each concert are $48,000. The additional costs per week are $150,000. How many weeks per year will this group need to be on tour if each wants to earn $60,000 per year?

Sometimes we need a simple formula from geometry or some other science to write the original equation.

▲ **EXAMPLE 3** An astronaut's space suit contains a small rectangular steel plate that supports the breathing control valve. The length of the rectangle is 3 millimeters more than double its width. Its perimeter is 108 millimeters. Find the width and length.

Solution

1. Understand the problem. We draw a sketch to assist us. The formula for the perimeter of a rectangle is $P = 2w + 2l$, where w = the width and l = the length. Since the length is compared to the width, we will begin with the width. Let w = the width. Then $2w + 3$ = the length.

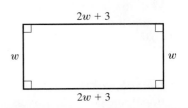

2. Write an equation.

$$P = 2w + 2l$$

$$108 = 2w + 2(2w + 3)$$

3. Solve the equation and state the answer.

$$108 = 2w + 4w + 6$$

$$108 = 6w + 6$$

$$102 = 6w$$

$$17 = w$$

Because $w = 17$, we have $2w + 3 = 2(17) + 3 = 37$. Thus, the rectangle is 17 millimeters wide and 37 millimeters long.

4. Check.

$$P = 2w + 2l$$

$$108 \overset{?}{=} 2(17) + 2(37)$$

$$108 \overset{?}{=} 34 + 74$$

$$108 = 108 \ \checkmark$$

▲ **Practice Problem 3** The perimeter of a triangular lawn is 162 meters. The length of the first side is twice the length of the second side. The length of the third side is 6 meters shorter than three times the length of the second side. Find the dimensions of the triangle.

Write an algebraic equation and use it to solve each problem.

1. Three-fifths of a number is -54. What is the number?

2. Five-sixths of a number is -60. What is the number?

Applications

3. ***Gym Membership Costs*** Allyson can pay for her gym membership on a monthly basis, but if she pays for an entire year's membership in advance, she'll receive a $50 discount. Her discounted bill for the year would then be $526. What is the monthly membership fee at her gym?

4. ***Parking Garage Fees*** The monthly parking fee at the Safety First Parking Garage is one and one-half times last year's monthly fee, but compact cars will now receive a $10 per month discount. If the owner of a compact car pays $98 per month this year, what was his monthly fee last year?

5. ***Computer Shipments*** Barbara Stormer's computer company ships its packages through Reliable Shipping Express. Reliable charges $3 plus $0.80 per pound to ship each package. If the charge for shipping a package to Idaho was $17.40, how much did the package weigh?

6. ***Parking Garage Fees*** The Greenville City Garage charges $5 for the first hour and then $2.75 for each additional hour. How long has a car been parked at this garage if the parking charge is $29.75?

7. ***Laundry Expenses*** Roberto and Maria Santanos spend approximately $11.75 per week to wash and dry their family's clothes at a local coin laundry. A new washer and dryer would cost them a total of $846. How many weeks will it take for the laundromat cost to equal the cost of a new washer and dryer?

8. ***Personal Banking*** Winchesters Community Bank charges its customers $6 per month plus 15¢ per check for the use of a standard checking account. Thomas had a checking account there for 8 months and was charged $53.10 in service charges. How many checks did he write during that period?

9. ***Guest Speaker Costs*** Mr. Ziglar is a famous motivational speaker. He travels across the country by jet and gives an average of eight presentations a week. He gives his speeches in auditoriums that are always full and hold an average of 2000 people each. He charges $15 per person at each of these speeches. His average expenses for each presentation, including renting the auditorium, are about $14,000. The cost of meals, motels, jet travel, and support personnel totals $85,000 per week. How many weeks will Mr. Ziglar need to travel if he wants to earn $129,000 per year?

10. ***Music Tour Costs*** The Three Tenors are planning a nationwide tour. They will travel across the country by jet and give four concerts a week. They will sing in concert halls that hold an average of 5000 people each. Concert tickets average $18 per person, and most concerts are totally sold out. The advance expenses for each concert are $55,000. The cost of meals, motels, jet travel, and support personnel totals $110,000 per week. How many weeks will the Three Tenors need to be on tour if each of them wants to earn $120,000 per year?

▲ **11.** ***Geometry*** A new Youth Opportunity Center is being built in Roxbury. The perimeter of the rectangular playing field is 340 yards. The length of the field is 6 yards less than triple the width. What are the dimensions of the playing field?

▲ **12.** ***Geometry*** Dave and Jane Wells have a new rectangular driveway. The perimeter of the driveway is 168 feet. The length is 12 feet longer than three times the width. What are the dimensions of the driveway?

▲ **13.** *Geometry* A vacant city lot is being turned into a neighborhood garden. The neighbors want to fence in a triangular section of the lot and plant flowers there. The longest side of the triangular section is 7 feet shorter than twice the shortest side. The third side is 6 feet longer than the shortest side. The perimeter is 59 feet. How long is each side?

▲ **14.** *Geometry* A leather coin purse has the shape of a triangle. Two sides are equal in length and the third side is 3 centimeters shorter than one and one-half times the length of the equal sides. The perimeter is 28.5 centimeters. Find the lengths of the sides.

Cumulative Review *Name the property that justifies each statement.*

15. **[1.1.2]** $57 + 0 = 57$

16. **[1.1.2]** $(2 \cdot 3) \cdot 9 = 2 \cdot (3 \cdot 9)$

17. **[1.2.3]** Evaluate. $7(-2) \div 7(-3) - 3$

18. **[1.3.3]** Evaluate. $(7 - 12)^3 - (-4) + 3^3$

Quick Quiz 2.4

1. Three-fourths of a number is -69. What is the number?

2. A triangle has a perimeter of 100 yards. The first side is twice the length of the second side. The third side is 8 yards longer than the second side. How long is each side?

3. The Manchester Airport Central Garage charges $6 to park for the first hour and then $3.50 an hour for any additional hours. Ben paid $65.50 to park there one day when he flew to Philadelphia. How many hours was he parked at the airport.

4. **Concept Check** Explain how you would set up an equation to solve the following problem. A driveway has a perimeter of 212 feet. The length of the driveway is 7 feet longer than four times the width. Find the width and the length of the driveway.

How are you doing with your homework assignments in Sections 2.1 to 2.4? Do you feel you have mastered the material so far? Do you understand the concepts you have covered? Before you go further in the textbook, take some time to do each of the following problems.

2.1

Solve for x.

1. $2x - 1 = 12x + 36$

2. $\dfrac{x-2}{4} = \dfrac{1}{2}x + 4$

3. $4(x - 3) = x + 2(5x - 1)$

4. $0.6x + 3 = 0.5x - 7$

2.2

5. Solve for y. $3x - 7y = 14$

6. Solve for a. $5ab - 2b = 16ab - 3(8 + b)$

7. Solve for r. $A = P + Prt$

8. Use your results from problem 7 to find r when $P = 100$, $t = 3$, and $A = 118$.

2.3

Solve for x.

9. $|3x - 2| = 7$

10. $|9 - x| + 2 = 5$

11. $\left|\dfrac{2x+3}{4}\right| = 2$

12. $|5x - 8| = |3x + 2|$

2.4

Use an algebraic equation to find a solution for each exercise.

▲ **13.** The length of a queen-size mattress is 20 inches longer than the width, and the perimeter is 280 inches. Find the dimensions of a queen-size mattress.

14. Eastern Bank charges its customers a flat fee of $6 per month for a checking account plus 12¢ for each check. The bank charged Jose $9.12 for his checking account last month. How many checks did he use?

15. Alan and Cindi pick up food donations for their local food bank from grocery stores. Alan picked up 80 more than half as many pounds of food as Cindi did. Together they picked up 455 pounds of food donations. How many pounds did each pick up?

▲ **16.** In Freeport, Maine, the north end of L. L. Bean has a triangular parking lot for bicycles. The longest side is 5 feet shorter than twice the shortest side. The third side is 9 feet longer than the shortest side. The perimeter is 62 feet. How long is each side?

Now turn to page SA-3 for the answers to each of these problems. Each answer also includes a reference to the objective in which the problem is first taught. If you missed any of these problems, you should stop and review the Examples and Practice Problems in the referenced objective. A little review now will help you master the material in the upcoming sections of the text.

1. _____
2. _____
3. _____
4. _____
5. _____
6. _____
7. _____
8. _____
9. _____
10. _____
11. _____
12. _____
13. _____
14. _____
15. _____
16. _____

Solving More-Involved Word Problems by Using an Equation

To solve some word problems, we might need to understand percents, simple interest, mixtures, or some other concept before we can use an algebraic equation as a model. We review several of these concepts in this section.

From arithmetic you know that to find a percent of a number we write the percent as a decimal and multiply the decimal by the number.

Thus, to find 36% of 85, we calculate $(0.36)(85) = 30.6$. If the number is not known, we can represent it by a variable.

Student Learning Objective

After studying this section, you will be able to:

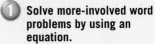 Solve more-involved word problems by using an equation.

EXAMPLE 1 The Wildlife Refuge Rangers tagged 144 deer. They estimate that they have tagged 36% of the deer in the refuge. If they are correct, approximately how many deer are in the refuge?

Solution

1. **Understand the problem.** Let n = the number of the deer in the refuge. Then $0.36n$ = 36% of the deer in the refuge.

2. **Write an equation.**

36%	of	the deer in the refuge	gives a total of	144 tagged deer
0.36	×	n	=	144

3. **Solve the equation and state the answer.**

$$0.36n = 144$$
$$\frac{0.36n}{0.36} = \frac{144}{0.36} \quad \text{Divide each side by 0.36.}$$
$$n = 400 \quad \text{Simplify.}$$

There are approximately 400 deer in the refuge.

4. **Check.** Is it true that 36% of 400 is 144?

$$(0.36)(400) \stackrel{?}{=} 144$$
$$144 = 144 \quad ✓$$

It checks. Our answer is correct.

Practice Problem 1 Technology Resources, Inc. sold 6900 computer workstations, a 15% increase in sales over the previous year. How many computer workstations were sold last year? (*Hint:* Let x = the amount of sales last year, then $0.15x$ = the increase in sales over last year.)

NOTE TO STUDENT: Fully worked-out solutions to all of the Practice Problems can be found at the back of the text starting at page SP-1

Adding two numbers yields a total. We can call one of the numbers x and the other number (total $- x$). We will use this concept in Example 2.

EXAMPLE 2 Bob's and Marcia's weekly salaries total $265. If they both went from part-time to full-time employment, their combined weekly income would be $655. Bob's salary would double, while Marcia's would triple. How much do they each make now?

Solution

1. **Understand the problem.** Let b = Bob's part-time salary.
   Since the total of the two part-time weekly salaries is $265, we can let $265 - b$ = Marcia's part-time salary.

2. **Write an equation.**

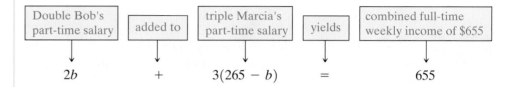

$$2b \qquad + \qquad 3(265 - b) \qquad = \qquad 655$$

3. **Solve the equation and state the answer.**

$$2b + 3(265 - b) = 655$$

$$2b + 795 - 3b = 655 \qquad \text{Remove parentheses.}$$

$$795 - b = 655 \qquad \text{Simplify.}$$

$$-b = -140 \qquad \text{Subtract 795 from each side.}$$

$$b = 140 \qquad \text{Multiply each side by } -1.$$

   If $b = 140$, then $265 - b = 265 - 140 = 125$. Thus, Bob's present part-time weekly salary is $140, and Marcia's present part-time weekly salary is $125.

4. **Check.** Do their present weekly salaries total $265? Yes: $140 + 125 = 265$. If Bob's income is doubled and Marcia's is tripled, will their new weekly salaries total $655? Yes: $2(140) + 3(125) = 280 + 375 = 655$. ✓

NOTE TO STUDENT: Fully worked-out solutions to all of the Practice Problems can be found at the back of the text starting at page SP-1

Practice Problem 2 Together Alicia and Heather sold forty-three cars at Prestige Motors last month. If Alicia doubles her sales and Heather triples her sales next month, they will sell 108 cars. How many cars did each person sell this month?

Simple interest is an income from investing money or a charge for borrowing money. It is computed by multiplying the amount of money borrowed or invested (called the *principal*) by the rate of interest and by the period of time it is borrowed or invested (usually measured in years unless otherwise stated). Hence

$$\text{interest} = \text{principal} \times \text{rate} \times \text{time}.$$
$$I = prt$$

All interest problems in this chapter involve simple interest.

EXAMPLE 3 Maria has a job as a financial advisor in a bank. She advised a customer to invest part of his money in a money market fund earning 12% simple interest and the rest in an investment fund earning 14% simple interest. The customer had $6000 to invest. If he earned $772 in interest in 1 year, how much did he invest in each fund?

Solution

1. *Understand the problem.* Let x = the amount of money invested at 12% interest. The other amount of money is (total $- x$).
   Let $6000 - x$ = the amount of money invested at 14% interest.

2. *Write an equation.*

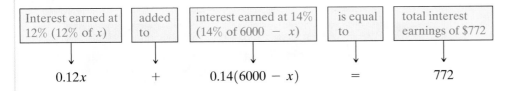

Interest earned at 12% (12% of x)	added to	interest earned at 14% (14% of 6000 $- x$)	is equal to	total interest earnings of $772
$0.12x$	$+$	$0.14(6000 - x)$	$=$	772

3. *Solve the equation and state the answer.*

$$0.12x + 0.14(6000 - x) = 772$$

$0.12x + 840 - 0.14x = 772$ Remove parentheses.

$840 - 0.02x = 772$ Add $0.12x$ to $-0.14x$.

$-0.02x = -68$ Subtract 840 from each side.

$$\frac{-0.02x}{-0.02} = \frac{-68}{-0.02}$$ Divide each side by -0.02.

$x = 3400$ Simplify.

If $x = 3400$, then $6000 - x = 6000 - 3400 = 2600$. Thus, $3400 was invested in the money market fund earning 12% interest, and $2600 was invested in the investment fund earning 14% interest.

4. *Check.* Do the two amounts of money total $6000?
   Yes: $3400 + $2600 = $6000.
   Does the total interest amount to $772?

$$(0.12)(3400) + (0.14)(2600) \overset{?}{=} 772$$

$$408 + 364 \overset{?}{=} 772$$

$$772 = 772 \checkmark$$

Our answers are correct.

Practice Problem 3 Tricia received an inheritance of $5500. She invested part of it at 8% simple interest and the remainder at 12% simple interest. At the end of the year she had earned $540. How much did Tricia invest at each interest amount?

Sometimes we encounter a situation in which two or more items are combined to form a mixture or solution. These types of problems are called **mixture problems.**

EXAMPLE 4 A small truck has a radiator that holds 20 liters. A mechanic needs to fill the radiator with a solution that is 60% antifreeze. He has 70% and 30% antifreeze solutions. How many liters of each should he use to achieve the desired mix?

Solution

1. **Understand the problem.** Let x = the number of liters of 70% antifreeze to be used.

   Since the total amount of solution must be 20 liters, we can use $20 - x$ for the other part. So $20 - x$ = the number of liters of 30% antifreeze to be used.

   In this problem a chart or table is very helpful. We will multiply the entry in column (A) by the entry in column (B) to obtain the entry in column (C).

	(A) Number of Liters of the Solution	(B) Percent Pure Antifreeze	(C) Number of Liters of Pure Antifreeze
70% antifreeze solution	x	70%	$0.70x$
30% antifreeze solution	$20 - x$	30%	$0.30(20 - x)$
Final 60% solution	20	60%	$0.60(20)$

2. **Write an equation.** Now we form an equation from the entries in column (C).

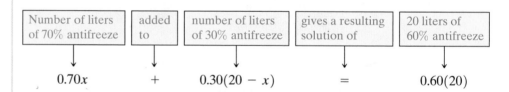

Number of liters of 70% antifreeze	added to	number of liters of 30% antifreeze	gives a resulting solution of	20 liters of 60% antifreeze
$0.70x$	$+$	$0.30(20 - x)$	$=$	$0.60(20)$

3. **Solve the equation and state the answer.**

$$0.70x + 0.30(20 - x) = 0.60(20)$$
$$0.70x + 6 - 0.30x = 12$$
$$6 + 0.40x = 12$$
$$0.40x = 6$$
$$\frac{0.40x}{0.40} = \frac{6}{0.40}$$
$$x = 15$$

If $x = 15$, then $20 - x = 20 - 15 = 5$. Thus, the mechanic needs 15 liters of 70% antifreeze solution and 5 liters of 30% antifreeze solution.

4. **Check.** Can you verify that this answer is correct?

NOTE TO STUDENT: *Fully worked-out solutions to all of the Practice Problems can be found at the back of the text starting at page SP-1*

Practice Problem 4 A jeweler wishes to prepare 200 grams of 80% pure gold from sources that are 68% pure gold and 83% pure gold. How many grams of each should he use?

Some problems involve the relationship distance = rate $\times$ time or $d = rt$.

EXAMPLE 5 Frank drove at a steady speed for 3 hours on the turnpike. He then slowed his traveling speed by 15 miles per hour on the secondary roads. The entire trip took 5 hours and covered 245 miles. What was his speed on the turnpike?

Solution

1. **Understand the problem.** Let x = speed on the turnpike in miles per hour. So $x - 15$ = speed on the secondary roads in miles per hour.

   Again, as in Example 4, a chart or table is very helpful. We will use a chart that records values of $(r)(t) = d$.

	Rate (miles per hour) r	Time (hours) t	Distance (miles) $(r)(t) = d$
Turnpike	x	3	$3x$
Secondary roads	$x - 15$	2	$2(x - 15)$
Entire trip	Not appropriate	5	245

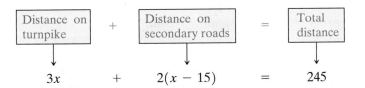

2. **Write an equation.** Using the distance entries (third column of the table), we have

$$3x \quad + \quad 2(x - 15) \quad = \quad 245$$

3. **Solve the equation and state the answer.**
$$3x + 2(x - 15) = 245$$
$$3x + 2x - 30 = 245$$
$$5x - 30 = 245$$
$$5x = 275$$
$$x = 55$$

   Thus, Frank traveled at an average speed of 55 miles per hour on the turnpike.

4. **Check.** Can you verify that this answer is correct?

Practice Problem 5 Wally drove for 4 hours at a steady speed. He slowed his speed by 10 miles per hour for the last part of the trip. The entire trip took 6 hours and covered 352 miles. How fast did he drive on each portion of the trip?

Write an algebraic equation for each problem and solve it.

1. **U.S. Population** The population of the United States in 2007 was estimated to be 301.1 million people. This was an increase of 21% from the population in 1990. What was the population of the United States in 1990? (*Source:* www.census.gov)

2. **National Debt** The U.S. national debt on January 24, 2008, was approximately $9.19 trillion. This was an increase of 61% from the national debt on January 24, 2000. What was the national debt on January 24, 2000? (*Source:* www.treasurydirect.gov)

3. **Consumer Purchases** Eastwing dormitory just purchased a new Sony color television on sale for $340. The sale price was 80% of the original price. What was the original price of the television set?

4. **Health Club Memberships** In 2006, the number of Americans who were members of a health club was 42.7 million. This was an increase of 106% from the number of health club members in 1990. How many Americans were health club members in 1990? (*Source:* www.emaxhealth.com)

5. **Unemployment Rates** Allentown's employment statistics show that 969 of its residents were unemployed last month. This is a decrease of 15% from the previous month. How many residents were unemployed in the previous month?

6. **Lyme Tick Disease** A wildlife expert at the Crane Wildlife Reserve has found fifteen deer with ticks that are carrying Lyme disease. She estimates that this number is about 60% of the total number of deer carrying infected ticks. Approximately how many such deer are there on the Reserve?

7. **Reforestation Program** In a reforestation program, 1400 seedling trees were planted. Twice as many spruce as hemlocks were planted. The number of balsams planted was twenty more than triple the number of hemlocks. How many of each type of seedling was planted?

8. **Summer Rental** Lynn and Judy are pooling their savings to rent a cottage in Maine for a week this summer. The rental cost is $950. Lynn's family is joining them, so she is paying a larger part of the cost. Her share of the cost is $250 less than twice Judy's. How much of the rental fee is each of them paying?

9. *Salary Increases* When Angela and Walker first started working for the supermarket, their weekly salaries totaled $500. Now during the last 25 years Walker has seen his weekly salary triple. Angela has seen her weekly salary become four times larger. Together their weekly salaries now total $1740. How much did they each make 25 years ago?

10. *Salary Increases* When Grace and Tony started as junior engineers at a manufacturing company, their weekly salaries totaled $1300. Now ten years later they have both become senior engineers. Grace has had her salary double. Tony has had his salary triple. Together their weekly salaries now total $3200. How much did they each make 10 years ago?

11. *Medicine Dosage* A hospital received a shipment of 8-milligram doses of a medicine. Each 8-milligram package was repacked into two smaller doses of unequal size and labeled packet A and packet B. The hospital then used 17 doses of packet A and 14 doses of packet B in one week. The hospital used a total of 127 milligrams of the medicine during that week. How many milligrams of the medicine are contained in each A packet? In each B packet?

12. *Cookie Sales* This year, two Girl Scout Troops together sold 460 boxes of cookies. Half of the Rockland troop's sales were Thin Mints and $\frac{2}{5}$ of the Harrisville troop's sales were Thin Mints. Together they sold 205 boxes of Thin Mints. How many boxes of cookies did each troop sell?

13. *Interest on a Loan* Rebecca and Peter borrowed $1500 at a simple interest rate of 9% for a period of 2 years. What was the interest?

14. *Interest on a Loan* Juanita and Carlos borrowed $4800 at a simple interest rate of 11% for a period of 2 years. What was the interest?

15. *College Fund Investment* The Vegas want to invest $5000 of their daughter's college fund for 18 months in a certificate of deposit that pays a simple interest rate of 3.1%. How much interest will they earn?

16. *Earned Interest* David had $4000 invested for one-fourth of a year at a simple interest rate of 6.1%. How much interest did he earn?

17. *Investment Income* Cynthia is a hardworking single mother on a very limited budget. She invests her savings only in safe investments because she is not in a position to take risks and lose any of her hard-earned money. She invested $6400 in two types of accounts for one year. The first type earned 5% simple interest, and the second type earned 8% simple interest. At the end of the year, Cynthia had earned $395 in interest. How much did she invest at each rate?

18. *Investment Income* The Johnsons' family business did well *this* year due to their investments *last* year. The business earned $6570 on an investment of $45,000 in mutual funds. There were two types of funds that Jenna Johnson invested in. The first was a pharmaceutical fund, which paid out simple interest of 13%. The second was a biotech fund, which paid out simple interest of 16%. How much did Jenna Johnson invest in each fund last year for her family's business?

19. *Retirement Fund Investment* Jim Jacobs decided to invest $18,000 of his retirement fund conservatively. He invested part of this money in a certificate of deposit that pays 3.5% simple interest and part in a fixed interest account that pays 2.2% simple interest. Last year he earned $552 in interest. How much did he invest in each type of account?

20. *Investment Income* Tori invested $8000 in money market funds. Part was invested at 5% simple interest, and the rest was invested at 7% simple interest. At the end of 1 year, Tori had earned $496 in interest. How much did she invest in each fund?

21. *Fat Content of Food* A chef has one cheese that contains 45% fat and another cheese that contains 20% fat. How many grams of each cheese should she use in order to obtain 30 grams of a cheese mixture that is 30% fat?

22. *Chemical Mixtures* Becky was awarded a college internship at a local pharmaceutical research corporation. She has been given her own corner of the laboratory to try her hand at pharmacology. One of her assignments is to mix two solutions, one that is 16% strength and the other that is 9% strength. How many milliliters of each should she use in order to obtain 350 milliliters of a 12% strength solution?

23. *Fat Content of Food* The meat department manager at a large food store wishes to mix some hamburger with 30% fat content and some hamburger that has 10% fat content in order to obtain 100 pounds of hamburger with 25% fat content. How much of each type of hamburger should she use?

24. *Cost of Tea* The manager of a gourmet coffee and tea store is mixing two green teas, one worth $7 per pound and the other worth $9 per pound. He wants to obtain 32 pounds of tea worth $8.50 per pound. How much of each tea should he use?

25. *Fertilizer Mix* A landscaping company needs 150 gallons of 18% fertilizer to fertilize the shrubs in an office park. They have in stock 25% fertilizer and 15% fertilizer. How much of each type should they mix together?

26. *Insecticide Manufacturing* Most mosquito repellents contain the ingredient deet. A manufacturer needs to produce 10-ounce spray cans of 40% deet. How much 100% deet and how much 25% deet should be mixed to produce each of these spray cans?

27. *Auto Travel* Susan drove for 4 hours on secondary roads at a steady speed. She completed her trip by driving 2 hours on an interstate highway. Her total trip was 250 miles. Her speed on the interstate highway portion of the trip was 20 miles per hour faster than her speed on the secondary roads. How fast did she travel on the secondary roads?

28. *Airplane Flight* Alice and Wendy flew a small plane for 930 miles. For the first 3 hours they flew at maximum speed. After refueling they finished the trip at a cruising speed that was 60 miles per hour slower than maximum speed. The entire trip took 5 hours of flying time. What is the maximum flying speed of the plane?

29. *Walking on a Treadmill* Yissania and Charlotte walked on treadmills at the gym for the same amount of time. Yissania walks at 5 miles per hour, and Charlotte walks at 4.2 miles per hour. If Yissania walked 0.6 miles farther than Charlotte, how long did they use the treadmills?

30. *Boating on a Lake* The Clarke family went sailing on a lake. Their boat averaged 6 kilometers per hour. The Rourke family took their outboard runabout for a trip on the lake for the same amount of time. Their boat averaged 14 kilometers per hour. The Rourke family traveled 20 kilometers farther than the Clarke family. How many hours did each family spend on their boat trip?

Cumulative Review *Evaluate.*

31. **[1.6.1]** $5a - 2b + c$ when $a = 1, b = -3, c = -4$

32. **[1.6.1]** $2x^2 - 3x + 1$ when $x = -2$

33. **[1.3.3]** $\dfrac{5 + 8(-2) + 2^4}{|2 - 7|}$

34. **[1.3.3]** $\dfrac{\sqrt{7^2 - 24}}{2^3(-1) + 7(4)}$

Quick Quiz 2.5

1. Eastview dormitory purchased a new 52-inch HD television for $2208 this year. This was a price that was 8% less than the one they purchased last year. How much was the television that they purchased last year?

2. A landscaping company needs 200 gallons of 45% fertilizer to fertilize the shrubs at North Shore Community College. They have in stock 50% fertilizer and 30% fertilizer. How much of each type should they mix together?

3. Lexi invested $4000 in two mutual funds. In one year part of her investment earned 5% simple interest and the other part earned 7% simple interest. At the end of that year Lexi had earned $250. How much did she invest at each rate?

4. **Concept Check** How would you set up an equation to solve the following problem? A new sports car sold for a certain amount of money. This year the price went up 12%. The new price is $39,200. What was the price the previous year?

Student Learning Objectives

After studying this section, you will be able to:

1 Determine whether one number is less than or greater than another number.

2 Graph linear inequalities in one variable.

3 Solve linear inequalities in one variable.

1 Determining Whether One Number Is Less Than or Greater Than Another Number

We briefly introduced inequalities in Chapter 1. Now we will discuss this concept more completely.

A **linear inequality** is a statement that describes how two numbers or linear expressions are related to one another. We can use a number line to visualize the concept of inequality.

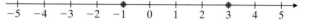

It is a mathematical property that -1 is less than 3. We write this in the following way:

$$-1 < 3.$$

Notice the position of these numbers on the number line. -1 is to the *left* of 3 on the number line. We say that a first number *is less than* a second number if the first number is *to the left* of the second number on a number line.

We could also say that 3 is greater than -1 (because it is *to the right* of -1 on the number line). We then write this in the following way:

$$3 > -1.$$

An inequality symbol can face either right or left, but its opening must face the larger number.

In general, if $a < b$, then it is also true that $b > a$.

EXAMPLE 1 Insert the proper symbol between the numbers.

(a) 8 _____ 6   (b) -2 _____ 3   (c) -4 _____ -2

(d) $\dfrac{1}{2}$ _____ $\dfrac{1}{3}$ (e) -0.033 _____ -0.0329

Solution

(a) $8 > 6$ because 8 is to the right of 6 on a number line.

(b) $-2 < 3$ because -2 is to the left of 3 on a number line.

(c) $-4 < -2$ because -4 is to the left of -2 on a number line.

(d) When comparing two fractions, rewrite them with a common denominator.

$\dfrac{1}{2}$ _____ $\dfrac{1}{3}$ Rewrite fractions with the same denominator.

$\dfrac{3}{6}$ _____ $\dfrac{2}{6}$ Compare the numerators: $3 > 2$.

$\dfrac{3}{6} > \dfrac{2}{6}$

Thus, $\dfrac{1}{2} > \dfrac{1}{3}$ because $\dfrac{3}{6} > \dfrac{2}{6}$.

(e) $-0.033 < -0.0329$ because $-0.0330 < -0.0329$.

Notice that since both numbers are negative, -0.0330 is to the left of -0.0329 on a number line.

NOTE TO STUDENT: *Fully worked-out solutions to all of the Practice Problems can be found at the back of the text starting at page SP-1*

Practice Problem 1   Insert the proper symbol between the two numbers.

(a) -1 _____ -2 **(b)** $\dfrac{2}{3}$ _____ $\dfrac{3}{4}$ **(c)** -0.561 _____ -0.5555

Numerical or algebraic expressions as well as numbers can be compared.

EXAMPLE 2   Insert the proper symbol between the expressions.

(a) $(5 - 8)$ _____ $(2 - 3)$ **(b)** $|1 - 7|$ _____ $|-4 - 12|$

Solution

(a) $(5 - 8)$ _____ $(2 - 3)$ Evaluate each expression and compare.
$$-3 < -1$$
Thus, $(5 - 8) < (2 - 3)$ because $-3 < -1$.

(b) $|1 - 7|$ _____ $|-4 - 12|$ Evaluate each expression and compare.
$$6 < 16$$
Thus, $|1 - 7| < |-4 - 12|$ because $6 < 16$.

Practice Problem 2   Insert $<$ or $>$ between the expressions.

(a) $(-8 - 2)$ _____ $(-3 - 12)$ **(b)** $|-15 + 8|$ _____ $|7 - 13|$

In addition to the $<$ and $>$ symbols, we will also encounter two other notations when we deal with inequalities. We briefly mentioned these in Chapter 1. If we want to say that a number x is greater than or equal to 5, we would write this using the following notation:

$$x \geq 5.$$

Likewise, if we want to say that a number x is less than or equal to 8, we would write it using the following notation:

$$x \leq 8.$$

The symbol $\leq$ means **less than or equal to,** and the symbol $\geq$ means **greater than or equal to.** We say that the inequality symbols $\leq$ and $\geq$ "contain the equals sign."

2 Graphing Linear Inequalities in One Variable

Inequality symbols are often used with variables. For example,

$$x > 4$$

means that x can be any number that is greater than 4. The variable x cannot equal 4. We can use a number line to graph this inequality.

On the number line, we shade the portion that is to the right of 4. Any point in the shaded portion will satisfy the inequality since all points to the right of 4 are greater than 4. The open circle at 4 means that x cannot be 4.

Let's look at the inequality $x < -1$. This means that x can be any number that is less than -1. To graph the inequality, we will shade all points to the left of -1 on a number line.

$$x < -1$$

Inequality symbols used with variables can include the equals sign. The inequality $x \geq -2$ means all numbers greater than or equal to -2. We graph this inequality as follows.

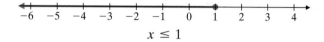

$$x \geq -2$$

The shaded circle at -2 means that the graph includes the point -2. -2 is a solution to the inequality.

Similarly, $x \leq 1$ is graphed as follows.

$$x \leq 1$$

EXAMPLE 3 Graph each inequality.

(a) $x < 0$ **(b)** $x \leq 0$ **(c)** $x > -5$ **(d)** $x \geq -5$ **(e)** $2 < x$

Solution

(a) $x < 0$

(b) $x \leq 0$

(c) $x > -5$

(d) $x \geq -5$

(e) We read an inequality starting with the variable. Thus, we read $2 < x$ as "x is greater than 2" and graph the expression accordingly.

Practice Problem 3 Graph each inequality.

(a) $x > 3.5$

(b) $x \leq -10$

(c) $x \geq -3$

(d) $-4 > x$

③ Solving Linear Inequalities in One Variable

Inequalities that have the same solution are said to be **equivalent.** Solving a first-degree inequality is similar to solving a first-degree equation. We use various properties of real numbers.

ADDITION AND SUBTRACTION PROPERTY FOR INEQUALITIES

For all real numbers a, b, and c, if $a < b$, then

$$a + c < b + c \quad \text{and} \quad a - c < b - c.$$

If the same number is added to or subtracted from both sides of an inequality, the result is an equivalent inequality.

The same number can be added to or subtracted from both sides of an inequality without affecting the direction of the inequality. (Any inequality symbol can be used. We used $<$ for convenience.)

EXAMPLE 4 Solve the inequality. Graph and check the solution. $x - 8 < 15$

Solution

$$x - 8 < 15$$

$x - 8 + 8 < 15 + 8$ Add +8 to each side.

$x < 23$ Simplify.

To check, choose any numerical value that lies in the region indicated by the red arrow on the number line. See whether a true statement results when you substitute it into the inequality. We will choose 22.5.

$$x - 8 < 15 \quad \text{Substitute 22.5 for } x \text{ in the original inequality.}$$

$$22.5 - 8 \overset{?}{<} 15$$

$$14.5 < 15 \ \checkmark$$

Practice Problem 4 Solve the inequality. Graph and check the solution.
$x + 2 > -12$

MULTIPLICATION OR DIVISION BY A POSITIVE NUMBER

For all real numbers a, b, and c when $c > 0$, if $a < b$, then

$$ac < bc \quad \text{and} \quad \frac{a}{c} < \frac{b}{c}.$$

If both sides of an inequality are multiplied or divided by the same *positive* number, the result is an equivalent inequality.

When we multiply or divide both sides of an inequality by a positive number, the direction of the inequality is not changed. That is, if $5x < -15$ and we divide both sides by 5, we obtain $x < -3$.

To check this solution, we choose the value -4. Is $5(-4) < -15$ true? Yes.

EXAMPLE 5 Solve $6x + 3 \le 2x - 5$. Graph your solution.

Solution

$6x + 3 - 3 \le 2x - 5 - 3$ Subtract 3 from each side.

$6x \le 2x - 8$ Simplify.

$6x - 2x \le 2x - 2x - 8$ Subtract $2x$ from each side.

$4x \le -8$ Simplify.

$\dfrac{4x}{4} \le \dfrac{-8}{4}$ Divide each side by 4. The inequality is not changed.

$\phantom{\dfrac{4x}{4}}x \le -2$

To check, we choose -2 and -3.

$$6x + 3 \leq 2x - 5$$
$$6(-2) + 3 \overset{?}{\leq} 2(-2) - 5$$
$$-9 \leq -9 \quad \checkmark$$

$$6x + 3 \leq 2x - 5$$
$$6(-3) + 3 \overset{?}{\leq} 2(-3) - 5$$
$$-15 \leq -11 \quad \checkmark$$

NOTE TO STUDENT: *Fully worked-out solutions to all of the Practice Problems can be found at the back of the text starting at page SP-1*

Practice Problem 5 Solve $8x - 8 \geq 5x + 1$. Graph your solution.

$$\xleftarrow{\quad}\underset{-1 \ \ 0 \ \ 1 \ \ 2 \ \ 3 \ \ 4 \ \ 5 \ \ 6 \ \ 7}{+\!+\!+\!+\!+\!+\!+\!+\!+}\xrightarrow{\quad}$$

When we multiply or divide both sides of an inequality by a *negative number*, the direction of the inequality is *reversed*. That is, if we divide both sides of $-3x < 21$ by -3, we obtain $x > -7$. If we divide both sides of $-4x \geq -16$ by -4, we obtain $x \leq 4$.

To check the solution of the first inequality, we choose the value 1. Is $-3(1) < 21$? Yes. To reverse the inequality symbol might seem to be an unusual move. Let's see what would happen if we did not reverse the symbol. If we did not, the solution would be $x < -7$. To check this solution, we choose -8. Is $-3(-8) < 21$ true? No, 24 is not less than 21.

MULTIPLICATION OR DIVISION BY A NEGATIVE NUMBER

For all real numbers a, b, and c when $c < 0$, if $a < b$, then

$$ac > bc \quad \text{and} \quad \frac{a}{c} > \frac{b}{c}.$$

If both sides of an inequality are multiplied or divided by the same *negative number and the inequality symbol is reversed*, the result is an equivalent inequality.

EXAMPLE 6 Solve. $-8x - 12 < -4(x - 4) + 8$

Solution

$-8x - 12 < -4x + 16 + 8$	Use the distributive property to remove the parentheses.
$-8x - 12 < -4x + 24$	Simplify.
$-8x + 4x - 12 < -4x + 4x + 24$	Add $4x$ to each side.
$-4x - 12 < 24$	Simplify.
$-4x - 12 + 12 < 24 + 12$	Add 12 to each side.
$-4x < 36$	Simplify.
$\dfrac{-4x}{-4} > \dfrac{36}{-4}$	Divide each side by -4 and reverse the inequality.
$x > -9$	Simplify.

Practice Problem 6 Solve. $2 - 12x > 7(1 - x)$

An inequality that contains all decimal terms is best handled by first multiplying both sides of the inequality by the appropriate power of 10.

EXAMPLE 7 Solve. $-0.3x + 1.0 \le 1.2x - 3.5$

Solution

$10(-0.3x + 1.0) \le 10(1.2x - 3.5)$	Multiply each side by 10.
$-3x + 10 \le 12x - 35$	
$-3x - 12x + 10 \le 12x - 12x - 35$	Subtract $12x$ from each side.
$-15x + 10 \le -35$	Simplify.
$-15x + 10 - 10 \le -35 - 10$	Subtract 10 from each side.
$-15x \le -45$	Simplify.
$\dfrac{-15x}{-15} \ge \dfrac{-45}{-15}$	Divide each side by -15 and reverse the direction of the inequality.
$x \ge 3$	

Practice Problem 7 Solve. $-0.8x + 0.9 \ge 0.5x - 0.4$

To solve an inequality that contains fractions, multiply both sides of the inequality by the LCD to clear the fractions.

EXAMPLE 8 Solve. $\dfrac{1}{7}(x + 5) > \dfrac{1}{5}(x + 1)$

Solution

$\dfrac{x}{7} + \dfrac{5}{7} > \dfrac{x}{5} + \dfrac{1}{5}$	Using the distributive property, remove the parentheses.
$35\left(\dfrac{x}{7}\right) + 35\left(\dfrac{5}{7}\right) > 35\left(\dfrac{x}{5}\right) + 35\left(\dfrac{1}{5}\right)$	Multiply each term by the LCD.
$5x + 25 > 7x + 7$	Simplify.
$5x + 25 - 25 > 7x + 7 - 25$	Subtract 25 from each side.
$5x > 7x - 18$	Simplify.
$5x - 7x > 7x - 7x - 18$	Subtract $7x$ from each side.
$-2x > -18$	Simplify.
$\dfrac{-2x}{-2} < \dfrac{-18}{-2}$	Divide each side by -2 and reverse the direction of the inequality.
$x < 9$	

Practice Problem 8 Solve.

$$\frac{1}{5}(x - 6) < \frac{1}{3}(x - 2)$$

TO THINK ABOUT: Multiplying by the LCD Another approach to solving the inequality in Example 8 would be to multiply each side of the inequality by the LCD before removing the parentheses. Try it. Think about the pros and cons of this approach. Choose the method you like best.

EXAMPLE 9 Lexi and her mother are using a public phone to make a long distance phone call from Honolulu, HI, to West Chicago, IL. The charge is $4.50 for the first minute and 85¢ for each additional minute. Any fractional part of a minute will be rounded up to the nearest whole minute. What is the maximum time that Lexi and her mother can talk if they have $15.55 in change to make the call?

Let x = the number of minutes they talk after the first minute. The cost must be less than or equal to $15.55. So we write the inequality

$$4.50 + 0.85x \leq 15.55$$
$$0.85x \leq 11.05 \quad \text{We subtract 4.50 from each side}$$
$$x \leq 13 \quad \text{We divide each side by 0.85}$$

Now we add the 13 minutes to the one minute that cost $4.50. This gives us 14 minutes. Thus the maximum amount of time they can talk is 14 minutes.

Practice Problem 9 Olivia and her mother are making a long distance phone call from Anchorage, AK, to Darien, CT. The charge is $3.50 for the first minute and 65¢ for each additional minute. Any fractional part of a minute will be rounded up to the nearest whole minute. What is the maximum time that Olivia and her mother can talk if they have $13.90 in change to make the call?

Solving inequalities is a very important skill. Take some extra time to review Examples 1–9. Carefully work out the solutions to Practice Problems 1–9. Turn to pages SP-5 and SP-6 and check to be sure you have done the Practice Problems correctly. Some extra time spent carefully studying the Examples and Practice Problems will make the homework exercises much easier to complete.

Verbal and Writing Skills

True or false?

1. The statement $6 < 8$ conveys the same information as $8 > 6$.

2. Adding $-5x$ to each side of an inequality reverses the direction of the inequality.

3. Dividing each side of an inequality by -4 reverses the direction of the inequality.

4. The graph of $x > -2$ is the set of all points to the right of -2 on the number line.

5. The graph of $x \leq 6$ does not include the point at 6 on the number line.

6. To solve the inequality $\frac{2}{3}x + \frac{3}{4} \geq \frac{1}{2}x - 4$, multiply each fraction by the LCD.

Insert the symbol $<$ or $>$ between each pair of numbers.

7. $6 \underline{\hspace{1cm}} -3$

8. $-15 \underline{\hspace{1cm}} 4$

9. $-7 \underline{\hspace{1cm}} -2$

10. $-5 \underline{\hspace{1cm}} -9$

11. $\frac{3}{4} \underline{\hspace{1cm}} \frac{2}{3}$

12. $\frac{5}{6} \underline{\hspace{1cm}} \frac{5}{7}$

13. $-\frac{2}{9} \underline{\hspace{1cm}} -\frac{3}{14}$

14. $-\frac{7}{16} \underline{\hspace{1cm}} -\frac{6}{13}$

15. $-3.4 \underline{\hspace{1cm}} -3.41$

16. $-2.69 \underline{\hspace{1cm}} -2.7$

17. $|3 - 7| \underline{\hspace{1cm}} |9 - 2|$

18. $|-8 + 2| \underline{\hspace{1cm}} |6 - 13|$

Graph each inequality.

19. $x \geq -2$

20. $x \geq -4$

21. $x < 15$

22. $x < 80$

Solve for x and graph your solution.

23. $2x - 7 \leq -5$

24. $3 + 5x \geq 18$

25. $3x - 7 > 9x + 5$

26. $2x + 5 > 4x - 5$

27. $0.5x + 0.1 < 1.1x + 0.7$

28. $1.7 - 0.6x \leq x + 0.1$

Solve for x.

29. $4x - 1 > 15$

30. $5x - 1 > 29$

31. $5x + 3 \leq 2x - 9$

32. $9x - 1 \leq 4x - 11$

33. $2x + \frac{5}{3} > \frac{2}{5}x - 1$

34. $2x + \frac{5}{2} > \frac{3}{2}x - 2$

35. $3x - 11 + 4(x + 8) < 0$

36. $2x - 11 + 3(x + 2) < 0$

37. $\frac{3}{5}x - (x + 2) \geq -2$

38. $-3(x + 1) - \frac{x}{2} + \frac{3}{2} < 0$

Mixed Practice *Solve for x.*

39. $0.4x + 1 \leq 2.6$

40. $-0.4x + 1.5 \geq 2.1 - x$

41. $0.1(x - 2) \geq 0.5x - 0.2$

42. $1.2 - 0.8x \leq 0.3(4 - x)$

43. $2 - \frac{1}{5}(x - 1) \geq \frac{2}{3}(2x + 1)$

44. $\frac{3}{4} + \frac{1}{2}(x - 7) \leq 1 - \frac{x}{4}$

45. $\frac{2x - 3}{5} + 1 \geq \frac{1}{2}x + 3$

46. $4 - \frac{3x - 1}{3} > \frac{x}{6} + \frac{7}{2}$

Applications *For exercises 47–52, describe the situation with a linear inequality and then solve the inequality.*

47. *Tip Income* Maria is a waitress and earns $4.50 per hour plus an average tip of $6 for every table served. How many tables must she serve to earn more than $72 for a 4-hour shift?

48. *Telemarketing* A phone solicitor selling long-distance services earns $7.75 per hour plus $25 for every new customer she signs up. How many customers must she sign up to earn more than $401.50 during the next 26 working hours?

49. *Telephone Rates* Rusty Slater is making a long-distance phone call to Orlando, Florida, from a pay phone. The operator informs him that the charge will be $3.95 for the first minute and 55¢ for each additional minute. Any fractional part of a minute used will be rounded up to the nearest whole minute. What is the maximum time Rusty can talk if he has $13.30 in change in his pocket?

50. *Aircraft Cargo Capacity* A small plane takes off with packages from Beverly Airport. Each package weighs 68.5 pounds. The plane has a carrying capacity for people and packages of 2395 pounds. The plane is carrying a pilot who weighs 180 pounds and a passenger who weighs 160 pounds. How many packages can be safely carried?

51. *Elevator Capacity* Molly and Denton from the computer services department are delivering several new computers to faculty offices using the college elevator. The elevator has a maximum capacity of 1100 pounds. Molly weighs 130 pounds, and Denton weighs 155 pounds. Each computer weighs 59 pounds. How many computers can Molly and Denton place on the elevator and then safely ride with the computers up to the next floor?

52. *Mailing Costs* DeWolf Associates sent out several boxes of literature by Priority Mail. They planned for a mailing budget of $8.46 per box. The post office charged $0.41 for the first ounce and $0.23 for each additional ounce. What was the most a box could weigh and still be mailed at a cost that did not exceed the budget?

Cumulative Review *Simplify.*

53. **[1.5.3]** $3xy(x + 2) - 4x^2(y - 1)$

54. **[1.5.2]** $\frac{2}{3}ab(6a - 2b + 9)$

55. **[1.4.3]** $\left(\dfrac{4x^2}{3yw^{-1}}\right)^3$

56. **[1.4.1]** $(-3a^0b^{-3}c^5)^{-2}$

Quick Quiz 2.6 Solve each inequality.

1. $5x + 7 > 3x - 9$

2. $-3(x + 4) < 6x - 8$

3. $\frac{1}{5}(x - 3) \geq \frac{1}{4}(x - 5) + 1$

4. **Concept Check** If you were to solve the inequalities $-3x < 9$ and $3x < 9$, in one case you would have to reverse the direction of the inequality and in the other case you would not. Explain how you can tell which case is which.

1 Graphing Compound Inequalities That Use *and*

Some inequalities consist of two inequalities connected by the word *and* or the word *or*. They are called **compound inequalities.** The solution of a compound inequality using the connective *and* includes all the numbers that make both parts true at the same time.

EXAMPLE 1 Graph the values of x where $7 < x$ and $x < 12$.

Solution We read the inequality starting with the variable. Thus, we graph all values of x, where x is greater than 7 and where x is less than 12. All such values must be between 7 and 12. Numbers that are greater than 7 and less than 12 can be written as $7 < x < 12$.

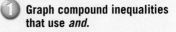

Practice Problem 1 Graph the values of x where $-8 < x$ and $x < -2$.

EXAMPLE 2 Graph the values of x where $-6 \leq x \leq 2$.

Solution Here we have that x is greater than or equal to -6 and that x is less than or equal to 2. We remember to include the points -6 and 2 since the inequality symbols contain the equals sign.

Practice Problem 2 Graph the values of x where $-1 \leq x \leq 5$.

EXAMPLE 3 Graph the values of x where $-8.5 \leq x < -1$.

Solution Note the shaded circle at -8.5 and the open circle at -1.

Practice Problem 3 Graph the values of x where $-10 < x \leq -5.5$.

EXAMPLE 4 Graph the salary range (*s*) of the full-time employees of Tentron Corporation. Each person earns at least $190 weekly, but not more than $800 weekly.

Solution "At least $190" means that the weekly salary of each person is greater than or equal to $190 weekly. We write $s \geq \$190$. "Not more than" means that the weekly salary of each person is less than or equal to $800. We write $s \leq \$800$. Thus, *s* may be between 190 and 800 and may include those endpoints.

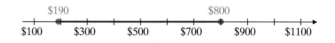

Practice Problem 4 Graph the weekly salary range of a person who earns at least $200 per week, but never more than $950 per week.

NOTE TO STUDENT: Fully worked-out solutions to all of the Practice Problems can be found at the back of the text starting at page SP-1

2 Graphing Compound Inequalities That Use *or*

The solution of a compound inequality using the connective *or* includes all the numbers that belong to either of the two inequalities.

EXAMPLE 5 Graph the region where *x* < 3 *or* *x* > 6.

Solution Notice that a solution to this inequality need not be in both regions at the same time.

Read the inequality as "*x* is less than 3 or *x* is greater than 6." Thus, *x* can be less than 3 or *x* can be greater than 6. This includes all values to the left of 3 as well as all values to the right of 6 on a number line. We shade these regions.

Practice Problem 5 Graph the region where *x* < 8 *or* *x* > 12.

EXAMPLE 6 Graph the region where *x* > −2 *or* *x* ≤ −5.

Solution Note the shaded circle at −5 and the open circle at −2.

Practice Problem 6 Graph the region where *x* ≤ −6 *or* *x* > 3.

EXAMPLE 7 Male applicants for the state police force in Fred's home state are ineligible for the force if they are shorter than 60 inches or taller than 76 inches. Graph the range of rejected applicants' heights.

Solution Each rejected applicant's height h will be less than 60 inches ($h < 60$) or will be greater than 76 inches ($h > 76$).

```
                              76
  ◄─┼─┼──────┼──────┼────────┼○─┼─►
   50 inches  60 inches  70 inches  80 inches
```

Practice Problem 7 Female applicants are ineligible if they are shorter than 56 inches or taller than 70 inches. Graph the range of rejected applicant's heights.

```
  ──┼──┼──┼──┼──┼──┼──┼──┼──┼──┼─►
     50 in.   60 in.   70 in.   80 in.   90 in.
```

Solving Compound Inequalities

When asked to solve a more complicated compound inequality for x, we normally solve each individual inequality separately.

EXAMPLE 8 Solve for x and graph the solution of $3x + 2 > 14$ *or* $2x - 1 < -7$.

Solution We solve each inequality separately.

$$
\begin{array}{ccc}
3x + 2 > 14 & or & 2x - 1 < -7 \\
3x > 12 & & 2x < -6 \\
x > 4 & & x < -3
\end{array}
$$

The solution is $x < -3$ *or* $x > 4$.

```
  ◄─┼──┼──○──┼──┼──┼──┼──┼──┼──○──┼─►
   −5 −4 −3 −2 −1  0  1  2  3  4  5
```

Practice Problem 8 Solve for x and graph the solution of $3x - 4 < -1$ *or* $2x + 3 > 13$.

```
  ──┼──┼──┼──┼──┼──┼──┼──┼──┼──┼─►
   −2 −1  0  1  2  3  4  5  6  7  8
```

EXAMPLE 9 Solve for x and graph the solution of $5x - 1 > -2$ *and* $3x - 4 < 8$.

Solution We solve each inequality separately.

$$
\begin{array}{ccc}
5x - 1 > -2 & and & 3x - 4 < 8 \\
5x > -1 & & 3x < 12 \\
x > -\dfrac{1}{5} & & x < 4
\end{array}
$$

The solution is the set of numbers between $-\frac{1}{5}$ and 4, not including the endpoints.

$$-\frac{1}{5} < x < 4$$

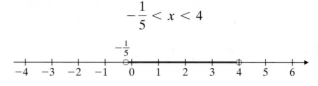

NOTE TO STUDENT: Fully worked-out solutions to all of the Practice Problems can be found at the back of the text starting at page SP-1

Practice Problem 9 Solve for x and graph the solution of $3x + 6 > -6$ and $4x + 5 < 1$.

EXAMPLE 10 Solve and graph. $2x + 5 \le 11$ and $-3x > 18$

Solution   We solve each inequality separately.

$$2x + 5 \le 11 \qquad and \qquad -3x > 18$$

$$2x \le 6 \qquad\qquad\qquad x < -\frac{18}{3}$$

$$x \le 3 \qquad\qquad\qquad x < -6$$

The solutions are $x < -6$ and $x \le 3$.

The only numbers that satisfy the statements $x \le 3$ and $x < -6$ at the same time are $x < -6$. Thus, $x < -6$ is the solution to the compound inequality.

Practice Problem 10 Solve and graph. $-2x + 3 < -7$ and $7x - 1 > -15$

EXAMPLE 11 Solve. $-3x - 2 < -5$ and $4x + 6 < -12$

Solution   We solve each inequality separately.

$$-3x - 2 < -5 \qquad and \qquad 4x + 6 < -12$$

$$-3x < -3 \qquad\qquad\qquad 4x < -18$$

$$\frac{-3x}{-3} > \frac{-3}{-3} \qquad\qquad \frac{4x}{4} < \frac{-18}{4}$$

$$x > 1 \qquad\qquad\qquad x < -4\frac{1}{2}$$

Now, clearly it is impossible for one number to be greater than 1 *and* at the same time be less than $-4\frac{1}{2}$.

Thus, there is *no solution*. We can express this by the notation $\varnothing$, which is the **empty set.** Or we can just state, "There is no solution."

Practice Problem 11 Solve. $-3x - 11 < -26$ and $5x + 4 < 14$

Graph the values of x that satisfy the conditions given.

1. $3 < x$ *and* $x < 8$

2. $5 < x$ *and* $x < 10$

3. $-4 < x$ *and* $x <$

5. $7 < x < 9$

6. $3 < x < 5$

7. $-2 < x \le \dfrac{1}{2}$

9. $x > 8$ *or* $x < 2$

10. $x \ge 2$ *or* $x \le 1$

11. $x \le -\dfrac{5}{2}$ *or* $x > 4$

12. $x <$

13. $x \le -10$ *or* $x \ge 40$

14. $x \le -6$ *or* $x \ge 2$

Solve for x and graph your results.

15. $2x + 3 \le 5$ *and* $x + 1 \ge -2$

16. $4x - 1 < 7$ *and* $x \ge -1$

17. $2x - 3 > 0$ *or* $x - 2 < -$

18. $x + 1 \ge 5$ *or* $x + 5 < 2.5$

19. $x < 8$ *and* $x > 10$

20. $x < 6$ *and* $x > 9$

Applications *Express as an inequality.*

21. *Toothpaste* A tube of toothpaste is not properly filled if the amount of toothpaste t in the tube is more than 11.2 ounces or fewer than 10.9 ounces.

22. *Clothing Standards* The width of a seam on a pair of blue jeans is unacceptable if it is narrower than 10 millimeters or wider than 12 millimeters.

23. *Interstate Highway Travel* The number of cars c driving over Interstate 91 during the evening hours in January was always at least 5000, but never more than 12,000.

24. *Campsite Capacity* The number of campers c at a campsite during the Independence Day weekend was always at least 490, but never more than 2000.

Temperature Conversion *Solve the following application problems by using the formula* $C = \dfrac{5}{9}(F - 32)$. *Round to the nearest tenth.*

25. When visiting Montreal this spring, Marcos had been advised that the temperature could range from $-20°C$ to $11°C$. Find an inequality that represents the range in Fahrenheit temperatures.

26. The temperature in Sao Paulo, Brazil, during January can range from $16°C$ to $24°C$. Find an inequality that represents the range in Fahrenheit temperatures.

ates *At one point in 2008, the exchange rate for converting American dollars into Japanese yen was* *— 5). In this equation, d is the number of American dollars, Y is the number of yen, and $5 represents a one-time* *ks sometimes charge for currency conversion. Use the equation to solve the following problems. (Round answers* *est cent.)*

x is traveling to Tokyo, Japan, for 2 weeks and been advised to have between 20,000 yen and 00 yen for spending money for each week he is re. Including the conversion charge, write an equality that represents the number of American ollars he will need to exchange at the bank for this -week period.

28. Carrie is traveling to Osaka, Japan, for 3 weeks. Her friend told her she should plan to have between 23,000 yen and 28,000 yen for spending money for each week she is there. Including the conversion charge, write an inequality that represents the number of American dollars she will need to exchange at the bank for the 3-week period.

ed Practice *Solve each compound inequality.*

$x - 3 > -5$ *and* $2x + 4 < 8$

30. $x + 3 < 7$ *and* $x - 2 < -3$

1. $-6x + 5 \geq -1$ *and* $2 - x \leq 5$

32. $8 - x \geq 6$ *and* $10x + 9 \geq -11$

33. $4x - 3 < -11$ *or* $7x + 2 \geq 23$

34. $5x + 1 < 1$ *or* $3x - 9 > 9$

35. $-0.3x + 1 \geq 0.2x$ *or* $-0.2x + 0.5 > 0.7$

36. $-0.3x - 0.4 \geq 0.1x$ *or* $0.2x + 0.3 \leq -0.4x$

37. $\dfrac{5x}{2} + 1 \geq 3$ *and* $x - \dfrac{2}{3} \geq \dfrac{4}{3}$

38. $\dfrac{5x}{3} - 2 < \dfrac{14}{3}$ *and* $3x + \dfrac{5}{2} < -\dfrac{1}{2}$

39. $2x + 5 < 3$ *and* $3x - 1 > -1$

40. $6x - 10 < 8$ *and* $2x + 1 > 9$

41. $2x - 3 \geq 7$ *and* $5x - 8 \leq 2x + 7$

42. $6x - 3 \geq 8x + 5$ *and* $x + 6 \geq 2$

To Think About *Solve the compound inequality.*

43. $\dfrac{1}{4}(x + 2) + \dfrac{1}{8}(x - 3) \leq 1$ *and* $\dfrac{3}{4}(x - 1) > -\dfrac{1}{4}$

44. $\dfrac{x - 4}{6} - \dfrac{x - 2}{9} \leq \dfrac{5}{18}$ *or* $-\dfrac{2}{5}(x + 3) < -\dfrac{6}{5}$

Cumulative Review *Solve for the specified variable.*

45. **[2.2.1]** Solve for x. $3y - 5x = 8$

46. **[2.2.1]** Solve for y. $7x + 6y = -12$

47. **[1.6.1]** Evaluate $x^2 + 5x - |x + 3|$ for $x = -2$.

48. **[1.6.1]** Evaluate $3x^3 - x^2 - \sqrt{8x + 9}$ for $x = 2$.

Quick Quiz 2.7 Find the values of x that satisfy the given conditions.

1. $3x + 2 < 8$ *and* $3x > -16$

2. $x > 5$ *and* $2x - 1 < 23$

3. $x - 7 \leq -15$ *or* $2x + 3 \geq 5$

4. **Concept Check** Explain why there are no values of x that satisfy these given conditions.

$$x + 8 < 3 \quad and \quad 2x - 1 > 5$$

2.8 ABSOLUTE VALUE INEQUALITIES

① Solving Absolute Value Inequalities of the Form $|ax + b| < c$

We begin by looking at $|x| < 3$. What does this mean? The inequality $|x| < 3$ means that x is less than 3 units from 0 on a number line. We draw a picture.

Student Learning Objectives

After studying this section, you will be able to:

① Solve absolute value inequalities of the form $|ax + b| < c$.

② Solve absolute value inequalities of the form $|ax + b| > c$.

This picture shows all possible values of x such that $|x| < 3$. We see that this occurs when $-3 < x < 3$. We conclude that $|x| < 3$ and $-3 < x < 3$ are equivalent statements.

> **DEFINITION OF $|x| < a$**
>
> If a is a positive real number and $|x| < a$, then $-a < x < a$.

EXAMPLE 1 Solve. $|x| \leq 4.5$

Solution The inequality $|x| \leq 4.5$ means that x is less than or equal to 4.5 units from 0 on a number line. We draw a picture.

Thus, the solution is $-4.5 \leq x \leq 4.5$.

Practice Problem 1 Solve and graph. $|x| < 2$

This same technique can be used to solve more complicated inequalities.

EXAMPLE 2 Solve and graph the solution of $|x + 5| \leq 10$.

Solution We want to find the values of x that make $-10 \leq x + 5 \leq 10$ a true statement. We need to solve the compound inequality.

To solve this inequality, we subtract 5 from each part.

$$-10 - 5 \leq x + 5 - 5 \leq 10 - 5$$
$$-15 \leq x \leq 5$$

Thus, the solution is $-15 \leq x \leq 5$. We graph this solution.

Practice Problem 2 Solve and graph the solution of $|x - 6| < 15$.
(*Hint:* Choose a convenient scale.)

EXAMPLE 3 Solve and graph the solution of $\left| x - \dfrac{2}{3} \right| \le \dfrac{5}{2}$.

Solution

$$-\frac{5}{2} \le x - \frac{2}{3} \le \frac{5}{2} \qquad \text{If } |x| < a, \text{ then } -a < x < a.$$

$$6\left(-\frac{5}{2}\right) \le 6(x) - 6\left(\frac{2}{3}\right) \le 6\left(\frac{5}{2}\right) \qquad \text{Multiply each part of the inequality by 6.}$$

$$-15 \le 6x - 4 \le 15 \qquad \text{Simplify.}$$

$$-15 + 4 \le 6x - 4 + 4 \le 15 + 4 \qquad \text{Add 4 to each part.}$$

$$-11 \le 6x \le 19 \qquad \text{Simplify.}$$

$$-\frac{11}{6} \le \frac{6x}{6} \le \frac{19}{6} \qquad \text{Divide each part by 6.}$$

$$-1\frac{5}{6} \le x \le 3\frac{1}{6} \qquad \text{Change to mixed numbers to facilitate graphing.}$$

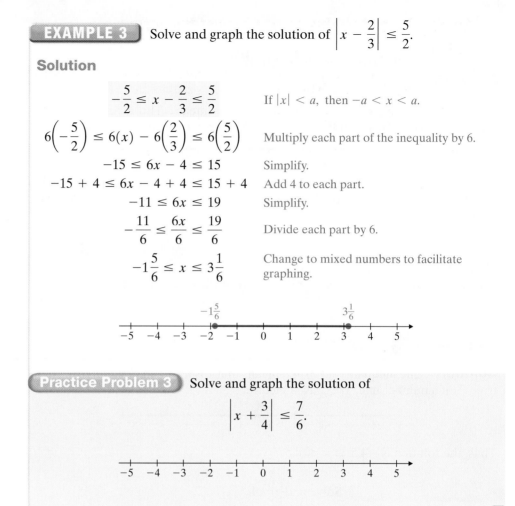

NOTE TO STUDENT: Fully worked-out solutions to all of the Practice Problems can be found at the back of the text starting at page SP-1

Practice Problem 3 Solve and graph the solution of
$$\left| x + \frac{3}{4} \right| \le \frac{7}{6}.$$

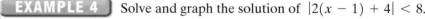

EXAMPLE 4 Solve and graph the solution of $|2(x - 1) + 4| < 8$.

Solution First we simplify the expression within the absolute value symbol.

$$|2x - 2 + 4| < 8$$

$$|2x + 2| < 8$$

$$-8 < 2x + 2 < 8 \qquad \text{If } |x| < a, \text{ then } -a < x < a.$$

$$-8 - 2 < 2x + 2 - 2 < 8 - 2 \qquad \text{Subtract 2 from each part.}$$

$$-10 < 2x < 6 \qquad \text{Simplify.}$$

$$\frac{-10}{2} < \frac{2x}{2} < \frac{6}{2} \qquad \text{Divide each part by 2.}$$

$$-5 < x < 3$$

Practice Problem 4 Solve and graph the solution of $|2 + 3(x - 1)| < 20$.

② Solving Absolute Value Inequalities of the Form $|ax + b| > c$

Now consider $|x| > 3$. What does this mean? This inequality $|x| > 3$ means that x is greater than 3 units from 0 on a number line. We draw a picture.

This picture shows all possible values of x such that $|x| > 3$. This occurs when $x < -3$ or when $x > 3$. (Note that a solution can be either in the region to the left of -3 on the number line or in the region to the right of 3 on the number line.) We conclude that the expression $|x| > 3$ and the expression $x < -3$ or $x > 3$ are equivalent statements.

> ### DEFINITION OF $|x| > a$
> If a is a positive real number and $|x| > a$, then $x < -a$ or $x > a$.

EXAMPLE 5   Solve and graph the solution of $|x| \geq 5\frac{1}{4}$.

Solution   The inequality $|x| \geq 5\frac{1}{4}$ means that x is more than $5\frac{1}{4}$ units from 0 on a number line. We draw a picture.

Thus, the solution is $x \leq -5\frac{1}{4}$ or $x \geq 5\frac{1}{4}$.

Practice Problem 5   Solve and graph the solution of $|x| > 2.5$.

This same technique can be used to solve more complicated inequalities.

EXAMPLE 6   Solve and graph the solution of $|x - 4| > 5$.

Solution   We want to find the values of x that make $x - 4 < -5$ or $x - 4 > 5$ a true statement. We need to solve the compound inequality.
   We will solve each inequality separately.

$$
\begin{array}{ccc}
x - 4 < -5 & or & x - 4 > 5 \\
x - 4 + 4 < -5 + 4 & & x - 4 + 4 > 5 + 4 \\
x < -1 & & x > 9
\end{array}
$$

Thus, the solution is $x < -1$ or $x > 9$. We graph the solution on a number line.

Practice Problem 6   Solve and graph the solution of $|x + 6| > 2$.

EXAMPLE 7   Solve and graph the solution of $|-3x + 6| > 18$.

Solution   By definition, we have the following compound inequality.

$$-3x + 6 > 18 \qquad\qquad or \qquad\qquad -3x + 6 < -18$$
$$-3x > 12 \qquad\qquad\qquad\qquad\qquad -3x < -24$$

$$\frac{-3x}{-3} < \frac{12}{-3} \longleftarrow \begin{array}{c}\text{Division by a negative}\\ \text{number reverses the}\\ \text{inequality sign.}\end{array} \longrightarrow \frac{-3x}{-3} > \frac{-24}{-3}$$

$$x < -4 \qquad\qquad\qquad\qquad\qquad x > 8$$

The solution is $x < -4$ *or* $x > 8$.

NOTE TO STUDENT: *Fully worked-out solutions to all of the Practice Problems can be found at the back of the text starting at page SP-1*

Practice Problem 7   Solve and graph the solution of $|-5x - 2| > 13$.

EXAMPLE 8   Solve and graph the solution of $\left|3 - \dfrac{2}{3}x\right| \geq 5$.

Solution   By definition, we have the following compound inequality.

$$3 - \frac{2}{3}x \geq 5 \qquad or \qquad 3 - \frac{2}{3}x \leq -5$$

$$3(3) - 3\left(\frac{2}{3}x\right) \geq 3(5) \qquad 3(3) - 3\left(\frac{2}{3}x\right) \leq 3(-5)$$

$$9 - 2x \geq 15 \qquad\qquad 9 - 2x \leq -15$$

$$-2x \geq 6 \qquad\qquad\qquad -2x \leq -24$$

$$\frac{-2x}{-2} \leq \frac{6}{-2} \qquad\qquad\qquad \frac{-2x}{-2} \geq \frac{-24}{-2}$$

$$x \leq -3 \qquad\qquad\qquad\qquad x \geq 12$$

The solution is $x \leq -3$ *or* $x \geq 12$.

Practice Problem 8   Solve and graph the solution of

$$\left|4 - \frac{3}{4}x\right| \geq 5.$$

EXAMPLE 9 When a new car transmission is built, the diameter d of the transmission must not differ from the specified standard s by more than 0.37 millimeter. The engineers express this requirement as $|d - s| \le 0.37$. If the standard s is 216.82 millimeters for a particular car, find the limits of d.

Solution

$$|d - s| \le 0.37$$
$$|d - 216.82| \le 0.37 \qquad \text{Substitute the known value of } s.$$
$$-0.37 \le d - 216.82 \le 0.37 \qquad \text{If } |x| \le a, \text{ then } -a \le x \le a.$$
$$-0.37 + 216.82 \le d - 216.82 + 216.82 \le 0.37 + 216.82$$
$$216.45 \le d \le 217.19$$

Thus, the diameter of the transmission must be at least 216.45 millimeters, but not greater than 217.19 millimeters.

Practice Problem 9 The diameter d of a transmission must not differ from the specified standard s by more than 0.37 millimeter. Solve to find the allowed limits of d for a truck transmission for which the standard s is 276.53 millimeters.

SUMMARY OF ABSOLUTE VALUE EQUATIONS AND INEQUALITIES

It may be helpful to review the key concepts of absolute value equations and inequalities that we have covered in Sections 2.3 and 2.8. For real numbers a, b, and c, where $a \ne 0$ and $c > 0$, we have the following:

Absolute value form of the equation or inequality	Equivalent form without the absolute value	Type of solution obtained	Graphed form of the solution on a number line
$\|ax + b\| = c$	$ax + b = c$ or $ax + b = -c$	Two distinct numbers: m and n	
$\|ax + b\| < c$	$-c < ax + b < c$	The set of numbers between the two numbers m and n: $m < x < n$	
$\|ax + b\| > c$	$ax + b < -c$ or $ax + b > c$	The set of numbers less than m or the set of numbers greater than n: $x < m$ or $x > n$	

Developing Your Study Skills

Problems with Accuracy

Strive for accuracy. The mistakes students make are often simple ones and not the result of a lack of understanding. Such mistakes are frustrating. A simple arithmetic or copying error can lead to an incorrect answer.

These five steps will help you to cut down on errors.

1. Work carefully and take your time. Do not rush through a problem just to get it done.

2. Concentrate on one problem at a time. Sometimes problems become mechanical, and your mind begins to wander. You can become careless and make a mistake.

3. Check your problem. Be sure that you copied it correctly from the book.

4. Check your computations from step to step. Check the solution in the problem. Does it work? Does it make sense?

5. Keep practicing new skills. Remember the old saying "Practice makes perfect." An increase in practice will result in an increase in accuracy. Many errors are due simply to lack of practice.

There is no magic formula for eliminating all errors, but these five steps will be a tremendous help in reducing them.

Solve and graph the solutions.

1. $|x| \le 8$

2. $|x| < 6$

3. $|x + 4.5| < 5$

4. $|x + 6| < 3.5$

Solve for x.

5. $|x - 3| \le 5$ **6.** $|x - 7| \le 10$ **7.** $|2x - 1| \le 5$ **8.** $|4x - 3| \le 9$

9. $|5x - 2| \le 4$ **10.** $|2x - 3| \le 1$ **11.** $|0.5 - 0.1x| < 1$ **12.** $|0.9 - 0.2x| < 2$

13. $\left|\frac{1}{4}x + 2\right| < 6$ **14.** $\left|\frac{1}{5}x + 1\right| < 5$ **15.** $\left|\frac{2}{3}(x - 2)\right| < 4$ **16.** $\left|\frac{3}{5}(x - 1)\right| < 3$

17. $\left|\frac{3x - 2}{4}\right| < 3$ **18.** $\left|\frac{5x - 3}{2}\right| < 4$

Solve for x.

19. $|x| > 5$ **20.** $|x| \ge 7$ **21.** $|x + 2| > 5$ **22.** $|x + 4| > 7$

23. $|x - 1| \ge 2$ **24.** $|x - 2| \ge 3$ **25.** $|4x - 7| \ge 9$ **26.** $|6x - 5| \ge 7$

27. $|6 - 0.1x| > 5$ **28.** $|0.4 - 0.2x| > 3$ **29.** $\left|\frac{1}{5}x - \frac{1}{10}\right| > 2$ **30.** $\left|\frac{1}{4}x - \frac{3}{8}\right| > 1$

31. $\left|\frac{1}{3}(x - 2)\right| < 5$ **32.** $\left|\frac{2}{5}(x - 2)\right| \le 4$

Mixed Practice

33. $|3x + 5| < 17$ **34.** $|2x + 3| < 5$ **35.** $|3 - 8x| > 19$ **36.** $|2 - 5x| > 2$

Applications

Manufacturing Standards *In a certain company, the measured thickness m of a helicopter blade must not differ from the standard s by more than 0.12 millimeter. The manufacturing engineer expresses this as* $|m - s| \le 0.12$.

37. Find the limits of m if the standard s is 18.65 millimeters.   **38.** Find the limits of m if the standard s is 17.48 millimeters.

Computer Chip Standards *A small computer microchip has dimension requirements. The manufacturing engineer has written the specification that the new length n of the chip can differ from the previous length p by only 0.05 centimeter or less. The equation is* $|n - p| \leq 0.05$.

39. Find the limits of the new length if the previous length was 9.68 centimeters.

40. Find the limits of the new length if the previous length was 7.84 centimeters.

Cumulative Review *Perform the correct order of operations to simplify.*

41. **[1.3.3]** $4^2 + (5 - 2)^3 \div (-9)$

42. **[1.2.3]** $(-4)(7) \div 2 + (-8) - 12$

In exercises 43 and 44 use $\pi \approx 3.14$. *Round answers to the nearest hundredth.*

▲ **43.** **[1.6.2]** *Geometry* The Outward Bound program in the United States is famous for teaching self-esteem and personal achievement to young people. One of its physical challenges is for a student to hang on to a rope 19 meters long and swing from one shore to another and then back. The rope swings through a circular arc, measuring $\frac{1}{8}$ of the circumference of a circle. How many meters does the end of the rope travel in one *round-trip* swing?

▲ **44.** **[1.6.2]** *Geometry* The rigging on a sailboat comes loose from the mast. The end of the wire rigging that is hanging down is 30 feet from the top of the mast. This end swings through a circular arc, measuring $\frac{1}{6}$ of the circumference of a circle. How many feet does the end of the wire travel in one *round-trip* swing?

Quick Quiz 2.8 Solve for *x*.

1. $\left|\frac{1}{2}x + \frac{1}{4}\right| < 6$

2. $|8x - 4| \leq 20$

3. $|5x + 2| > 7$

4. **Concept Check** Explain what happens when you try $|7x + 3| < -4$

Putting Your Skills to Work: Use Math to Save Money

SO YOU'RE READY TO LIVE ON YOUR OWN?

Mariam has decided that it is time to move out of her parents' house and rent her first apartment. She is planning on rooming with Jen, her college roommate. They have found the apartment of their dreams; it is close to school, has a fitness center and a pool, and seems to have a lot of young people. Mariam and Jen are still in school, but with their part-time jobs, they are sure they can easily pay the $950 monthly rent between them.

Mariam talks it over with her parents and finds that it is not that simple. She discovers that the apartment complex requires a first and last month's deposit, that they need to pay their own electricity and cable, and that the apartment is unfurnished, so they need furniture. Her parents agree to loan her the deposit, and she can take any old furniture from the basement. Her mother also suggests renter's insurance because Mariam and Jen expect to have computers, stereos, and other expensive electronic gadgets.

Look at the following costs they need to consider:

Monthly bill associated with the apartment	Monthly cost
Rent	$950
Electricity	$200/if bills are averaged over a year
Cable TV, phone, internet	$90.00
Laundry	$60.00 ($30.00 each)
Renter's insurance	$300.00 ($150 each)

1. How much might the monthly apartment costs be, based on the above information?

2. Mariam had expected her costs to be half the $950, or $475. Based on the previous information, what would her new expected monthly cost be? How much more is that per month than she planned to pay?

3. Mariam's part-time job is at an Italian restaurant, where she waits on tables. If she works an average of 15 full shifts per month, earning a take-home salary of about $130 each shift, can she pay for these expenses?

Making It Personal for You

4. How much can you afford per month for apartment costs? How much would you pay in rent, utilities, and insurance?

5. Can you afford to move into an apartment on your own or with a roommate? If your roommate decides to move out, can you afford the bills until you get a new one?

Facts You Should Know

Before you rent take the following actions:

- Check your lease for rent-increase clauses, damage costs, and other unusual charges.
- Compare apartments with utilities included to apartments without. You may find one is more cost efficient for you, based on your lifestyle.
- Check the neighborhood. Is it a safe place to live?
- A standard lease is one year. What happens if you decide to move out earlier?
- Try to consider all costs (rent, utilities, laundry, insurance, commuting, food, furnishings) before you sign any papers.
- Have someone in the know read your lease before you sign. Even if the landlord promises something, if it is not in writing, you can't depend on it.

Topic	Procedure	Examples
Solving first-degree equations, p. 59.	1. Remove any grouping symbols in the proper order. 2. If fractions exist, multiply all terms on both sides by the LCD of all the fractions. 3. If decimals exist, multiply all terms on both sides by a power of 10. 4. Combine like terms if possible. 5. Add or subtract terms on both sides of the equation to get all terms with the variable on one side of the equation. 6. Add or subtract a value on both sides of the equation to get all terms not containing the variable on the other side of the equation. 7. Divide both sides of the equation by the coefficient of the variable. 8. Simplify the solution (if possible). 9. Check the solution.	Solve for x. $$\frac{1}{3}(2x - 3) + \frac{1}{2}(x + 1) = 3$$ $$\frac{2}{3}x - 1 + \frac{1}{2}x + \frac{1}{2} = 3$$ $$6\left(\frac{2}{3}x\right) - 6(1) + 6\left(\frac{1}{2}x\right) + 6\left(\frac{1}{2}\right) = 6(3)$$ $$4x - 6 + 3x + 3 = 18$$ $$7x - 3 = 18$$ $$7x = 21$$ $$x = 3$$ **Check.** $\frac{1}{3}[2(3) - 3] + \frac{1}{2}(3 + 1) \overset{?}{=} 3$ $$\frac{1}{3}(3) + \frac{1}{2}(4) \overset{?}{=} 3$$ $$1 + 2 = 3 \checkmark$$
Equations and formulas with more than one variable, p. 66.	If an equation or formula has more than one variable, you can solve for a particular variable by using the procedure for solving linear equations. Remember that your goal is to get all terms containing the desired variable on one side of the equation and all other terms on the opposite side of the equation.	Solve for r. $A = P(1 + rt)$ $$A = P + Prt$$ $$A - P = Prt$$ $$\frac{A - P}{Pt} = \frac{Prt}{Pt}$$ $$\frac{A - P}{Pt} = r$$
Absolute value equations, p. 72.	To solve an equation that involves an absolute value, we rewrite the absolute value equation as two separate equations without the absolute value. We solve each equation. If $|ax + b| = c$ where $c > 0$, then $ax + b = c$ or $ax + b = -c$.	Solve for x. $\|4x - 1\| = 17$ $4x - 1 = 17$ *or* $4x - 1 = -17$ $4x = 17 + 1$ $4x = -17 + 1$ $4x = 18$ $4x = -16$ $x = \frac{18}{4}$ $x = \frac{-16}{4}$ $x = \frac{9}{2}$ $x = -4$

Solving word problems, pp. 78 and 85.	1. *Understand the problem.*
	   (a) Read the word problem carefully to get an overview.
	   (b) Determine what information you need to solve the problem.
	   (c) Draw a sketch. Label it with the known information. Determine what needs to be found.
	   (d) Choose a variable to represent one unknown quantity.
	   (e) If necessary, represent other unknown quantities in terms of that same variable.
	 2. *Write an equation.*
	   (a) Look for key words to help you translate the words into algebraic symbols.
	   (b) Use a relationship given in the problem or an appropriate formula in order to write an equation.
	 3. *Solve the equation and state the answer.*
	 4. *Check.*
	   (a) Check the solution in the original equation.
	   (b) Be sure the solution to the equation answers the question in the word problem. You may need some additional calculations if it does not.
		Fred and Linda invested $7000 for 1 year. Part was invested at 3% simple interest, and part was invested at 5% simple interest. If they earned a total of $320 in interest, how much did they invest at each rate?
		 1. *Understand the problem.*
		 Let $x =$ the amount invested at 3% simple interest. Then $7000 - x =$ the amount invested at 5% simple interest.
		 2. *Write an equation.* $$0.03x + 0.05(7000 - x) = 320$$ **3.** *Solve the equation and state the answer.* $$0.03x + 350 - 0.05x = 320$$ $$350 - 0.02x = 320$$ $$-0.02x = -30$$ $$x = 1500$$ $$7000 - x = 5500$$ Therefore, $1500 was invested at 3% simple interest and $5500 at 5% simple interest.
		 4. *Check.*
		 $0.03(1500) \; + \; 0.05(7000 - 1500) \overset{?}{=} 320$
		 $45 \quad + \quad 275 \quad \overset{?}{=} 320$
		 $320 \quad = 320 \checkmark$

(Continued on next page)

Topic	Procedure	Examples
Solving linear inequalities, pp. 97 and 98.	1. If $a < b$, then for all real numbers a, b, and c, $$a + c < b + c \text{ and } a - c < b - c.$$ 2. If $a < b$, then for all real numbers a, b, and c when $c > 0$, $$ac < bc \quad \text{and} \quad \frac{a}{c} < \frac{b}{c}.$$ Multiplying or dividing both sides of an inequality by a positive number does **not** reverse the inequality. 3. If $a < b$, then for all real numbers a, b, and c when $c < 0$, $$ac > bc \quad \text{and} \quad \frac{a}{c} > \frac{b}{c}.$$ Multiplying or dividing both sides of an inequality **reverses the direction** of the inequality symbol.	Solve and graph. $$3(2x - 4) + 1 \geq 7$$ $$6x - 12 + 1 \geq 7$$ $$6x - 11 \geq 7$$ $$6x \geq 18$$ $$x \geq 3$$ Solve and graph. $$\frac{1}{4}(x + 3) \leq \frac{1}{3}(x - 2)$$ $$\frac{1}{4}x + \frac{3}{4} \leq \frac{1}{3}x - \frac{2}{3}$$ $$3x + 9 \leq 4x - 8$$ $$-1x + 9 \leq -8$$ $$-1x \leq -17$$ $$\frac{-1x}{-1} \geq \frac{-17}{-1}$$ $$x \geq 17$$
Solving compound inequalities containing and, p. 103.	The solution is the desired region containing all values of x that meet both conditions.	Graph the values of x satisfying $x + 6 > -3$ *and* $2x - 1 < -4$. $x + 6 - 6 > -3 - 6$ *and* $2x - 1 + 1 < -4 + 1$ $\quad\quad x > -9 \quad$ *and* $\quad\quad x < -1.5$
Solving compound inequalities containing or, p. 104.	The solution is the desired region containing all values of x that meet either of the two conditions.	Graph the values of x satisfying $-3x + 1 \leq 7$ *or* $3x + 1 \leq -11$. $-3x + 1 - 1 \leq 7 - 1$ *or* $3x + 1 - 1 \leq -11 - 1$ $\quad -3x \leq 6 \quad$ *or* $\quad\quad 3x \leq -12$ $\quad \frac{-3x}{-3} \geq \frac{6}{-3} \quad$ *or* $\quad \frac{3x}{3} \leq \frac{-12}{3}$ $\quad\quad x \geq -2 \quad$ *or* $\quad\quad x \leq -4$
Solving absolute value inequalities involving $<$ or $\leq$, p. 109.	Let a be a positive real number. If $|x| < a$, then $-a < x < a$. If $|x| \leq a$, then $-a \leq x \leq a$.	Solve and graph. $$|3x - 2| < 19$$ $$-19 < 3x - 2 < 19$$ $$-19 + 2 < 3x - 2 + 2 < 19 + 2$$ $$-17 < 3x < 21$$ $$-\frac{17}{3} < \frac{3x}{3} < \frac{21}{3}$$ $$-5\frac{2}{3} < x < 7$$

Topic	Procedure	Examples
Solving absolute value inequalities involving > or ≥, p. 111.	Let a be a positive real number. If $\|x\| > a$, then $x < -a$ or $x > a$. If $\|x\| \geq a$, then $x \leq -a$ or $x \geq a$.	Solve and graph. $\left\| \frac{1}{3}(x - 2) \right\| \geq 2$ $\frac{1}{3}(x - 2) \leq -2 \qquad or \qquad \frac{1}{3}(x - 2) \geq 2$ $\frac{1}{3}x - \frac{2}{3} \leq -2 \qquad\qquad \frac{1}{3}x - \frac{2}{3} \geq 2$ $x - 2 \leq -6 \qquad\qquad\qquad x - 2 \geq 6$ $x \leq -6 + 2 \qquad\qquad\qquad x \geq 6 + 2$ $x \leq -4 \qquad or \qquad x \geq 8$

Chapter 2 Review Problems

Solve for x.

1. $7x - 3 = -5x - 18$

2. $8 - 2(x + 3) = 24 - (x - 6)$

3. $5(x - 2) + 4 = x + 9 - 2x$

4. $x - \frac{4}{3} = \frac{11}{12} + \frac{3}{4}x$

5. $\frac{1}{9}x - 1 = \frac{1}{2}\left(x + \frac{1}{3}\right)$

6. $\frac{x - 4}{2} - \frac{1}{5} = \frac{7x + 1}{20}$

7. $5x = 3(1.6x - 4.2)$

8. $1.2x - 1 = 2(1.6x + 1.5)$

9. Solve for y. $6x - 11y = 8$

10. Solve for a. $P = \frac{1}{2}ab$

11. Solve for a. $2(3ax - 2y) - 6ax = -3(ax + 2y)$

12. Solve for b: $\frac{1}{2}a + 3b = \frac{2}{3}(2b - 1)$

13. (a) Solve for F: $C = \frac{5F - 160}{9}$

   (b) Now find F when $C = 10°$.

▲ **14.** (a) Solve for W: $P = 2W + 2L$

   (b) Now find W when $P = 100$ meters and $L = 20.5$ meters.

Solve for x.

15. $|2x - 7| = 9$ **16.** $|5x + 2| = 7$ **17.** $|3 - x| = |5 - 2x|$ **18.** $|x + 8| = |2x - 4|$

19. $\left|\dfrac{1}{4}x - 3\right| = 8$ **20.** $|4 - 7x| = 25$ **21.** $|2x - 8| + 7 = 12$ **22.** $|0.2x - 1| + 1.2 = 2.3$

Solve each problem.

▲ **23.** *Geometry* Jessica wants to fence her rectangular vegetable garden. The length of the garden is 3 feet longer than twice its width. The perimeter is 42 feet. What are the length and width of the garden?

24. *Education* The number of men attending Western Tech is 200 less than twice the number of women. The number of students at the school is 280. How many men attend? How many women attend?

25. *Car Rental Costs* Rent-It-Right rents compact cars for $38 per day plus 15¢ per mile. Lucia rented a car for 3 days and was charged $150. How many miles did she drive?

26. *Taxi Costs* Stephanie recently stayed at a hotel in Worcester. She took a taxi to the airport. The cab driver said the cost of the trip was $14.75. The cabs in Worcester charge $2.50 for the first $\frac{1}{5}$ of a mile and $0.35 for each additional $\frac{1}{5}$ of a mile. How many miles is it from the hotel to the airport?

27. *Withholding from Monthly Paycheck* Alice's employer withholds from her monthly paycheck $102 for federal and state taxes and for retirement. She noticed that the amount withheld for her state income tax is $13 more than that withheld for retirement. The amount withheld for federal income tax is three times the amount withheld for the state tax. How much is withheld monthly for federal tax? State tax? Retirement?

28. *Raffle Tickets* Emma, Jackson, and Nicholas have been selling raffle tickets to raise money for the math club. Emma sold 5 less than twice as many tickets as Nicholas. Jackson sold 10 more than twice as many tickets as Nicholas. Together the three sold 180 tickets. How many did each student sell?

29. *Education* Valleyview College has 15% more students this year than five years ago. There are 2415 enrolled. How many students were enrolled five years ago?

30. *Auto Manufacturing* An auto manufacturer wants to make 260,000 sedans a year. Some will be two-door sedans; the rest will be four-door. They also want to make three times as many four-door sedans as two-door sedans. How many of each type should be manufactured?

31. *Investments* Huang invested $9000 in mutual funds and bonds. The mutual fund earned 11% simple interest. The bonds earned 6% simple interest. At the end of one year, he had earned $815 in interest. How much had he invested at each rate?

32. *Chemical Mixtures* To make a weak solution of 24 liters of 4% acid, a lab technician will use some premixed solutions: one is 2% acid and the other is 5% acid. How many liters of each type should he use to obtain the desired solution?

33. *Coffee Costs* A local coffee specialty shop wants to obtain 30 pounds of a mixture of coffee beans costing $4.40 per pound. They have a mixture costing $4.25 per pound and a mixture costing $4.50 per pound. How much of each should be used?

34. *Education* When Eastern Slope Community College opened, the number of students (full-time and part-time) was 380. Since then the number of full-time students has doubled, and the number of part-time students has tripled. There are now 890 students at the school. How many of the present students are full-time? Part-time?

Solve for x.

35. $7x + 8 < 5x$

36. $9x + 3 < 12x$

37. $4x - 1 < 3(x + 2)$

38. $3(3x - 2) < 4x - 16$

39. $\frac{7}{8}x - \frac{1}{4} > \frac{1}{2}$

40. $\frac{5}{3} - x \geq -\frac{1}{6}x + \frac{5}{6}$

41. $\frac{1}{3}(x - 2) < \frac{1}{4}(x + 5) - \frac{5}{3}$

42. $\frac{1}{3}(x + 2) > 3x - 5(x - 2)$

43. $7x - 6 \leq \frac{1}{3}(-2x + 5)$

Graph the values of x that satisfy the conditions given.

44. $-3 \leq x < 2$

45. $-4 < x \leq 5$

46. $-8 \leq x \leq -4$

47. $-9 \leq x \leq -6$

48. $x < -2$ *or* $x \geq 5$

49. $x < -3$ *or* $x \geq 6$

50. $x > -5$ *and* $x < -1$

51. $x > -8$ *and* $x < -3$

52. $x + 3 > 8$ *or* $x + 2 < 6$

Solve for x.

53. $x - 2 > 7$ *or* $x + 3 < 2$

54. $x + 3 > 8$ *and* $x - 4 < -2$

55. $-1 < x + 5 < 8$

56. $0 \leq 5 - 3x \leq 17$

57. $2x - 7 < 3$ *and* $5x - 1 \geq 8$

58. $4x - 2 < 8$ *or* $3x + 1 > 4$

Solve for x.

59. $|x + 7| < 15$

60. $|x + 9| < 18$

61. $\left|\frac{1}{2}x + 2\right| < \frac{7}{4}$

62. $\left|\frac{1}{5}x + 3\right| < \frac{11}{5}$

63. $|2x - 1| \geq 9$

64. $|3x - 1| \geq 2$

65. $|4(x + 1)| \geq 3$

66. $|2(x - 5)| \geq 2$

67. *Telephone Charges* Greg Salzman is making a long-distance phone call to Chicago, Illinois, from a pay phone. The operator informs him that the charge will be $3.95 for the first minute and 65¢ for each additional minute. Any fractional part of a minute used will be rounded up to the nearest whole minute. What is the maximum time Greg can talk if he has $13.05 in change in his pocket?

68. *Airplane Capacity* A small plane takes off with packages from Manchester Airport. Each package weighs 77.5 pounds. The plane has a carrying capacity for people and packages of 1765 pounds. The plane is carrying a pilot who weighs 170 pounds and a passenger who weighs 200 pounds. How many packages can be safely carried?

69. *Landscaping Costs* Emmanuel Vargas wants to order bark mulch from a landscaping company. He does not want to spend more than $250 for the order. The mulch costs $28 per cubic yard and there is a $40 delivery charge. Only whole cubic yards can be ordered. What is the maximum number of cubic yards of mulch he can order?

70. *Post Office Costs* The Ramsey County Government Office sent out several large envelopes of literature by Priority Mail. They planned for a mailing budget of $4.60 per envelope. The post office charges $0.41 for the first ounce and $0.28 for each additional ounce. What was the most that an envelope could weigh and still be mailed at a cost that did not exceed the budget?

71. *Manufacturing Requirements* The Trubolt Manufacturing Company ships some of its bolts in small plastic boxes. For shipping purposes, it is best for each box to weigh less than 14 ounces. Each plastic box weighs 1.5 ounces and each bolt weighs 2.5 ounces. What is the maximum number of bolts the company should package in each box?

72. *Census* The Census Bureau has projected that the population of Nevada in the year 2025 may be as large as 2,854,000 or as small as 2,312,000. A group of real estate offices in Nevada has projected that the larger number should be 6% higher. They feel that the smaller number should be 4% higher. Assuming the Nevada real estate offices are correct, write an inequality that expresses the revised population projections for the year 2025. Use the variable x to represent the population of Nevada in the year 2025. (*Source: U.S. Census Bureau*)

Mixed Practice *Solve for x.*

73. $4 - 7x = 3(x + 3)$

74. Solve for B. $H = \dfrac{3}{4}B - 16$

Use an algebraic equation to find a solution.

75. *Chemical Mixtures* A technician needs 100 grams of an alloy that is 80% pure copper (Cu). She has one alloy that is 77% pure Cu and another that is 92% pure Cu. How many grams of each alloy should she use to obtain 80% pure Cu?

Solve for x. Graph your solution.

76. $7x + 12 < 9x$

77. $\dfrac{2}{3}x - \dfrac{5}{6}x - 3 \le \dfrac{1}{2}x - 5$

Solve for x. Graph your solution.

78. $-2 \le x + 1 \le 4$

79. $2x + 3 < -5$ or $x - 2 > 1$

Solve for x.

80. $|2x - 7| + 4 = 5$

81. $\left| \dfrac{2}{3}x - \dfrac{1}{2} \right| \le 3$

82. $|2 - 5x - 4| > 13$

How Am I Doing? Chapter 2 Test

Remember to use your Chapter Test Prep Video CD to see the worked-out solutions to the test problems you want to review.

Solve for x.

1. $5x - 8 = -6x - 10$

2. $3(7 - 2x) = 14 - 8(x - 1)$

3. $\frac{1}{3}(-x + 1) + 4 = 4(3x - 2)$

4. $0.5x + 1.2 = 4x - 3.05$

5. Solve for n. $L = a + d(n - 1)$

6. Solve for b. $A = \frac{1}{2}bh$

7. Use your answer for problem 6 to evaluate b when $A = 15$ cm^2 and $h = 10$ cm.

8. Solve for r. $H = \frac{1}{2}r + 3b - \frac{1}{4}$

Solve for x.

9. $|5x - 2| = 37$

10. $\left|\frac{1}{2}x + 3\right| - 2 = 4$

Use an algebraic equation to find a solution.

▲ **11.** A triangle has a perimeter of 69 meters. The length of the second side is twice the length of the first side. The third side is 5 meters longer than the first side. How long is each side?

12. Mercy Hospital's electric bill for September was $2489. This is a decrease of 5% from August's electric bill. What was the hospital's electric bill in August?

13. Linda needs 10 gallons of solution that is 60% antifreeze. She has a solution that is 90% antifreeze and another that is 50% antifreeze. How much of each should she use?

14. Lon Triah invested $5000 at a local bank. Part was invested at 6% simple interest and the remainder at 10% simple interest. At the end of one year, Lon had earned $428 interest. How much was invested at each rate?

Solve and graph.

15. $5 - 6x < 2x + 21$

$$\overset{\longrightarrow}{\underset{-5\ -4\ -3\ -2\ -1\ \ 0\ \ 1\ \ 2\ \ 3}{\quad}}$$

16. $-\frac{1}{2} + \frac{1}{3}(2 - 3x) \geq \frac{1}{2}x + \frac{5}{3}$

$$\overset{\longrightarrow}{\underset{-6\ -5\ -4\ -3\ -2\ -1\ \ 0\ \ 1\ \ 2}{\quad}}$$

Find the values of x that satisfy the given conditions.

17. $-11 < 2x - 1 \leq -3$

18. $x - 4 \leq -6$ *or* $2x + 1 \geq 3$

Solve each absolute value inequality.

19. $|7x - 3| \leq 18$

20. $|3x + 1| > 7$

1. _____

2. _____

3. _____

4. _____

5. _____

6. _____

7. _____

8. _____

9. _____

10. _____

11. _____

12. _____

13. _____

14. _____

15. _____

16. _____

17. _____

18. _____

19. _____

20. _____

Cumulative Test for Chapters 1–2

1. _____

2. _____

3. _____

4. _____

5. _____

6. _____

7. _____

8. _____

9. _____

10. _____

11. _____

12. _____

13. _____

14. _____

15. _____

16. _____

17. _____

18. _____

19. _____

20. _____

This test is made up of problems from Chapters 1 and 2.

1. Consider the set of numbers
$$\left\{-12, -3, -\frac{5}{8}, 0, \frac{1}{4}, 2.16, 2.333\ldots, 2.9614371823\ldots, 3\right\}.$$
List all the rational numbers.

2. Name the property that justifies the statement $7 + (6 + 3) = (7 + 6) + 3$.

3. Evaluate. $\sqrt{100} + 4(3 - 5)^3 - (-20)$

4. Simplify. $\left(-\frac{2}{3}x^4y^{-2}z^0\right)(6x^{-1}y^6z^2)$

5. Simplify. $\dfrac{6a^{-1}b^3}{-18a^5b}$

6. Evaluate for $x = -2, y = 1$: $2x^2 + 3xy - y^2$

▲ 7. Find the area of a circle with a radius of 7 inches. Use $\pi \approx 3.14$.

8. Simplify. $2x - [6x - 3(x + 5y)]$

9. Solve for x.
$$\frac{1}{4}(x + 5) - \frac{5}{3} = \frac{1}{3}(x - 2)$$

10. Solve for b.
$$h = \frac{2}{3}(b + d)$$

▲ 11. A triangle has a perimeter of 112 inches. The second side is 15 inches longer than the first. The third side is 7 inches shorter than double the first. How long is each side?

12. Hamid paid $68 for a power saw that was on sale. This sale price was 85% of the original price of the saw. What was the original price of the saw?

13. Hector needs 9 gallons of a solution that is 70% antifreeze. He will combine a solution that is 80% antifreeze with a solution that is 50% antifreeze to obtain the desired 9 gallons. How many gallons of each should he use?

14. Wendy invested $6500 for one year. She invested part at 12% simple interest and part at 10% simple interest. She earned $690 in interest. How much did she invest at each rate?

Solve and graph.

15. $-4 - 3x < -2x + 6$

⊢─┼─┼─┼─┼─┼─┼─┼─┼─┼─→

16. $\frac{1}{3}(x + 2) \le \frac{1}{5}(x + 6)$

⊢─┼─┼─┼─┼─┼─┼─┼─┼─┼─→

Find the values of x that satisfy the given conditions.

17. $-13 < 4x - 5 < 3$

18. $x + 5 \le -4$ *or* $2 - 7x \le 16$

Solve each absolute value inequality.

19. $\left|\frac{1}{2}x + 2\right| \le 8$

20. $|3x - 4| > 11$

124

CHAPTER
3

Successful fruit agriculture involves more than just planting trees and waiting for the fruit to ripen for harvest. In addition to concerns over weather, pests, and disease, yield per acre is a variable that the farmer needs to consider. In this chapter you will develop your knowledge of functions to investigate some of these issues that might be used in managing a cherry orchard.

Equations and Inequalities in Two Variables and Functions

3.1   GRAPHING LINEAR EQUATIONS WITH TWO UNKNOWNS   126

3.2   SLOPE OF A LINE   135

3.3   GRAPHS AND THE EQUATIONS OF A LINE   143

       HOW AM I DOING? SECTIONS 3.1–3.3   153

3.4   LINEAR INEQUALITIES IN TWO VARIABLES   154

3.5   CONCEPT OF A FUNCTION   160

3.6   GRAPHING FUNCTIONS FROM EQUATIONS
       AND TABLES OF DATA   169

       CHAPTER 3 ORGANIZER   184

       CHAPTER 3 REVIEW PROBLEMS   186

       HOW AM I DOING? CHAPTER 3 TEST   191

       CUMULATIVE TEST FOR CHAPTERS 1–3   193

Student Learning Objectives

After studying this section, you will be able to:

1 Graph a linear equation in two variables.

2 Use *x*- and *y*-intercepts to graph a linear equation.

3 Graph horizontal and vertical lines.

4 Graph a linear equation using different scales.

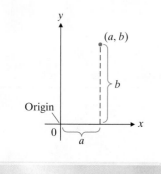

Graphing Calculator

Graphing Ordered Pairs

You can use a graphing calculator to graph ordered pairs. Most graphing calculators can plot statistical data. Use your calculator's statistical plot feature to plot the points $(3, 2)$; $(0, -4)$; $(-2, -1)$; $(-5, -4)$; and $(-3, 4)$. The display should be similar to the one below. Be sure to use an appropriate window.

Display:

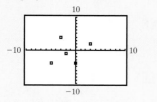

Graphs are often used to show the relationships among sets of data. You may be familiar with graphs that are used in applied mathematics, science, and business.

The following is a graph that could be found in a local newspaper. It shows the daily low temperature for the first 30 days of January 2009 and compares this to the normal (average) low temperature, as well as to the record low temperature for that month in that region of the country.

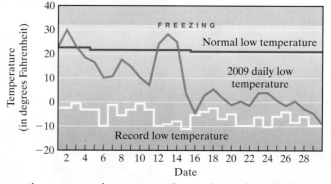

In mathematics we can also use graphs to show the relationships among the variables in an equation. To do so, we will use two number lines. For convenience we construct two real number lines—one horizontal and one vertical—that intersect to form a **rectangular coordinate system.** The horizontal line is the **x-axis.** The vertical line is the **y-axis.** They intersect at the **origin.**

An **ordered pair** of real numbers (a, b) represents a point on the rectangular coordinate system. See the graph in the left margin. To graph an ordered pair, we begin at the origin. The first coordinate is the *x*-coordinate, and the second coordinate is the *y*-coordinate. To locate the point described by the ordered pair (a, b), where $a, b > 0$, we move *a* units to the right of the origin along the *x*-axis. Then we move *b* units up parallel to the *y*-axis. Since (a, b) is an ordered pair, order is important. That is, if $a \neq b$, then $(a, b) \neq (b, a)$.

The next figure shows the graphs of $(3, 4)$ and $(4, 3)$.

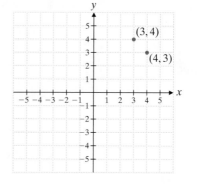

1 Graphing a Linear Equation in Two Variables

We now define a *linear equation in two variables.*

A **linear equation in two variables** is an equation that can be written in the form $Ax + By = C$, where A, B, and C are real numbers and A and B are not both zero. This form is called the **standard form** of a linear equation in two variables.

A **solution** of an equation in two variables is an ordered pair of real numbers that *satisfies* the equation. In other words, when we substitute the values of the coordinates into the equation, we get a true statement. For example, $(6, 4)$ is a solution to $4x - 3y = 12$. The ordered pair $(6, 4)$ means that $x = 6$ and $y = 4$.

$$4x - 3y = 12$$
$$4(6) - 3(4) \stackrel{?}{=} 12$$
$$24 - 12 \stackrel{?}{=} 12$$
$$12 = 12 \checkmark$$

To graph the equation, we could graph all its solutions. However, this would be impossible, since there are an infinite number of solutions. It is a mathematical property that the graph of an equation of the form $Ax + By = C$, where A, B, and C are constants, is a straight line. Hence, to graph the equation we graph three ordered pair solutions and connect them with a straight line. (The third ordered pair solution is used to check the line.)

EXAMPLE 1   Graph the equation $y = -3x + 2$.

Solution   We choose three values of x and then substitute them into the equation to find the corresponding values of y. Let's choose $x = -1$, $x = 1$, and $x = 2$.

For $x = -1$, $y = -3(-1) + 2 = 5$, so the first point, or solution, is $(-1, 5)$.

For $x = 1$, $y = -3(1) + 2 = -1$, and for $x = 2$, $y = -3(2) + 2 = -4$.

We can condense this information by using a table.

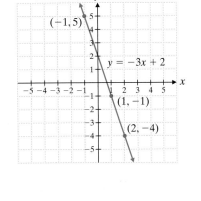

x	y
-1	5
1	-1
2	-4

Practice Problem 1   Graph the equation $y = -4x + 2$.

PRACTICE PROBLEM 1

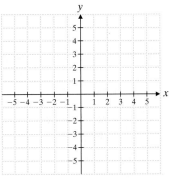

NOTE TO STUDENT: *Fully worked-out solutions to all of the Practice Problems can be found at the back of the text starting at page SP-1*

2 Using x- and y-intercepts to Graph a Linear Equation

We can usually graph a straight line by using the x- and y-intercepts. A straight line that is not vertical or horizontal has these two intercepts.

> The **x-intercept** of a line is the point where the line crosses the x-axis (that is, where $y = 0$). It is described by an ordered pair of the form $(a, 0)$.
>
> The **y-intercept** of a line is the point where the line crosses the y-axis (that is, where $x = 0$). It is described by an ordered pair of the form $(0, b)$.

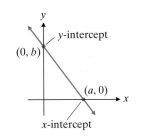

EXAMPLE 2   Find the x-intercept, the y-intercept, and one additional ordered pair that satisfies the equation. Then graph the equation $4x - 3y = -12$.

Solution

Find the x-intercept by using $y = 0$.

$$4x - 3(0) = -12$$
$$4x = -12$$
$$x = -3$$

The x-intercept is $(-3, 0)$.

Find the y-intercept by using $x = 0$.

$$4(0) - 3y = -12$$
$$-3y = -12$$
$$y = 4$$

The y-intercept is $(0, 4)$.

We can now pick any value of x or y to find our third point. Let's pick $y = 2$.

$$4x - 3(2) = -12$$
$$4x - 6 = -12$$
$$4x = -12 + 6$$
$$4x = -6$$
$$x = -\frac{6}{4} = -\frac{3}{2}$$

PRACTICE PROBLEM 2

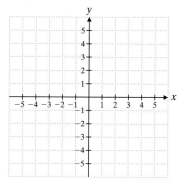

Hence, the third point on the line is $\left(-\frac{3}{2}, 2\right)$. A table and the graph of the equation is shown below.

x	y
−3	0
0	4
$-\frac{3}{2}$	2

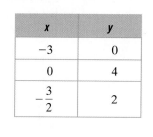

Practice Problem 2 Find the x-intercept, the y-intercept, and one additional ordered pair that satisfies the equation. Then graph the equation $3x - 2y = -6$.

Graphing Calculator

Graphing a Line

You can graph a line given in the form $y = mx + b$ using a graphing calculator. For example, to graph the equation in Example 2, first rewrite the equation by solving for y.

$$4x - 3y = -12$$
$$-3y = -4x - 12$$
$$y = \frac{4}{3}x + 4$$

Enter the right-hand side of the resulting equation in the Y = editor of your calculator and graph. Choose an appropriate window to show all the intercepts. The following window is $[-10, 10]$ by $[-10, 10]$.

Display:

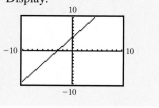

③ Graphing Horizontal and Vertical Lines

Let's look at the standard form of a linear equation, $Ax + By = C$, when $B = 0$.

$$Ax + (0)y = C$$
$$Ax = C$$
$$x = \frac{C}{A}$$

Notice that when we solve for x we get $x = \frac{C}{A}$, which is a constant. For convenience we will rename it a. The equation then becomes

$$x = a.$$

What does this mean? The equation $x = a$ means that for any value of y, x is a. The graph is a vertical line.

What happens to $Ax + By = C$ when $A = 0$?

$$(0)x + By = C$$
$$By = C$$
$$y = \frac{C}{B}$$

We will rename the constant $\frac{C}{B}$ as b. The equation then becomes

$$y = b.$$

What does this mean? The equation $y = b$ means that for any value of x, y is b. The graph is a horizontal line.

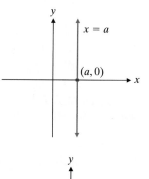

The graph of the equation $x = a$, where a is any real number, is a **vertical line** through the point $(a, 0)$.

The graph of the equation $y = b$, where b is any real number, is a **horizontal line** through the point $(0, b)$.

EXAMPLE 3   Simplify and graph each equation.

(a) $x = -3$ **(b)** $2y - 4 = 0$

Solution

(a) The equation $x = -3$ means that for any value of y, x is -3. The graph of $x = -3$ is a vertical line 3 units to the left of the origin.

(b) The equation $2y - 4 = 0$ can be simplified.

$$2y - 4 = 0$$
$$2y = 4$$
$$y = 2$$

The equation $y = 2$ means that, for any value of x, y is 2. The graph of $y = 2$ is a horizontal line 2 units above the x-axis.

Practice Problem 3   Simplify and graph each equation.

(a) $x = 4$ **(b)** $3y + 12 = 0$

PRACTICE PROBLEM 3

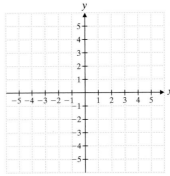

④ Graphing a Linear Equation Using Different Scales for the Axes

By common convention, each tick mark on a graph's axis indicates 1 unit, so we don't need to use a marked scale on each axis. But sometimes a different scale is more appropriate. This new scale must then be clearly labeled on each axis.

EXAMPLE 4   A company's finance officer has determined that the monthly cost in dollars for leasing a photocopier is $C = 100 + 0.002n$, where n is the number of copies produced in a month in excess of a specified number. Graph the equation using $n = 0$, $n = 30,000$, and $n = 60,000$. Let the n-axis be the horizontal axis.

Solution   For each value of n we obtain C.

When $n = 0$,
then $C = 100 + 0.002(0) = 100 + 0 = 100$.
When $n = 30,000$,
then $C = 100 + 0.002(30,000)$
$= 100 + 60 = 160$.
When $n = 60,000$,
then $C = 100 + 0.002(60,000)$
$= 100 + 120 = 220$.

The table of values is shown next.

n	C
0	100
30,000	160
60,000	220

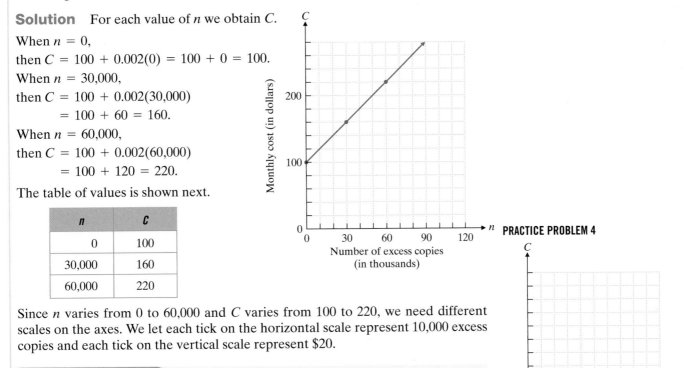

PRACTICE PROBLEM 4

Since n varies from 0 to 60,000 and C varies from 100 to 220, we need different scales on the axes. We let each tick on the horizontal scale represent 10,000 excess copies and each tick on the vertical scale represent $20.

Practice Problem 4   The cost of a product in dollars is given by $C = 300 + 0.15n$, where n is the number of products produced. Graph the equation using an appropriate scale. Use $n = 0$, $n = 1000$, and $n = 2000$.

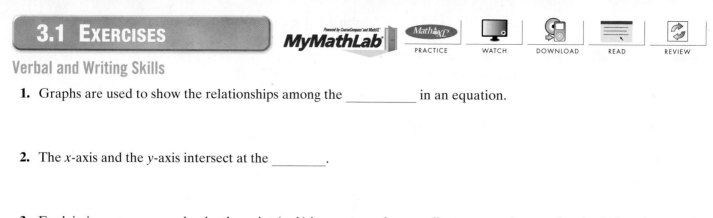
Verbal and Writing Skills

1. Graphs are used to show the relationships among the _____ in an equation.

2. The *x*-axis and the *y*-axis intersect at the _____.

3. Explain in your own words why the point (a, b) in a rectangular coordinate system is an *ordered* pair. In other words, what is the importance of the word *ordered* when we say it is an ordered pair? Give an example.

4. $(5, 1)$ is a solution to the equation $2x - 3y = 7$. What does this mean?

Find the missing coordinate.

5. $(-2, \_\_\_\_)$ is a solution of $y = 3x - 7$.

6. $(-3, \_\_\_\_)$ is a solution of $y = 4 - 3x$.

7. $\left(\_\_\_\_, \dfrac{1}{4}\right)$ is a solution of $5x + 12y = -17$.

8. $\left(\_\_\_\_, \dfrac{1}{2}\right)$ is a solution of $-x + 2y = -7$.

Graph each equation.

9. $y = 2x - 3$

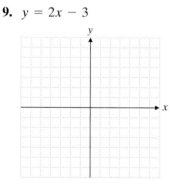

10. $y = 3x + 2$

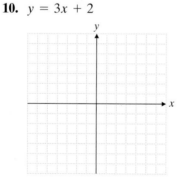

11. $y = 4 - 2x$

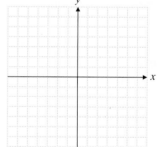

12. $y = -5x - 2$

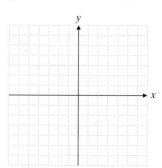

13. $y = \dfrac{2}{3}x - 4$

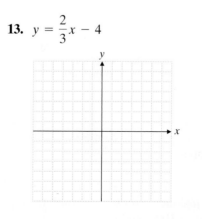

14. $y = \dfrac{5}{2}x + 1$

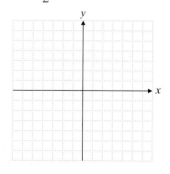

Simplify the equation if possible. Find the x-intercept, the y-intercept, and one or two additional ordered pairs that are solutions to the equation. Then graph the equation.

15. $2y - 3x = 6$

16. $2y + 5x = 10$

17. $2x - y = 6$

18. $4x - y = -4$

19. $-4x - 3y = 6$

20. $5x - 2y = -12$

21. $5y - 4 = 3x - 4$

22. $4x + 6y + 2 = 2$

Simplify the equation if possible. State whether the equation represents a horizontal or a vertical line. Then graph the equation.

23. $x = -5$

24. $x = 2$

25. $4x - 16 = 0$

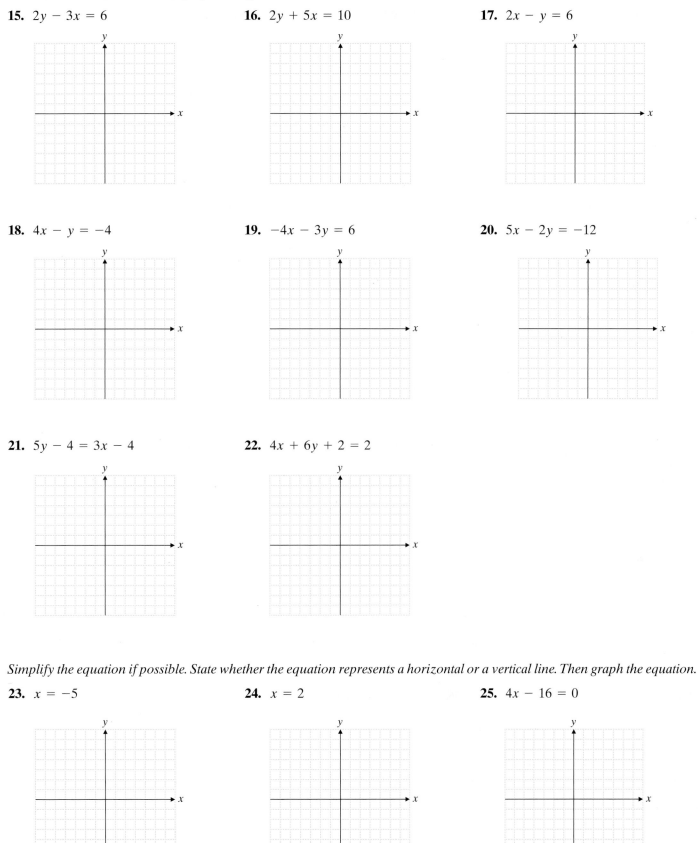

26. $2x - 3 = 3x$

27. $2y + 8 = 0$

28. $5y + 6 = 2y$

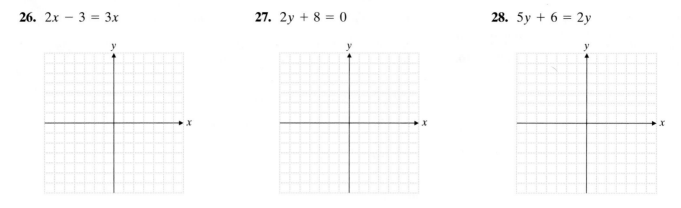

Mixed Practice *Simplify each equation if possible. Then graph the equation by any appropriate method.*

29. $y = -1.5x + 2$

30. $y = 0.5x + 4$

31. $2x + 5y = -5$

32. $4x - 3y = 6$

33. $5x + y + 4 = 8x$

34. $5x - 4y - 4 = 4x$

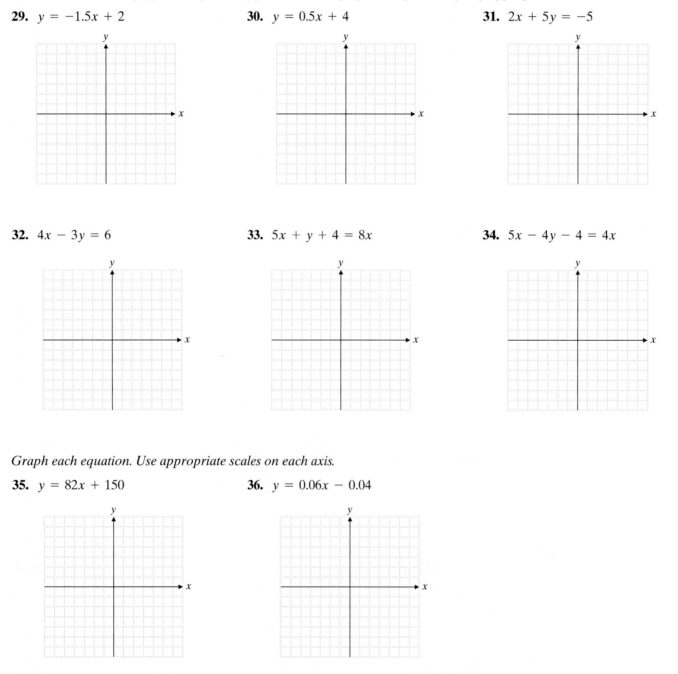

Graph each equation. Use appropriate scales on each axis.

35. $y = 82x + 150$

36. $y = 0.06x - 0.04$

To Think About

Income of Men Versus Women *The following graph shows the median weekly earnings of men and women in the United States during the period 1997 to 2007. Use the graph to answer exercises 37–42.*

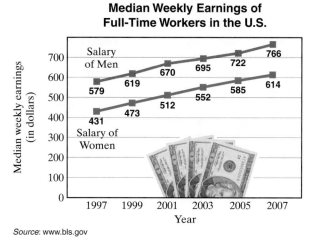

Median Weekly Earnings of Full-Time Workers in the U.S.

Source: www.bls.gov

37. During what 2-year period did the greatest increase in the median weekly earnings of men occur?

38. During what 2-year period did the greatest increase in the median weekly earnings of women occur?

39. In what year did the median weekly earnings of men and women have the largest difference?

40. In what year did the median weekly earnings of men and women have the smallest difference?

41. What was the percent of increase in earnings for women during the period 1997 to 2007? Round your answer to the nearest tenth of a percent.

42. What was the percent of increase in earnings for men during the period 1997 to 2007? Round your answer to the nearest tenth of a percent.

Applications

43. ***Baseball*** If a baseball is thrown vertically upward by Paul Frydrych when he is standing on the ground, the velocity of the baseball V (in feet per second) after T seconds is $V = 120 - 32T$.

(a) Find V for $T = 0, 1, 2, 3,$ and 4.

(b) Graph the equation, using T as the horizontal axis.

(c) What is the significance of the negative value of V when $T = 4$?

44. ***Gasoline Storage Tank*** A full storage tank on the Robinson family farm contains 900 gallons of gasoline. Gasoline is then pumped from the tank at a rate of 15 gallons per minute. The equation $G = 900 - 15m$ describes the number of gallons of gasoline G in the tank m minutes after the pumping began.

(a) Find G for $m = 0, 10, 20, 30,$ and 60.

(b) Graph the equation, using m as the horizontal axis.

(c) What happens when $m = 61$?

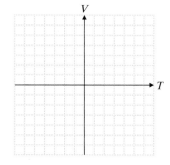

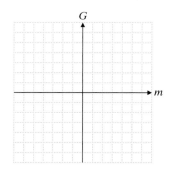

Cumulative Review

45. **[1.3.3]** Evaluate. $36 \div (8 - 6)^2 + 3(-4)$

46. **[2.6.3]** Solve for x. $3(x - 6) + 2 \le 4(x + 2) - 21$

47. **[2.4.1.]** *Balloon Giveaway* A novelty company is giving away balloons in a shopping mall. There are twice as many red balloons as green balloons. There are three times as many blue balloons as red balloons. There are half as many white balloons as there are yellow balloons. There are half as many yellow balloons as there are red balloons. There are 130 white balloons. How many balloons of each color are being given away?

48. **[2.5.1]** *Commission Sales* At Greenland Realty, a salesperson receives a commission of 7% on the first $100,000 of the selling price of a house and 3% on the amount that exceeds $100,000. Ray Peterson received a commission of $9100 for selling a house. What was the selling price of the house?

Quick Quiz 3.1 Graph each of the following. Plot at least three points.

1. $y = -\dfrac{2}{3}x + 4$

2. $7y - 5 = 16$

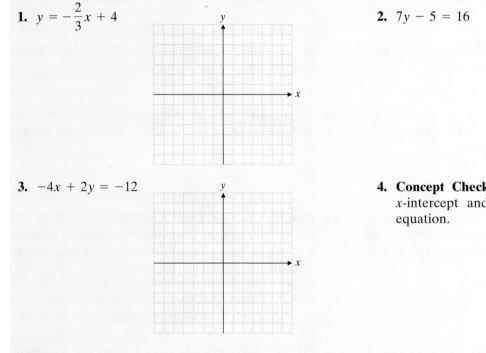

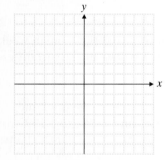

3. $-4x + 2y = -12$

4. **Concept Check** Explain how you would find the x-intercept and the y-intercept for the following equation.

$$7x + 3y = -14$$

3.2 SLOPE OF A LINE

1 Finding the Slope If Two Points Are Known

The concept of slope is one of the most useful in mathematics and has many practical applications. For example, a carpenter needs to determine the slope (or pitch) of a roof. (You may have heard someone say that a roof has a 5 : 12 pitch.) Road engineers must determine the proper slope (or grade) of a roadbed. If the slope is steep, you feel as if you're driving almost straight up. Simply put, slope is a measure of steepness. That is, slope measures the ratio of the vertical change (*rise*) to the horizontal change (*run*).

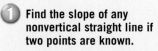

Mathematically, we define slope of a line as follows:

> The **slope of a straight line** containing points (x_1, y_1) and (x_2, y_2) is
>
> $$\text{slope} = m = \frac{y_2 - y_1}{x_2 - x_1} \qquad x_2 \neq x_1.$$

Student Learning Objectives

After studying this section, you will be able to:

1 Find the slope of any nonvertical straight line if two points are known.

2 Determine whether two lines are parallel or perpendicular by comparing their slopes.

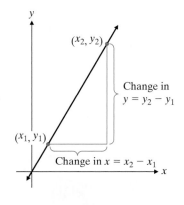

In the sketch, we see that the rise is $y_2 - y_1$ and the run is $x_2 - x_1$.

EXAMPLE 1 Find the slope of the line passing through $(-2, -3)$ and $(1, -4)$.

Solution Identify the *y*-coordinates and the *x*-coordinates for the points $(-2, -3)$ and $(1, -4)$.

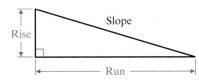

Use the formula.

$$\text{slope} = m = \frac{y_2 - y_1}{x_2 - x_1}$$

$$= \frac{-4 - (-3)}{1 - (-2)}$$

$$= -\frac{1}{3}$$

Practice Problem 1 Find the slope of the line passing through $(-6, 1)$ and $(-5, -2)$.

NOTE TO STUDENT: Fully worked-out solutions to all of the Practice Problems can be found at the back of the text starting at page SP-1

Notice that it does not matter which ordered pair we label (x_1, y_1) and which we label (x_2, y_2) as long as we subtract the *x*-coordinates in the same order that we subtract the *y*-coordinates. Let's redo Example 1.

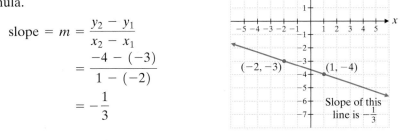

135

Slopes

Using a graphing calculator, graph

$$y_1 = 3x + 1,$$

$$y_2 = \frac{1}{3}x + 1,$$

$$y_3 = -3x + 1, \text{ and}$$

$$y_4 = -\frac{1}{3}x + 1$$

on the same set of axes. How is the coefficient of x in each equation related to the slope of the line? Will the graph of the line $y = -2x + 3$ slope upward or downward? How do you know? Verify using your calculator.

The Mount Washington Cog Railway in New Hampshire has a train track that in some places rises 37 feet for every 100 feet horizontally. This is a slope of $\frac{37}{100}$.

1. Lines sloping upward to the right have positive slopes.
2. Lines sloping downward to the right have negative slopes.

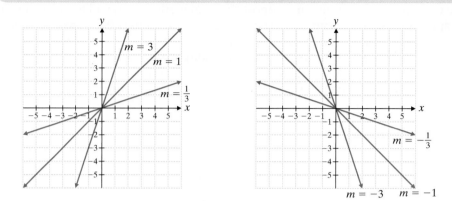

What is the slope of a horizontal line? Let's look at the equation $y = 4$. This equation $y = 4$ is the horizontal line 4 units above the x-axis. It means that for any x, y is 4. The slope is

$$m = \frac{y_2 - y_1}{x_2 - x_1} = \frac{4 - 4}{x_2 - x_1} = \frac{0}{x_2 - x_1} = 0.$$

In general, for all horizontal lines, $y_2 - y_1 = 0$. Hence, the slope is 0.

What is the slope of a vertical line? The equation of the vertical line 4 units to the right of the y-axis is $x = 4$. It means that for any y, x is 4. The slope is

$$m = \frac{y_2 - y_1}{x_2 - x_1} = \frac{y_2 - y_1}{4 - 4} = \frac{y_2 - y_1}{0}$$

Because division by zero is not defined, we say that a vertical line has no slope or the slope is undefined.

The slope of a horizontal line is 0. The slope of a vertical line is undefined.

EXAMPLE 2 Find the slope, if possible, of the line passing through each pair of points.

(a) $(1.6, 2.3)$ and $(-6.4, 1.8)$

(b) $\left(\frac{5}{3}, -\frac{1}{2}\right)$ and $\left(\frac{2}{3}, -\frac{1}{4}\right)$

Solution

(a) $m = \dfrac{1.8 - 2.3}{-6.4 - 1.6} = \dfrac{-0.5}{-8.0} = 0.0625$

(b) $m = \dfrac{-\dfrac{1}{4} - \left(-\dfrac{1}{2}\right)}{\dfrac{2}{3} - \dfrac{5}{3}} = \dfrac{-\dfrac{1}{4} + \dfrac{2}{4}}{-\dfrac{3}{3}} = \dfrac{\dfrac{1}{4}}{-1} = -\dfrac{1}{4}$

Practice Problem 2 Find the slope, if possible, of the line through each pair of points.

(a) $(1.8, -6.2)$ and $(-2.2, -3.4)$

(b) $\left(\frac{1}{5}, -\frac{1}{2}\right)$ and $\left(\frac{4}{15}, -\frac{3}{4}\right)$

When dealing with practical situations, such as the grade of a road or the pitch of a roof, we can find the slope by using the formula

$$\text{slope} = \frac{\text{rise}}{\text{run}}.$$

EXAMPLE 3 Find the pitch of a roof as shown in the sketch. Use only positive numbers in your calculation.

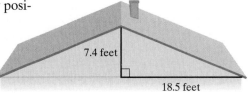

$$\text{slope} = \frac{\text{rise}}{\text{run}} = \frac{7.4}{18.5} = 0.4.$$

Solution This could also be expressed as the fraction $\frac{2}{5}$. A builder might refer to this as a *pitch* (slope) of 2:5.

Practice Problem 3 Find the slope of a river that drops 25.92 feet vertically over a horizontal distance of 1296 feet. (*Hint:* Use only positive numbers. In everyday use, the slope of a river or road is always considered to be a positive value.)

2 Determining Whether Two Lines Are Parallel or Perpendicular

We can tell a lot about a line by looking at its slope. A positive slope tells us that the line rises from left to right. A negative slope tells us that the line falls from left to right. What might be true of the slopes of parallel lines? Determine the slope of each line in the following graphs, and compare the slopes of the parallel lines.

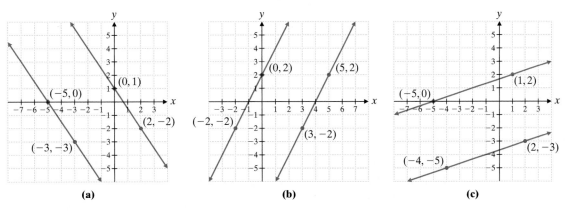

(a)   (b)   (c)

Since parallel lines are lines that never intersect, their slopes must be equal.

PARALLEL LINES

Two different lines with slopes m_1 and m_2 are *parallel* if $m_1 = m_2$.

Now take a look at perpendicular lines. Determine the slope of each line in the following graphs, and compare the slopes of the perpendicular lines. What do you notice?

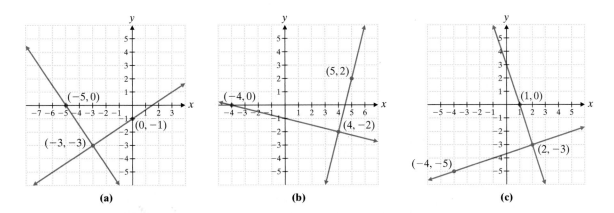

(a)   (b)   (c)

By definition, two lines are perpendicular if they intersect at right angles. (A right angle is an angle of 90°.) We would expect that one slope would be positive and the other slope negative.

In fact, it is true that if the slope of a line is 5, then the slope of a line that is perpendicular to it is $-\frac{1}{5}$.

> ## PERPENDICULAR LINES
> Two lines with slopes m_1 and m_2 are *perpendicular* if $m_1 = -\dfrac{1}{m_2}$ $(m_1, m_2 \neq 0)$.

EXAMPLE 4 Find the slope of a line that is perpendicular to the line l that passes through $(4, -6)$ and $(-3, -5)$.

Solution The slope of line l is

$$m_l = \frac{-5 - (-6)}{-3 - 4} = \frac{-5 + 6}{-7} = \frac{1}{-7} = -\frac{1}{7}.$$

The slope of a line perpendicular to line l must have a slope of 7.

Practice Problem 4 If a line l passes through $(5, 0)$ and $(6, -2)$, what is the slope of a line h that is perpendicular to l?

It is helpful to remember that if two lines are perpendicular, their slopes are negative reciprocals. One slope will be positive and the other negative. If one line had a slope of $-\frac{3}{7}$, the slope of a line perpendicular to it would have a slope of $\frac{7}{3}$.

Now we will use our knowledge of slopes to determine if three points lie on the same line.

EXAMPLE 5 Without plotting any points, show that the points $A(-5, -1)$, $B(-1, 2)$, and $C(3, 5)$ lie on the same line.

Solution First we find the slope of the line segment from A to B and the slope of the line segment from B to C.

$$m_{AB} = \frac{2 - (-1)}{-1 - (-5)} = \frac{2 + 1}{-1 + 5} = \frac{3}{4}$$

$$m_{BC} = \frac{5 - 2}{3 - (-1)} = \frac{3}{3 + 1} = \frac{3}{4}$$

Since the slopes are equal, we must have one line or two parallel lines. But the line segments have a point (B) in common, so all three points lie on the same line.

Practice Problem 5 Without plotting the points, show that $A(1, 5)$, $B(-1, 1)$, and $C(-2, -1)$ lie on the same line.

Graphing Calculator

Perpendicular Lines

In order to have perpendicular lines appear perpendicular to each other on a graphing calculator, you must use a "square" window setting. In other words, the actual unit length along both axes must be the same. You can produce square windows on most calculators with a few keystrokes.

Using a graphing calculator, graph the following equations on one coordinate system.

$$y_1 = 1.5625x + 1.8314$$
$$y_2 = -0.64x - 1.3816$$

Do the graphs appear to represent perpendicular lines? Are the lines really perpendicular? Why or why not?

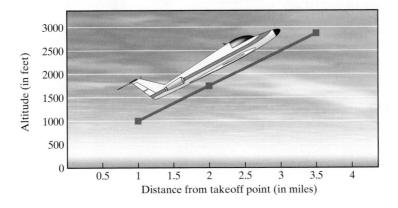

At takeoff an aircraft climbs into the sky at a certain rate of speed. This measurement of a slope is often called the *rate of climb*.

EXAMPLE 6 A gulfstream jet takes off from Orange County Airport in California. At 1 mile from the takeoff point, the jet is 1000 feet above the ground and begins a specified rate of climb. At 2 miles from the takeoff point, it is 1750 feet above the ground. At 3.5 miles from the takeoff point, it is 2865 feet above the ground. Look at the graph at the top of the page. Does it appear that the jet traveled in a straight line from the 1-mile point to the 3.5-mile point?

Solution From the 1-mile point to the 2-mile point, the jet traveled 750 feet upward over 1 horizontal mile. From the 2-mile point to the 3.5-mile point, the jet traveled 1115 feet upward over 1.5 horizontal miles. Thus, the first slope is 750 feet per mile, and the second slope is $743\frac{1}{3}$ feet per mile. These slopes are not the same, so the jet has not traveled in a straight line.

Practice Problem 6 On the return flight, the jet is at an altitude of 4850 feet when it is 4.5 miles from the airport. Its altitude is 3650 feet when it is 3.5 miles from the airport and 1010 feet when it is 1.3 miles from the airport. Does it appear that the jet is descending in a straight line?

Developing Your Study Skills

Getting Organized for an Exam

Studying adequately for an exam requires careful preparation. Begin early so that you will be able to spread your review over several days. Even though you may still be learning new material at this time, you can be reviewing concepts previously learned in the chapter. Giving yourself plenty of time for review will take the pressure off. You need this time to process what you have learned and to tie concepts together.

Adequate preparation enables you to feel confident and to think clearly with less anxiety.

Verbal and Writing Skills

1. Slope measures _____ change (rise) versus _____ change (run).

2. A positive slope indicates that the line slopes _____ to the right.

3. The slope of a horizontal line is _____.

✓ 4. Two different lines are parallel if their slopes are _____.

5. Does the line passing through $(-3, -7)$ and $(-3, 5)$ have a slope? Give a reason for your answer.

6. Let $(x_1, y_1) = (-6, -3)$ and $(x_2, y_2) = (-4, 5)$. Find $\dfrac{y_2 - y_1}{x_2 - x_1}$ and $\dfrac{y_1 - y_2}{x_1 - x_2}$. Are the results the same? Why or why not?

Find the slope, if possible, of the line passing through each pair of points.

7. $(2, 2)$ and $(6, -6)$

✓ 8. $(2, -1)$ and $(6, 3)$

9. $\left(\dfrac{3}{2}, 4\right)$ and $(-2, 0)$

10. $(6, 1)$ and $\left(0, \dfrac{1}{3}\right)$

11. $(6.8, -1.5)$ and $(5.6, -2.3)$

12. $(-2, 5.2)$ and $(4.8, -1.6)$

13. $\left(\dfrac{3}{2}, -2\right)$ and $\left(\dfrac{3}{2}, \dfrac{1}{4}\right)$

14. $\left(\dfrac{7}{3}, -6\right)$ and $\left(\dfrac{7}{3}, \dfrac{1}{6}\right)$

15. $(-7, -3)$ and $(10, -3)$

16. $(4, 12)$ and $(-5, 12)$

17. $\left(-\dfrac{1}{3}, -2\right)$ and $(1, -4)$

18. $\left(5, \dfrac{1}{2}\right)$ and $(2, 0)$

Applications

19. **Snowboarding** Find the slope (grade) of a snowboard recreation hill that rises 48 feet vertically over a horizontal distance of 80 feet.

20. **Grade of a Driveway** Find the grade of a driveway that rises 3.2 feet vertically over a horizontal distance of 80 feet. (The grade of a road or driveway is usually expressed as a percent.)

21. ***Rock Formation*** Find the slope (pitch) of a perfectly smooth rock formation that rises 35.7 feet vertically over a horizontal distance of 142.8 feet.

22. ***Pitch of a Roof*** Find the slope (pitch) of a roof that rises 3.15 feet vertically over a horizontal distance of 10.50 feet.

23. ***Slope of Street*** Baldwin Street in Dunedin, New Zealand, is the world's steepest residential street. The slope of this street is 0.35. How many feet vertically does it rise over a horizontal distance of 120 feet?

24. ***Slope of Roller Coaster*** The El Toro Roller Coaster at Six Flags Great Adventure in New Jersey has the steepest drop of any wooden roller coaster in the world. The slope of the drop is 4, or $\frac{4}{1}$. The horizontal distance of the drop is 44 feet. How many feet does it fall vertically?

Find the slope of a line parallel to the line that passes through the following points.

25. $(6, 7)$ and $(24, 3)$

26. $(35, -3)$ and $(5, 9)$

27. $(5.5, 2)$ and $(5, 4)$

28. $(4, 0)$ and $(3.8, 2)$

29. $\left(-9, \frac{1}{2}\right)$ and $(-6, 5)$

30. $\left(1, \frac{5}{2}\right)$ and $\left(\frac{1}{3}, 2\right)$

Find the slope of a line perpendicular to the line that passes through the following points.

31. $(8, 12)$ and $(3, 9)$

32. $(3, 9)$ and $(7, 15)$

33. $\left(3, \frac{1}{2}\right)$ and $\left(2, -\frac{3}{2}\right)$

34. $\left(\frac{1}{4}, -1\right)$ and $\left(\frac{5}{4}, \frac{1}{2}\right)$

35. $(-8.4, 0)$ and $(0, 4.2)$

36. $(0, -5)$ and $(-2, 0)$

To Think About

37. A line k passes through the points $(-3, -9)$ and $(1, 11)$. A second line h passes through the points $(-2, -13)$ and $(2, 7)$. Is line k parallel to line h? Why or why not?

38. A line k passes through the points $(4, 2)$ and $(-4, 4)$. A second line h passes through the points $(-8, 1)$ and $(8, -3)$. Is line k parallel to line h? Why or why not?

39. Show that $ABCD$ is a parallelogram if the four vertices are $A(2, 1)$, $B(-1, -2)$, $C(-7, -1)$, and $D(-4, 2)$. (*Hint:* A parallelogram is a four-sided figure with opposite sides parallel.)

40. Do the points $A(-1, -2)$, $B(2, -1)$, and $C(8, 1)$ lie on a straight line? Explain.

41. **Handicapped Ramp** Most new buildings are required to have a ramp for the handicapped that has a maximum vertical rise of 5 feet for every 60 feet of horizontal distance.

    (a) What is the value of the slope of a ramp for the handicapped?

    (b) If the builder constructs a new building in which the ramp has a horizontal distance of 24 feet, what is the maximum height of the doorway above the level of the parking lot where the ramp begins?

    (c) What is the shortest possible length of the ramp if the architect redesigns the building so that the doorway is 1.7 feet above the parking lot?

42. **Rate of Climb of Aircraft** A small Cessna plane takes off from Hyannis Airport. When the plane is 1 mile from the airport, it is flying at an altitude of 3000 feet. When the plane is 2 miles from the airport, it is flying at an altitude of 4300 feet. Round your answers to the following questions to the nearest tenth.

    (a) If the plane continues flying at the same slope (the same rate of climb), what will its altitude be when it is 4.8 miles from the airport?

    (b) If the plane continues flying at the same rate of climb, how many miles from the airport will it be when it reaches an altitude of 6000 feet?

    (c) A Lear jet leaves the airport at the same time and has the same altitude (3000 feet) as the Cessna when each plane is 1 mile from the airport. When the jet is 1.8 miles from the airport, it is flying at an altitude of 4040 feet. Is the jet being flown at the same rate of climb as the Cessna?

Cumulative Review *Evaluate.*

43. **[1.3.3]** $\dfrac{5 + 3\sqrt{9}}{|2 - 9|}$

44. **[1.3.3]** $2(3 - 6)^3 + 20 \div (-10)$

Simplify.

45. **[1.4.3]** $\dfrac{-15x^6y^3}{-3x^{-4}y^6}$

46. **[1.5.3]** $8x(x - 1) - 2(x + y)$

Quick Quiz 3.2

1. Find the slope of the line passing through $(-7, -2)$ and $\left(3, \dfrac{1}{2}\right)$.

2. Find the slope of the line passing through $(-8, -3)$ and $(5, -3)$.

3. Find the slope of a line that is *perpendicular* to the line passing through $(19, 4)$ and $(-3, -7)$.

4. **Concept Check** How would you find the grade (slope) of a driveway that rises 8.5 feet vertically over a horizontal distance of 120 feet?

1 Using the Slope–Intercept Form of the Equation of a Line

Recall that the standard form of the equation of a line is $Ax + By = C$. Although the standard form tells us that the graph is a straight line, it reveals little about the line. A more useful form of the equation is the **slope–intercept form.** The slope–intercept form immediately reveals the slope of the line and where it intersects the y-axis. This is important information that will help us graph the line.

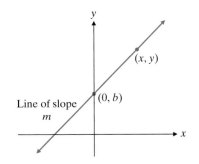

We will use the definition of slope to derive the equation. Let a nonvertical line with slope m cross the y-axis at some point $(0, b)$. Choose any other point on the line and label it (x, y). By the definition of slope, $\dfrac{y_2 - y_1}{x_2 - x_1} = m$. But here, $x_1 = 0$, $y_1 = b$, $x_2 = x$, and $y_2 = y$. So

$$\frac{y - b}{x} = m.$$

Now we solve this equation for y.

$$y - b = mx$$
$$y = mx + b$$

SLOPE–INTERCEPT FORM

The **slope–intercept form** of the equation of a line is $y = mx + b$, where m is the slope and $(0, b)$ is the y-intercept.

Write the Equation of a Line Given Its Slope and y-intercept

EXAMPLE 1 Write an equation of the line with slope $-\frac{2}{3}$ and y-intercept $(0, 5)$.

Solution
$$y = mx + b$$
$$y = \left(-\frac{2}{3}\right)x + (5) \quad \text{Substitute } -\frac{2}{3} \text{ for } m \text{ and } 5 \text{ for } b.$$
$$y = -\frac{2}{3}x + 5$$

Practice Problem 1 Write an equation of the line with slope 4 and y-intercept $\left(0, -\frac{3}{2}\right)$.

Write the Equation of a Line Given the Graph We can write an equation of a line if we are given its graph since we can determine the y-intercept and the slope from the graph.

Student Learning Objectives

After studying this section, you will be able to:

1 Use the slope–intercept form of the equation of a line.

2 Use the point–slope form of the equation of a line.

3 Write the equation of the line passing through a given point that is parallel or perpendicular to a given line.

Graphing Calculator

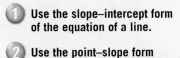

Exploring y-intercepts

Using a graphing calculator, graph
$$y_1 = 2x,$$
$$y_2 = 2x + 1,$$
$$y_3 = 2x - 1, \text{ and}$$
$$y_4 = 2x + 2$$

on the same set of axes. Where does each graph cross the y-axis? What effect does b have on the graph of $y = mx + b$? What would the graph of the line $y = 2x - 5$ look like? Use your graphing calculator to verify your conclusion.

Graphing Calculator

Exploring Slopes

Using a graphing calculator, graph

$$y_1 = x + 1,$$
$$y_2 = 3x + 1, \text{ and}$$
$$y_3 = \frac{1}{3}x + 1$$

on the same set of axes. What effect does m have on the graph of $y = mx + b$? What would the graph of the line $y = \frac{1}{2}x + 1$ look like?

PRACTICE PROBLEM 2

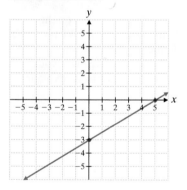

EXAMPLE 2 Find the slope and y-intercept of the line whose graph is shown. Then use these to write an equation of the line.

Solution Looking at the graph, we can see that the y-intercept is at $(0, 5)$. That is, $b = 5$. If we can identify the coordinates of another point on the line, we will have two points and we can determine the slope. Another point on the line is $(3, 3)$.

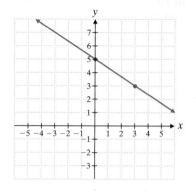

Thus,

$$(x_1, y_1) = (0, 5),$$
$$(x_2, y_2) = (3, 3), \text{ and}$$
$$m = \frac{y_2 - y_1}{x_2 - x_1} = \frac{3 - 5}{3 - 0} = -\frac{2}{3}.$$

We can now write the equation of the line in slope–intercept form.

$$y = mx + b$$
$$y = -\frac{2}{3}x + 5 \quad \text{Substitute } -\frac{2}{3} \text{ for } m \text{ and 5 for } b.$$

Practice Problem 2 Find the slope and y-intercept of the line whose graph is shown in the margin on the left. Then use these to write an equation of this line.

Use the Slope–Intercept Form to Graph an Equation We have just seen that given the graph of a line, we can determine its equation by identifying the y-intercept and finding the slope. We can also draw the graph of an equation without plotting points if we can write the equation in slope–intercept form and then locate the y-intercept on the graph.

EXAMPLE 3 Find the slope and the y-intercept. Then sketch the graph of the equation $28x - 7y = 21$.

Solution First we will change the standard form of the equation into slope–intercept form. This is a very important procedure. Be sure that you understand each step.

$$28x - 7y = 21$$
$$-7y = -28x + 21$$
$$\frac{-7y}{-7} = \frac{-28x}{-7} + \frac{21}{-7}$$

$$\overbrace{y = 4x}^{\text{slope}} + (-3) \qquad y = mx + b$$
$$\underbrace{}_{\text{gives } y\text{-intercept}}$$

Thus, the slope is 4, and the y-intercept is $(0, -3)$.

To sketch the graph, begin by plotting the point where the graph crosses the y-axis, $(0, -3)$. Plot the point. Now look at the slope. The slope, m, is 4 or $\frac{4}{1}$. This means there is a rise of 4 for every run of 1. From the point $(0, -3)$ go up 4 units and to the right 1 unit to locate a second point on the line. Draw the straight line that contains these two points, and you have the graph of the equation $28x - 7y = 21$.

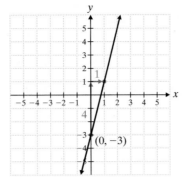

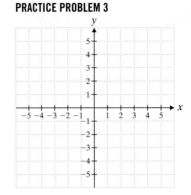

Practice Problem 3 Find the slope and the y-intercept. Then sketch the graph of the equation $3x - 4y = -8$.

NOTE TO STUDENT: Fully worked-out solutions to all of the Practice Problems can be found at the back of the text starting at page SP-1

2 Using the Point–Slope Form of the Equation of a Line

What happens if we know the slope of a line and a point on the line that is not the y-intercept? Can we write the equation of the line? By the definition of slope, we have the following:

$$m = \frac{y - y_1}{x - x_1}$$

$$m(x - x_1) = y - y_1$$

That is, $y - y_1 = m(x - x_1)$.

This is the point–slope form of the equation of a line.

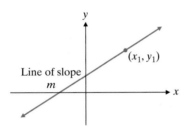

Line of slope m

(x_1, y_1)

POINT–SLOPE FORM

The **point–slope form** of the equation of a line is $y - y_1 = m(x - x_1)$, where m is the slope and (x_1, y_1) are the coordinates of a known point on the line.

Write the Equation of a Line Given Its Slope and One Point on the Line

EXAMPLE 4 Find an equation of the line that has slope $-\frac{3}{4}$ and passes through the point $(-6, 1)$. Express your answer in standard form.

Solution Since we don't know the y-intercept, we can't use the slope–intercept form easily. Therefore, we use the point–slope form.

$$y - y_1 = m(x - x_1)$$

$$y - 1 = -\frac{3}{4}[x - (-6)] \quad \text{Substitute the given values.}$$

$$y - 1 = -\frac{3}{4}x - \frac{9}{2} \quad \text{Simplify. (Do you see how we did this?)}$$

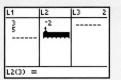
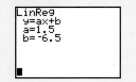
$$4y - 4(1) = 4\left(-\frac{3}{4}x\right) - 4\left(\frac{9}{2}\right) \quad \text{Multiply each term by the LCD 4.}$$

$$4y - 4 = -3x - 18 \qquad \text{Simplify.}$$

$$3x + 4y = -18 + 4 \qquad \text{Add } 3x + 4 \text{ to each side.}$$

$$3x + 4y = -14 \qquad \text{Add like terms.}$$

The equation in standard form is $3x + 4y = -14$.

Practice Problem 4 Find an equation of the line that passes through $(5, -2)$ and has a slope of $\frac{3}{4}$. Express your answer in standard form.

Write the Equation of a Line Given Two Points on the Line We can use the point–slope form to find the equation of a line if we are given two points. Carefully study the following example. Be sure you understand each step. You will encounter this type of problem frequently.

EXAMPLE 5 Find an equation of the line that passes through $(3, -2)$ and $(5, 1)$. Express your answer in slope–intercept form.

Solution First we find the slope.

$$m = \frac{y_2 - y_1}{x_2 - x_1} = \frac{1 - (-2)}{5 - 3} = \frac{1 + 2}{2} = \frac{3}{2}$$

Now we substitute the value of the slope and the coordinates of either point into the point–slope equation. Let's use $(5, 1)$.

$$y - y_1 = m(x - x_1)$$

$$y - 1 = \frac{3}{2}(x - 5) \qquad \text{Substitute } m = \frac{3}{2} \text{ and } (x_1, y_1) = (5, 1).$$

$$y - 1 = \frac{3}{2}x - \frac{15}{2} \qquad \text{Remove parentheses.}$$

$$y = \frac{3}{2}x - \frac{15}{2} + 1 \qquad \text{Add 1 to each side of the equation.}$$

$$y = \frac{3}{2}x - \frac{15}{2} + \frac{2}{2} \qquad \text{Add the two fractions.}$$

$$y = \frac{3}{2}x - \frac{13}{2} \qquad \text{Simplify.}$$

Practice Problem 5 Find an equation of the line that passes through $(-4, 1)$ and $(-2, -3)$. Express your answer in slope–intercept form.

Before we go further, we want to point out that these various forms of the equation of a straight line are just that—*forms* for convenience. We are *not* using different equations each time, nor should you simply try to memorize the different variations without understanding when to use them. They can easily be derived from the definition of slope, as we have seen. And remember, you can *always* use the definition of slope to find the equation of a line. You may find it helpful to review Examples 4 and 5 for a few minutes before going ahead to Example 6. It is important to see how each example is different.

3 Writing the Equation of a Parallel or Perpendicular Line

Let us now look at parallel and perpendicular lines. If we are given the equation of a line and a point not on the line, we can find the equation of a second line that passes through the given point and is parallel or perpendicular to the first line. We can do this because we know that the slopes of parallel lines are equal and that the slopes of perpendicular lines are negative reciprocals of each other.

We begin by finding the slope of the given line. Then we use the point–slope form to find the equation of the second line. Study each step of the following example carefully.

EXAMPLE 6 Find an equation of the line passing through the point $(-2, -4)$ and parallel to the line $2x + 5y = 8$. Express the answer in standard form.

Solution   First we need to find the slope of the line $2x + 5y = 8$. We do this by writing the equation in slope–intercept form.

$$5y = -2x + 8$$
$$y = -\frac{2}{5}x + \frac{8}{5}$$

The slope of the given line is $-\frac{2}{5}$. Since parallel lines have the same slope, the slope of the unknown line is also $-\frac{2}{5}$. Now we substitute $m = -\frac{2}{5}$ and the coordinates of the point $(-2, -4)$ into the point–slope form of the equation of a line.

$y - y_1 = m(x - x_1)$	
$y - (-4) = -\dfrac{2}{5}[x - (-2)]$	Substitute.
$y + 4 = -\dfrac{2}{5}(x + 2)$	Simplify.
$y + 4 = -\dfrac{2}{5}x - \dfrac{4}{5}$	Remove parentheses.
$5y + 5(4) = 5\left(-\dfrac{2}{5}x\right) - 5\left(\dfrac{4}{5}\right)$	Multiply each term by the LCD 5.
$5y + 20 = -2x - 4$	Simplify.
$2x + 5y = -4 - 20$	Add $2x - 20$ to each side.
$2x + 5y = -24$	Simplify.

$2x + 5y = -24$ is an equation of the line passing through the point $(-2, -4)$ and parallel to the line $2x + 5y = 8$.

Practice Problem 6 Find an equation of the line passing through $(4, -5)$ and parallel to the line $5x - 3y = 10$. Express the answer in standard form.

Some extra steps are needed if the desired line is to be perpendicular to the given line. Carefully note the approach in Example 7.

EXAMPLE 7 Find an equation of the line that passes through the point $(2, -3)$ and is perpendicular to the line $3x - y = -12$. Express the answer in standard form.

Solution To find the slope of the line $3x - y = -12$, we rewrite it in slope–intercept form.

$$-y = -3x - 12$$

$$y = 3x + 12$$

This line has a slope of 3. Therefore, the slope of a line perpendicular to this line is the negative reciprocal $-\frac{1}{3}$.

Now substitute the slope $m = -\frac{1}{3}$ and the coordinates of the point $(2, -3)$ into the point–slope form of the equation.

$$y - y_1 = m(x - x_1)$$

$$y - (-3) = -\frac{1}{3}(x - 2) \qquad \text{Substitute.}$$

$$y + 3 = -\frac{1}{3}(x - 2) \qquad \text{Simplify.}$$

$$y + 3 = -\frac{1}{3}x + \frac{2}{3} \qquad \text{Remove parentheses.}$$

$$3y + 3(3) = 3\left(-\frac{1}{3}x\right) + 3\left(\frac{2}{3}\right) \qquad \text{Multiply each term by the LCD 3.}$$

$$3y + 9 = -x + 2 \qquad \text{Simplify.}$$

$$x + 3y = 2 - 9 \qquad \text{Add } x - 9 \text{ to each side.}$$

$$x + 3y = -7 \qquad \text{Simplify.}$$

$x + 3y = -7$ is an equation of the line that passes through the point $(2, -3)$ and is perpendicular to the line $3x - y = -12$.

NOTE TO STUDENT: Fully worked-out solutions to all of the Practice Problems can be found at the back of the text starting at page SP-1

Practice Problem 7 Find an equation of the line that passes through $(-4, 3)$ and is perpendicular to the line $6x + 3y = 7$. Express the answer in standard form.

Developing Your Study Skills

Making a Friend in the Class

Try to make a friend in your class. You may find that you enjoy sitting together and drawing support and encouragement from one another. Exchange phone numbers so you can call each other whenever you get stuck while doing your homework. Set up convenient times to study together on a regular basis, to do homework, and to review for exams.

You must not depend on a friend or fellow student to tutor you, do your work for you, or in any way be responsible for your learning. However, you will learn from one another as you seek to master the course. Studying with a friend and comparing notes, methods, and solutions can be very helpful. And it can make learning mathematics a lot more fun!

Verbal and Writing Skills

1. You are given two points that lie on a line. Explain how you would find an equation of the line.

2. Suppose $y = -\frac{2}{7}x + 5$. What can you tell about the graph by looking at the equation?

Write an equation of the line with the given slope and the given y-intercept. Leave the answer in slope–intercept form.

3. Slope $\frac{3}{4}$, y-intercept $(0, -9)$

✔ **4.** Slope $-\frac{2}{3}$, y-intercept $(0, 5)$

Write an equation of the line with the given slope and the given y-intercept. Express the answer in standard form.

5. Slope $\frac{3}{4}$, y-intercept $\left(0, \frac{1}{2}\right)$

6. Slope $\frac{5}{6}$, y-intercept $\left(0, \frac{1}{3}\right)$

Find the slope and y-intercept of each of the following lines. Then use these to write an equation of the line.

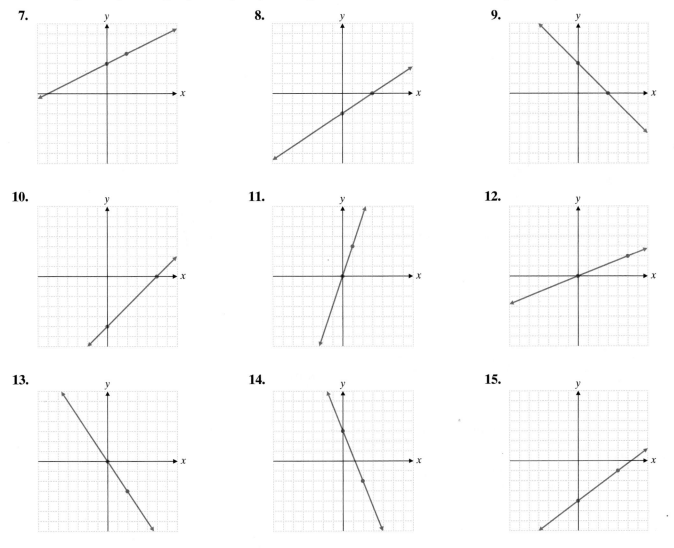

Write each equation in slope–intercept form. Then identify the slope and the y-intercept for each line.

16. $2x - y = 12$

17. $x - y = 5$

18. $2x - 3y = -8$

19. $5x - 4y = -20$

20. $\frac{1}{3}x + 2y = 7$

21. $2x + \frac{3}{4}y = -3$

For each equation find the slope and the y-intercept. Use these to graph the equation.

22. $y = 3x + 4$

23. $y = \frac{1}{2}x - 3$

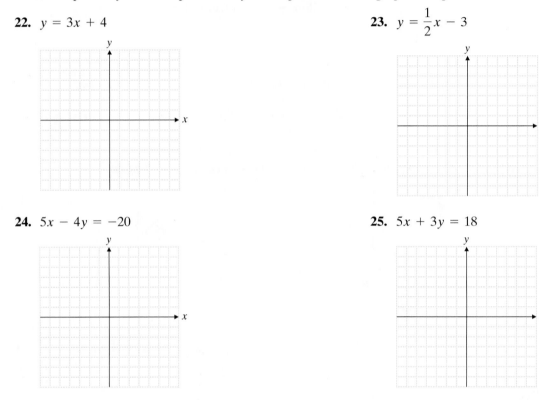

24. $5x - 4y = -20$

25. $5x + 3y = 18$

Find an equation of the line that passes through the given point and has the given slope. Express your answer in slope–intercept form.

26. $(6, 5), m = \frac{1}{3}$

27. $(4, 4), m = -\frac{1}{2}$

28. $(8, 0), m = -3$

29. $(-7, -2), m = 5$

30. $(0, -1), m = -\frac{5}{3}$

31. $(6, 0), m = -\frac{1}{5}$

Find an equation of the line passing through the pair of points. Write the equation in slope–intercept form.

32. $(5, -3)$ and $(1, -4)$

33. $(-4, -1)$ and $(3, 4)$

34. $\left(\frac{7}{6}, 1\right)$ and $\left(-\frac{1}{3}, 0\right)$

35. $\left(\frac{1}{2}, -3\right)$ and $\left(\frac{7}{2}, -5\right)$

36. $(4, 8)$ and $(-3, 8)$

37. $(12, -3)$ and $(7, -3)$

Find an equation of the line satisfying the conditions given. Express your answer in standard form.

38. Parallel to $3x - y = -5$ and passing through $(-1, 0)$

39. Parallel to $5x - y = 4$ and passing through $(-2, 0)$

40. Parallel to $2y + x = 7$ and passing through $(-5, -4)$

41. Parallel to $x = 3y - 8$ and passing through $(5, -1)$

42. Perpendicular to $y = 5x$ and passing through $(4, -2)$

43. Perpendicular to $2y = -3x$ and passing through $(6, -1)$

44. Perpendicular to $x - 4y = 2$ and passing through $(3, -1)$

45. Perpendicular to $x + 7y = -12$ and passing through $(-4, -1)$

To Think About *Without graphing, determine whether the following pairs of lines are (a) parallel, (b) perpendicular, or (c) neither parallel nor perpendicular.*

46. $5x - 6y = 19$
$6x + 5y = -30$

47. $-3x + 5y = 40$
$5y + 3x = 17$

48. $y = \dfrac{2}{3}x + 6$
$-2x - 3y = -12$

49. $y = -\dfrac{3}{4}x - 2$
$6x + 8y = -5$

50. $y = \dfrac{3}{7}x - \dfrac{1}{14}$
$14y + 6x = 3$

51. $y = \dfrac{5}{6}x - \dfrac{1}{3}$
$6x + 5y = -12$

Optional Graphing Calculator Problems *If you have a graphing calculator, use it to graph each pair of equations. Do the graphs appear to be parallel?*

52. $y = -2.39x + 2.04$ and $y = -2.39x - 0.87$

53. $y = 1.43x - 2.17$ and $y = 1.43x + 0.39$

Applications

Cost of Homes *The median sale price of single-family homes in the United States has been increasing steadily. The increase can be approximated by a linear equation of the form $y = mx + b$. The U.S. Census Bureau reported that in 1995 the median sale price of a single-family home in the United States was $133,900. In 2005, the median sale price of a single-family home was $240,900. We can record the data as follows:*

Number of Years Since 1995	Price of Home in Thousands of Dollars
0	133.9
10	240.9

Source: www.census.gov

Use the table of values on p. 151 for exercises 54–57.

54. Using these two ordered pairs, find the equation $y = mx + b$ where x is the number of years since 1995 and y is the median sale price of a single-family home in thousands of dollars.

55. Use the equation obtained in exercise 54 to find the expected median sale price of a single-family home in the year 2015 (20 years after 1995).

56. Graph the equation using the data for 1995, 2005, and 2015.

57. Use your graph to estimate the median cost of a home in 2010 (15 years after 1995).

Median Sale Price of Single-Family Homes

Cost of homes (in thousands of dollars)

Number of years since 1995

Source: www.census.gov

Cumulative Review *Solve for x.*

58. **[2.1.1]** $11 - (x + 2) = 7(3x + 6)$

59. **[2.1.1]** $0.3x + 0.1 = 0.27x - 0.02$

60. **[2.1.1]** $70 + 70(0.01x) + 3 = 82.10$

61. **[2.2.1]** $\dfrac{5}{4} - \dfrac{3}{4}(2x + 1) = x - 2$

Quick Quiz 3.3

1. Write the following equation in slope–intercept form. Then identify the slope and y-intercept. $-3x + 7y = -9$

2. Find the standard form of the equation of the line that passes through $(-3, 2)$ and $(-4, -7)$.

3. Find an equation of the line with slope $-\frac{3}{4}$ and y-intercept $(0, -5)$. Write the answer in slope–intercept form.

4. **Concept Check** How would you find an equation of the horizontal line that passes through the point $(-4, -8)$?

How are you doing with your homework assignments in Sections 3.1 to 3.3? Do you feel you have mastered the material so far? Do you understand the concepts you have covered? Before you go further in the textbook, take some time to do each of the following problems.

1. Find the value of a if $(a, 6)$ is a solution to $5x + 2y = -12$.

Graph each equation.

2. $y = -\dfrac{1}{2}x + 5$

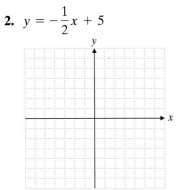

3. $5x + 3y = -15$

4. $4y + 6x = -8 + 9x$

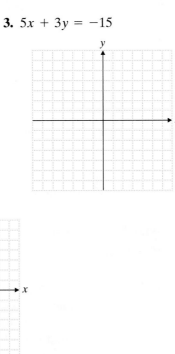

5. Find the slope of the line passing through the points $(-2, 3)$ and $(-1, -6)$.

6. Find the slope of a line parallel to the line that passes through $\left(\dfrac{2}{3}, 4\right)$ and $\left(\dfrac{5}{6}, -2\right)$.

7. Find the slope of a line perpendicular to the line passing through the points $(5, 0)$ and $(-13, -4)$.

8. A Colorado road has a slope of 0.13. How many feet does it rise vertically over a horizontal distance of 500 feet?

9. Find the slope and the y-intercept of $5x + 3y = 9$.

10. Write an equation of the line of slope -2 that passes through $(7, -3)$.

11. Write an equation of the line passing through $(-1, -2)$ and perpendicular to $3x - 5y = 10$.

12. Find an equation of the line that passes through $(-3, -10)$ and $(1, 2)$.

Now turn to page SA-6 for the answers to each of these problems. Each answer also includes a reference to the objective in which the problem is first taught. If you missed any of these problems, you should stop and review the Examples and Practice Problems in the referenced objective. A little review now will help you master the material in the upcoming sections of the text.

1. _____

2. _____

3. _____

4. _____

5. _____

6. _____

7. _____

8. _____

9. _____

10. _____

11. _____

12. _____

① **Graphing a Linear Inequality in Two Variables**

A linear inequality in two variables is similar to a linear equation in two variables. However, in place of the = sign, there appears instead one of the following four inequality symbols: $<, >, \leq, \geq$.

LINEAR INEQUALITY IN TWO VARIABLES

A **linear inequality in two variables** is an inequality that can be written

$$\text{as} \quad Ax + By > C \quad \text{or} \quad Ax + By < C$$
$$\text{or} \quad Ax + By \geq C \quad \text{or} \quad Ax + By \leq C$$

where A, B, and C are real numbers and A and B are not both zero.

The graph of this type of linear inequality is a half-plane that lies on one side of a straight line. It will also include the boundary line if the inequality contains the $\leq$ or the $\geq$ symbol.

PROCEDURE FOR GRAPHING LINEAR INEQUALITIES

1. Replace the inequality symbol by an equals sign. This equation will be the boundary for the desired region.

2. Graph the line obtained in step 1. Use a dashed line if the original inequality contains a $<$ or $>$ symbol. Use a solid line if the original inequality contains a $\leq$ or $\geq$ symbol.

3. Choose a test point that does not lie on the boundary line. Substitute the coordinates into the original inequality. If you obtain an inequality that is true, shade the region on the side of the line containing the test point. If you obtain a false inequality, shade the region on the side of the line opposite the test point.

If the boundary line does not pass through $(0, 0)$, that is usually a good test point to use.

EXAMPLE 1 Graph $y < 2x + 3$.

Solution

1. The boundary line is $y = 2x + 3$.
2. We graph $y = 2x + 3$ using a dashed line because the inequality contains $<$.
3. Since the line does not pass through $(0, 0)$, we can use it as a test point. Substituting $(0, 0)$ into $y < 2x + 3$, we have the following:

$$0 < 2(0) + 3$$
$$0 < 3$$

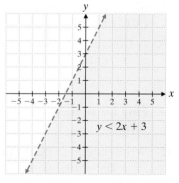

This inequality is true. We therefore shade the region on the same side of the line as $(0, 0)$. See the sketch. The solution is the shaded region *not including* the dashed line.

PRACTICE PROBLEM 1

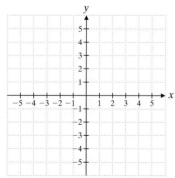

 Graph $y > 3x + 1$.

EXAMPLE 2   Graph $3x + 2y \geq 4$.

Solution

1. The boundary line is $3x + 2y = 4$.
2. We graph the boundary line with a solid line because the inequality contains $\geq$.
3. Since the line does not pass through $(0, 0)$, we can use it as a test point. Substituting $(0, 0)$ into $3x + 2y \geq 4$ gives the following:

$$3(0) + 2(0) \geq 4$$
$$0 + 0 \geq 4$$
$$0 \geq 4$$

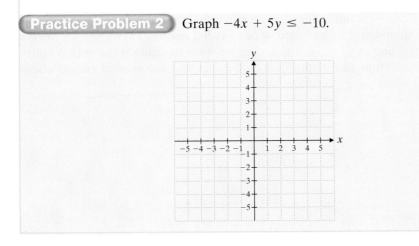

This inequality is false. We therefore shade the region on the side of the line opposite $(0, 0)$. See the sketch. The solution is the shaded region *including* the boundary line.

Practice Problem 2   Graph $-4x + 5y \leq -10$.

Graphing Calculator

 **Graphing Linear Inequalities**

You can graph a linear inequality like $y \leq 3x + 1$ on a graphing calculator. On some calculators you can enter the expression for the boundary directly into the $Y =$ editor of your graphing calculator and then select the appropriate direction for shading. Other calculators may require using a Shade command to shade the region. In general, most calculator displays do not distinguish between dashed and solid boundaries.

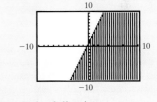

Graph the following:

$$y < 3.45x - 1.232$$

EXAMPLE 3   Graph $4x - y < 0$.

Solution

1. The boundary line is $4x - y = 0$.
2. We graph the boundary line with a dashed line.
3. Since $(0, 0)$ lies on the line, we cannot use it as a test point. We must choose another point not on the line. We try to pick some point that is *not* close to the dashed boundary line. Let's pick $(-1, 5)$. Substituting $(-1, 5)$ into $4x - y < 0$ gives the following:

$$4(-1) - 5 < 0$$
$$-4 - 5 < 0$$
$$-9 < 0$$

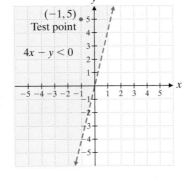

This is true. Therefore, we shade the side of the boundary line that contains $(-1, 5)$. See the sketch. The solution is the shaded region above the dashed line but *not including* the dashed line.

Practice Problem 3   Graph $3y + x < 0$.

PRACTICE PROBLEM 3

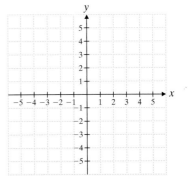

② Graphing a Linear Inequality for Which the Coefficient of One Variable Is Zero

EXAMPLE 4   Simplify and graph $3y \geq -12$.

Solution   First we will need to simplify the original inequality by dividing each side by 3.

$$3y \geq -12$$

$$\frac{3y}{3} \geq \frac{-12}{3}$$

$$y \geq -4$$

This inequality is equivalent to

$$0x + y \geq -4.$$

We find that if $y \geq -4$, any value of x will make the inequality true. Thus, x can be any value at all and still be included in our shaded region. Therefore, we draw a solid horizontal line at $y = -4$. The region we want to shade is the region above the line. The solution to our problem is the line and the shaded region above the line.

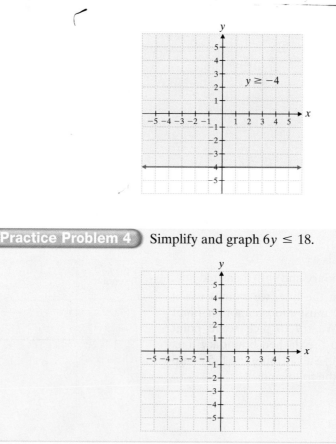

$y \geq -4$

NOTE TO STUDENT: *Fully worked-out solutions to all of the Practice Problems can be found at the back of the text starting at page SP-1*

Practice Problem 4   Simplify and graph $6y \leq 18$.

Verbal and Writing Skills

1. Explain when to use a dashed line as a boundary when graphing a linear inequality.

✓ 2. Explain when to use a solid line as a boundary when graphing a linear inequality.

3. When graphing $x > 5$, should the region to the left of the line $x = 5$ be shaded or should the region to the right of the line $x = 5$ be shaded?

✓ 4. When graphing $y < -6$ should the region below the line $y = -6$ be shaded or should the region above the line $y = -6$ be shaded?

5. When we graph the inequality $3x - 2y \geq 0$, why can't we use $(0, 0)$ as a test point? If we test the point $(-4, 2)$, do we obtain a false statement or a true one?

✓ 6. When we graph the inequality $4x - 3y \geq 0$, why can't we use $(0, 0)$ as a test point? If we test the point $(6, -5)$, do we obtain a false statement or a true one?

$$4(6) - 3(-5) \geq 0$$
$$24 \ \angle 15$$

Graph each region.

7. $y > -2x + 4$

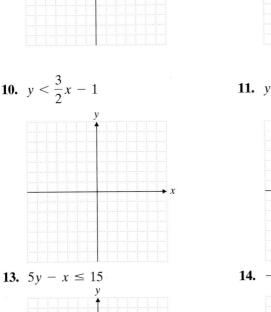

8. $y > -3x + 2$

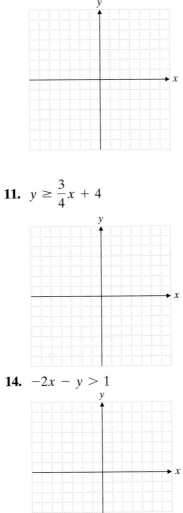

9. $y < \dfrac{2}{3}x - 2$

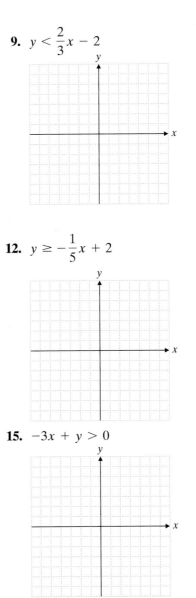

10. $y < \dfrac{3}{2}x - 1$

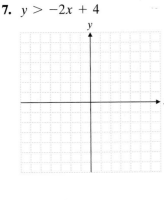

11. $y \geq \dfrac{3}{4}x + 4$

12. $y \geq -\dfrac{1}{5}x + 2$

13. $5y - x \leq 15$

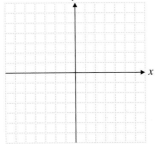

14. $-2x - y > 1$

15. $-3x + y > 0$

16. $x + y < 0$ **17.** $5x - 2y \geq 0$ **18.** $x - 3y \geq 0$

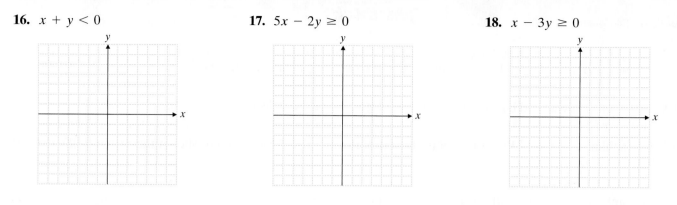

Simplify and graph each inequality in a rectangular coordinate system.

19. $x > -4$ **20.** $x < 3$ **21.** $y \leq -1$

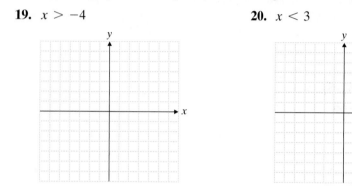
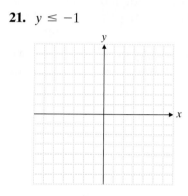

22. $y \geq -1$ **23.** $-8x \leq -12$ **24.** $-5x \leq -10$

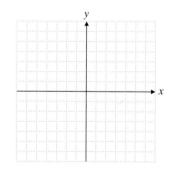
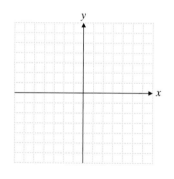
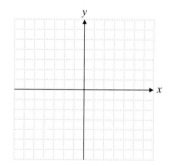

25. $4y \geq 2$ **26.** $3y + 2 > 0$

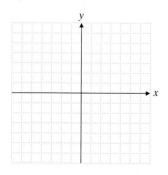
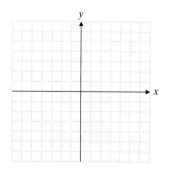

Cumulative Review

27. **[1.6.1]** Evaluate $2x^2 - 5x + 4$ when $x = -3$.

28. **[1.6.2]** Evaluate $A = \frac{1}{2}a(b + c)$ when $a = 4.0, b = 1.5$, and $c = 7.5$.

▲ **29.** **[1.2.2]** *Floor Tiles* A commons area at a summer camp measures 20 feet by 18 feet. Dan and Marsha Perkins want to cover the room with floor tiles. A package of 1-foot-square tiles contains thirty-two tiles. How many packages will they need to purchase?

Quick Quiz 3.4

1. The point $(3,-4)$ lies below the line $y = 3x + 4$. Now suppose we want to graph the region where $y > 3x + 4$. Test the point $(3,-4)$ in the inequality. Should the graph of the region be shaded above the line or below the line?

2. Graph $y < -\frac{3}{5}x + 3$.

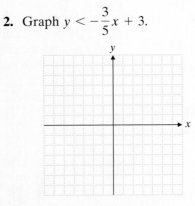

3. Graph $3x - 5y \leq -10$.

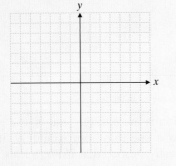

4. **Concept Check** Explain how you would graph the following region. $4y - 8 < 0$

— PIZZA PLUS —

Size of Pizza	Diameter of Pizza	Price of Pizza
Small	5"	$4.75
Medium	10"	$7.00
Large	15"	$9.25
Party Size	20"	$11.50

1 Describing a Relation and Determining Its Domain and Range

Whenever we collect and study data, we look for relationships. If we can determine a relationship between two sets of data, we can make predictions about the data. Let's begin with a simple finite example.

A local pizza parlor offers four different sizes of pizza and prices them as shown in the table on the left. By looking at the table, we can see that the price depends on the diameter of the pizza. The larger the size, the more expensive it will be. This appears to be an increasing relation. We can use a set of ordered pairs to show the correspondence between the diameter of the pizza and the price.

$$\{(5 \text{ in.}, \$4.75), \quad (10 \text{ in.}, \$7.00), \quad (15 \text{ in.}, \$9.25), \quad (20 \text{ in.}, \$11.50)\}$$

We can graph the ordered pairs to get a better picture of the relationship.

Following convention, we will assign the **independent variable** to the horizontal axis and the **dependent variable** to the vertical axis. Note that the dependent variable is price because price *depends* on size.

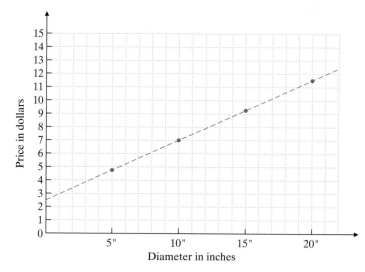

Just as we suspected, the relation is increasing. That is, the graph goes up as we move from left to right. We draw a line that approximately fits the data. This allows us to analyze the data more easily. Notice that if Pizza Plus decides to come out with a size between the small pizza and the medium pizza, it will probably be priced around $6. Thus, we can see that there is a relation between the diameter of a pizza and its price and that we can express this relation as a table of values, as a set of ordered pairs, or in a graph.

Mathematicians have found it most useful to define a relation in terms of ordered pairs.

RELATION

A **relation** is any set of ordered pairs.

All the first items of each ordered pair in a relation can be grouped as a set called the **domain** of the relation. The domain of the pizza example is the set of sizes (diameters measured in inches) {5, 10, 15, 20}. These are all the possible values of the independent variable *diameter*. The domain is the input. It is the starting value.

All the second items of each ordered pair in a relation can be grouped as a set called the **range** of the relation. The range of the pizza example is the set of prices {$4.75, $7.00, $9.25, $11.50}. These are the corresponding values of the dependent variable *price*. The range is the output. The output is the result that comes out once you pick a diameter.

EXAMPLE 1   The information in the following table can be found in most almanacs.

Look at the data for the men's 100-meter race. Is there a relation between any two sets of data in this table? If so, describe the relation as a table of values, a set of ordered pairs, and a graph.

Olympic Games: 100-Meter Race for Men		
Year	**Winning runner, Country**	**Time**
1900	Francis W. Jarvis, USA	11.0 s
1912	Ralph Craig, USA	10.8 s
1924	Harold Abrahams, Great Britain	10.6 s
1936	Jesse Owens, USA	10.3 s
1948	Harrison Dillard, USA	10.3 s
1960	Armin Harg, Germany	10.2 s
1972	Valery Borzov, USSR	10.14 s
1984	Carl Lewis, USA	9.99 s
1996	Donovan Bailey, Canada	9.84 s
2000	Maurice Greene, USA	9.87 s
2004	Justin Gatlin, USA	9.85 s

Source: The World Almanac.

Solution   A useful relation might be the correspondence between the year the event occurred and the time in which the race was won. Let's see how this looks in a table, as a set of ordered pairs, and on a graph. Because we will choose the year as the independent variable, we will list it first in the table, as is customary.

Table.

Year	1900	1912	1924	1936	1948	1960	1972	1984	1996	2000	2004
Time in Seconds	11.0	10.8	10.6	10.3	10.3	10.2	10.14	9.99	9.84	9.87	9.85

Ordered pairs.

{(1900, 11.0), (1912, 10.8), (1924, 10.6), (1936, 10.3), (1948, 10.3), (1960, 10.2), (1972, 10.14), (1984, 9.99), (1996, 9.84), (2000, 9.87), (2004, 9.85)}

Graph. To save space, we draw the graph so that the time values on the vertical axis range from 9 seconds to 11 seconds rather than from 0 seconds to 11 seconds. We indicate a break like this in the scale on a vertical axis with the symbol ⌇ on the axis. On a horizontal axis we use the symbol ⌁. By looking at the graph, we can see that the winning time usually decreases each year. This is a decreasing relation most of the time. What might we expect the winning time to be in 2008? Can we expect the time to decrease indefinitely? Why or why not?

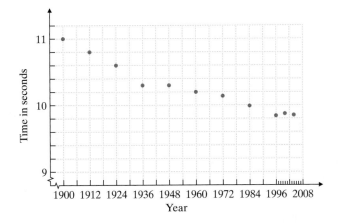

NOTE TO STUDENT: Fully worked-out solutions to all of the Practice Problems can be found at the back of the text starting at page SP-1

Practice Problem 1

The data for the women's 100-meter race in the Olympic Games for selected years are given in the table on the right. Describe the relation between the year and the winning time in a table, as a set of ordered pairs, and as a graph.

	Olympic Games: 100-Meter Race for Women	
Year	**Winning runner, Country**	**Time**
1936	Helen Stephens, USA	11.5 s
1948	Francina Blankers-Koen, Netherlands	11.9 s
1960	Wilma Rudolph, USA	11.0 s
1972	Renate Stecher, East Germany	11.07 s
1984	Evelyn Ashford, USA	10.97 s
1996	Gail Devers, USA	10.94 s
2000	Ekaterina Thanar, Greece	11.12 s
2004	Yulia Nesterenko, Belarus	10.93 s

Source: The World Almanac.

Year								
Time in Seconds								

Determining Whether a Relation Is a Function

The two relations we have just discussed have a special characteristic. Each value in the domain is matched with exactly one value in the range. This is true of our pizza example. One size of pizza does not have several different prices. This is also true of the relation between the year and winning time in the 100-meter race for men. No year has several different winning times.

Looking at this issue in terms of ordered pairs, we say that no two different ordered pairs have the same first coordinate. We will list each set of ordered pairs so that you can verify this characteristic.

$\{(5 \text{ in.}, \$4.75), (10 \text{ in.}, \$7.00), (15 \text{ in.}, \$9.25), (20 \text{ in.}, \$11.50)\}$

$\{(1900, 11.0), (1912, 10.8), (1924, 10.6), (1936, 10.3), (1948, 10.3), (1960, 10.2),$
$(1972, 10.14), (1984, 9.99), (1996, 9.84), (2000, 9.87), (2004, 9.85)\}$

All such relations with this special property are called *functions*. Each input has only one output.

FUNCTION

A **function** is a relation in which no two different ordered pairs have the same first coordinate.

Notice that if we reverse the order of coordinates in the ordered pairs in Example 1, the resulting relation is not a function. We can see this readily if we list the ordered pairs.

$\{(11.0, 1900), (10.8, 1912), (10.6, 1924), (10.3, 1936), (10.3, 1948), (10.2, 1960),$
$(10.14, 1972), (9.99, 1984), (9.84, 1996), (9.87, 2000), (9.85, 2004)\}$

Two pairs, (10.3, 1936) and (10.3, 1948), have the same first coordinate. This relation is not a function.

EXAMPLE 2 Give the domain and range of each relation. Indicate whether the relation is a function.

(a) $g = \{(2, 8), (2, 3), (3, 7), (5, 12)\}$

(b) Individuals' Incomes and Taxes

Income in Dollars	14,000	18,000	24,500	33,000	50,000	50,000
Income Tax in Dollars	2350	2800	2900	3750	1350	7980

(c) Women's Tibia Bone Lengths and Heights

Length of Tibia Bone in Centimeters	33	34	35	36
Height of the Woman in Centimeters	151	154	156	159

Source: National Center for Health Statistics.

Solution Recall that the domain of a function consists of all the possible values of the independent variable or input. In a set of ordered pairs, this is the first item in each ordered pair.

The range of a function consists of the corresponding values of the dependent variable. In a set of ordered pairs, this is the second item or output in each ordered pair.

(a) Domain = $\{2, 3, 5\}$

Range = $\{8, 3, 7, 12\}$

(2, 8) and (2, 3) have the same first coordinate. Thus, g is not a function.

(b) Domain = $\{14{,}000, 18{,}000, 24{,}500, 33{,}000, 50{,}000\}$

Range = $\{2350, 2800, 2900, 3750, 1350, 7980\}$

(50,000, 1350) and (50,000, 7980) have the same first coordinate. This relation is not a function.

(c) Domain = $\{33, 34, 35, 36\}$

Range = $\{151, 154, 156, 159\}$

No two different ordered pairs have the same first coordinate. This relation is a function.

Practice Problem 2 Give the domain and range of the given relation. Indicate whether the relation is a function.

Car Performance

Horsepower	158	161	163	160	161
Top Speed (Mph)	98.6	89.2	101.4	102.3	94.9

If we are looking at the graph of a function, it will never have two different ordered pairs with the same *x*-value. We will examine some graphs in exercises 17–25.

Often we can tell a lot about the graph of a relation. Any graph that is not a function will have at least one region in which a vertical line will cross the graph more than once.

VERTICAL LINE TEST

If a vertical line can intersect the graph of a relation more than once, the relation is not a function. If no such line can be drawn, then the relation is a function.

EXAMPLE 3 Determine whether each of the following is a graph of a function.

(a) 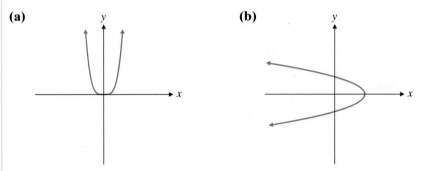 (b)

Solution

(a) This is a function. No vertical line will pass through more than one ordered pair on the curve.
(b) This is not a function. A vertical line could pass through two points on the curve.

Practice Problem 3 Determine whether each of following is a graph of a function.

(a) 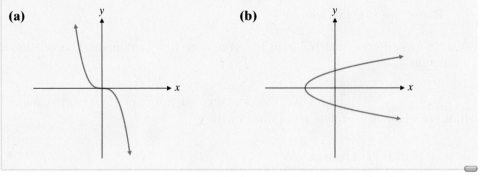 (b)

NOTE TO STUDENT: Fully worked-out solutions to all of the Practice Problems can be found at the back of the text starting at page SP-1

③ Evaluating a Function Using Function Notation

We can use **function notation** to indicate the relationship between the ordered pairs in a function. Looking back at our pizza example, we see that the diameter of a pizza determines its price. Since price is a function of diameter and a 5-inch-diameter pizza costs $4.75 we can write

$$f(5) = 4.75.$$

Some equations describe functions. You are already familiar with linear equations. Let's look at one such equation.

$$y = 3x - 2$$

In this equation, y is a function of x. That is, for each value of x in the domain, there is a unique value of y in the range. We can use function notation when we determine the function value y for specific values of x. Let $x = 0$. Then we can write the following.

$$f(x) = 3x - 2$$
$$f(0) = 3(0) - 2 \quad \text{Substitute 0 for } x.$$
$$f(0) = -2 \qquad \text{Evaluate.}$$

We see that when x is 0, $f(x)$ is -2. That is, the value of the function is -2 when x is 0. We can write this ordered pair solution as $(x, f(x)) = (0, f(0)) = (0, -2)$.

Note: The notation $f(x)$ does not mean that f is multiplied by x. It means that for any specific value of x, there is only one value for y. $f(x)$ is read as "f of x." Although we commonly use f as the function name, we can also use other variables, like g and h.

It may help you to understand the idea of a function by imagining a "function machine." An item from the domain enters as input into the machine—the function—and a member of the range results as output.

Function, f

x

$f(x)$

EXAMPLE 4   If $f(x) = 4 - 3x$, find each of the following.

(a) $f(2)$ **(b)** $f(-1)$ **(c)** $f\left(\dfrac{1}{3}\right)$

Solution   In each case we replace x by the specific value from the domain.

(a) $f(2) = 4 - 3(2)$
$$= 4 - 6$$
$$f(2) = -2$$

(b) $f(-1) = 4 - 3(-1)$
$$= 4 + 3$$
$$f(-1) = 7$$

(c) $f\left(\dfrac{1}{3}\right) = 4 - 3\left(\dfrac{1}{3}\right)$
$$= 4 - 1$$
$$f\left(\dfrac{1}{3}\right) = 3$$

Practice Problem 4   If $f(x) = 2x^2 - 8$, find each of the following.

(a) $f(-3)$ **(b)** $f(4)$ **(c)** $f\left(\dfrac{1}{2}\right)$

Verbal and Writing Skills

1. Explain the difference between a relation and a function.

2. Explain the difference between the domain and the range of a relation.

3. What are the three ways you can describe a function?

4. Write the ordered pair for $f(-5) = 8$ and identify the x- and y-values.

What are the domain and range of each relation? Is the relation a function?

5. $D = \{(0, 0), (5, 13), (7, 11), (5, 0)\}$

6. $C = \left\{\left(\frac{1}{2}, 5\right), (-3, 7), \left(\frac{3}{2}, 5\right), \left(\frac{1}{2}, -1\right)\right\}$

7. $F = \{(85, -12), (16, 4), (-102, 4), (62, 48)\}$

8. $E = \{(40, 10), (-18, 27), (38, 10), (57, -15)\}$

9. Women's Dress Sizes

USA	6	8	10	12	14
France	38	40	42	44	46

10. Women's Dress Sizes

France	40	42	44	46	48
Britain	14	15	16	17	18

11. Average Monthly Fahrenheit Temperature: Pago Pago, Samoa

Month	Jan.	Feb.	Mar.	Apr.	May	June	July	Aug.	Sept.	Oct.	Nov.	Dec.
Temperature	81	81	81	81	80	80	79	79	79	80	80	81

Source: United Nations Statistics Division

12. Highest Waterfalls of the World

Country	Venezuela	South Africa	Norway	Norway	Peru	Zimbabwe
Height in Feet	3212	3110	2625	2540	2532	2499

Source: www.infoplease.com

13. Tallest Buildings, USA

City	Chicago	New York	New York	Chicago	Chicago	New York
Height in Feet	1454	1350	1250	1136	1127	1046

Source: World Almanac

14. Metric Conversion

Miles	1	2	3	4	5
Kilometers	1.61	3.22	4.83	6.44	8.05

15. Mariner's Speed Conversion

Knots	10	20	30	40	50
Miles per Hour	11.51	23.02	34.53	46.04	57.55

16. Metric Conversion

Yards	1	2	3	4	5	6
Meters	0.91	1.83	2.74	3.66	4.57	5.48

Determine whether the graph represents a function.

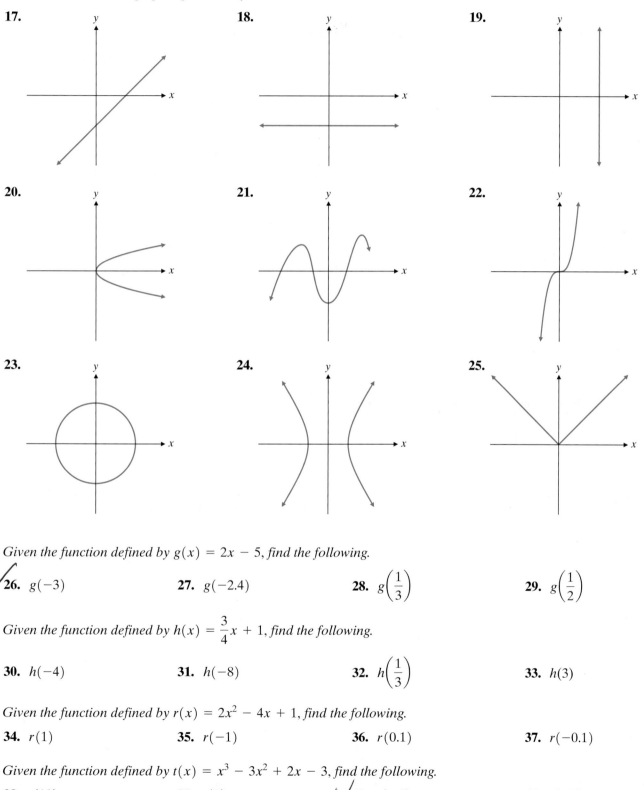

17.

18.

19.

20.

21.

22.

23.

24.

25.

Given the function defined by $g(x) = 2x - 5$, *find the following.*

26. $g(-3)$ **27.** $g(-2.4)$ **28.** $g\left(\dfrac{1}{3}\right)$ **29.** $g\left(\dfrac{1}{2}\right)$

Given the function defined by $h(x) = \dfrac{3}{4}x + 1$, *find the following.*

30. $h(-4)$ **31.** $h(-8)$ **32.** $h\left(\dfrac{1}{3}\right)$ **33.** $h(3)$

Given the function defined by $r(x) = 2x^2 - 4x + 1$, *find the following.*

34. $r(1)$ **35.** $r(-1)$ **36.** $r(0.1)$ **37.** $r(-0.1)$

Given the function defined by $t(x) = x^3 - 3x^2 + 2x - 3$, *find the following.*

38. $t(10)$ **39.** $t(4)$ **40.** $t(-1)$ **41.** $t(-2)$

42. If $f(x) = \sqrt{x + 10}$, find $f(-6)$.

43. If $f(x) = \sqrt{2 - x}$, find $f(-2)$.

44. If $g(x) = |x^2 + 5|$, find $g(-4)$.

45. If $g(x) = |6 - x^2|$, find $g(2)$.

To Think About *Find the range of the function for the given domain.*

46. $f(x) = x + 3$ Domain $= \{-2, -1, 0, 1, 2\}$

47. $g(x) = x^2 + 3$ Domain $= \{-2, -1, 0, 1, 2\}$

Find the domain of the function for the given range. Be sure to find all possible values.

48. $h(x) = \frac{1}{2}x + 3$ Range $= \left\{0, 2, \frac{7}{2}, 9\right\}$

49. $d(x) = 4 - \frac{1}{3}x$ Range $= \left\{0, \frac{4}{3}, 4, 5\right\}$

Cumulative Review

50. [2.3.1] Solve. $|3x - 2| = 1$

51. [2.8.1] Solve. $|x - 5| \le 3$

Quick Quiz 3.5

1. What is the domain and range of the following relation?

$$\{(9, 3), (-2, 3), (4, 5), (-4, 5)\}$$

2. If $p(x) = \left|-\frac{3}{4}x + 2\right|$, find $p(-8)$.

3. If $f(x) = -2x^3 + 4x^2 - x + 4$, find $f(-2)$.

4. Concept Check Explain how you can determine from a graph whether or not the graph represents a function.

3.6 GRAPHING FUNCTIONS FROM EQUATIONS AND TABLES OF DATA

1 Graphing a Function from an Equation

Frequently, we are given a function in the form of an equation and are asked to graph it. Each value of the function $f(x)$, often labeled y, corresponds to a value in the domain, often labeled x. This correspondence is the ordered pair (x, y) or $(x, f(x))$. The graph of the function is the graph of the ordered pairs.

If a function can be written in the form $f(x) = mx + b$, it is called a **linear function.** The graph of a linear function is a straight line.

If we can describe a real-life relationship with a function in the form of an equation, a table of values, and/or a graph, we can determine characteristics of the relationship and make predictions.

Student Learning Objectives

After studying this section, you will be able to:

1 Graph a function from an equation.

2 Graph a function from a table of data.

EXAMPLE 1 A salesperson earns $15,000 a year plus a 20% commission on her total sales. Express her annual income in dollars as a function of her total sales. Determine values of the function for total sales of $0, $25,000, and $50,000. Graph the function and determine whether the function is increasing or decreasing.

Solution Since we want to express income as a function of sales, let's see how income is determined.

$$\text{income} = \$15,000 + 20\% \text{ of total sales}$$

Income depends on total sales. Thus, total sales is the independent variable, and income is the dependent variable.

Let x = the amount of total sales in dollars and
$i(x)$ = income in dollars.
$i(x) = 15,000 + 0.20x$ Change 20% to 0.20.
$i(0) = 15,000 + 0.20(0) = 15,000$
$i(25,000) = 15,000 + 0.20(25,000) = 20,000$
$i(50,000) = 15,000 + 0.20(50,000) = 25,000$

We can put this information in a table of values.

x (dollars)	i(x) (dollars)
0	15,000
25,000	20,000
50,000	25,000

To facilitate the graphing, we will use a scale in thousands. Thus, we modify the table to make our task of graphing easier.

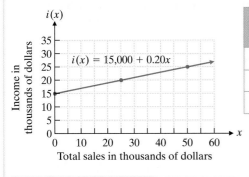

x (thousands of dollars)	i(x) (thousands of dollars)
0	15
25	20
50	25

The function is increasing.

PRACTICE PROBLEM 1

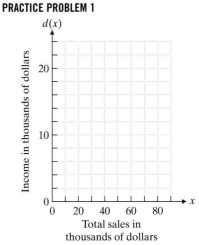

Practice Problem 1 A salesperson earns $10,000 a year plus a 15% commission on her total sales. Express her annual income in dollars as a function d of her total sales x. Use sales amounts of $0, $40,000, and $80,000 to graph her salary.

NOTE TO STUDENT: Fully worked-out solutions to all of the Practice Problems can be found at the back of the text starting at page SP-1

Thus far, most of our work has been with linear functions. Let's look at some other functions and their graphs. We will begin with the **absolute value function** $f(x) = |x|$.

EXAMPLE 2   Graph $f(x) = |x|$.

Solution   Let's find the function values for five values of the independent variable x. Because of the nature of the function, we will choose both negative and positive values for x. The table of values and the resulting graph are as follows:

x	f(x)
−2	2
−1	1
0	0
1	1
2	2

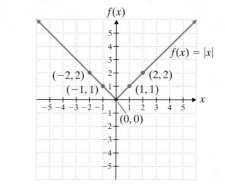

Notice that the graph is **symmetric about the y-axis.** That is, the graph on one side of the y-axis is a mirror image of the graph on the other side of the y-axis. This means that if the point (a, b) is on the graph, then the point $(-a, b)$ is also on the graph.

Practice Problem 2   Graph the function $h(x) = |x + 2|$. Use integer values for x from -4 to 0.

PRACTICE PROBLEM 2

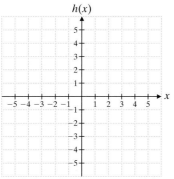

TO THINK ABOUT: Example 2 Follow-up   How are the graphs of $f(x) = |x|$ and $h(x) = |x + 2|$ the same? How are they different? What would the graph of $k(x) = |x - 2|$ look like?

Now let's take a look at another function, $p(x) = x^2$. What happens when $x = 2$? When $x = -2$? Notice that when x is 2 or -2, the value of the function is 4. The ordered pairs are $(-2, 4)$ and $(2, 4)$. This graph is also symmetric about the y-axis as we shall see in the next example.

EXAMPLE 3   Graph: **(a)** $p(x) = x^2$ **(b)** $q(x) = (x + 2)^2$

Solution   For each function we will choose both negative and positive values of x. Since these are *not* linear functions, we use a curved line to connect the points. The tables of values and the resulting graphs are as follows.

(a)

x	p(x)
−2	4
−1	1
0	0
1	1
2	4

(b)

x	q(x)
−4	4
−3	1
−2	0
−1	1
0	4
1	9

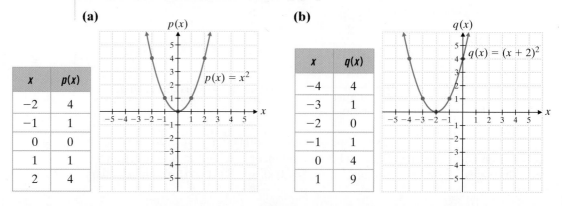

Practice Problem 3   Graph: **(a)** $r(x) = (x - 2)^2$ **(b)** $s(x) = x^2 + 2$

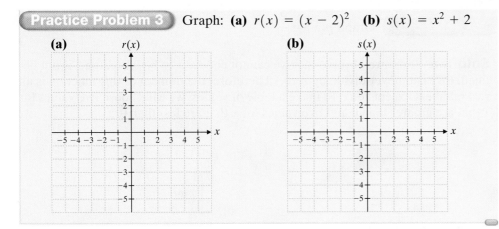

(a) $r(x)$ **(b)** $s(x)$

TO THINK ABOUT: Example 3 Follow-up   How are the graphs of each of the previous functions the same? How are they different? Describe how changing $p(x) = x^2$ to $q(x) = (x + 5)^2$ and to $s(x) = x^2 + 5$ would affect the graph of the function.

Another interesting graph is the graph of the function $g(x) = x^3$. A quick look at $g(x)$ when x is 2 and -2 reveals that $g(x)$ is *not* symmetric about the y-axis. That is, there are different function values for 2 and -2. Let's see.

$$g(x) = x^3$$

$$g(2) = 2^3 \qquad\qquad g(-2) = (-2)^3$$
$$\qquad = 8 \qquad\qquad\qquad\qquad = -8$$

EXAMPLE 4   Graph: **(a)** $g(x) = x^3$ **(b)** $h(x) = x^3 + 1$

Solution   We will pick five values for x, find the corresponding function values, and plot the five points to assist us in sketching the graph.

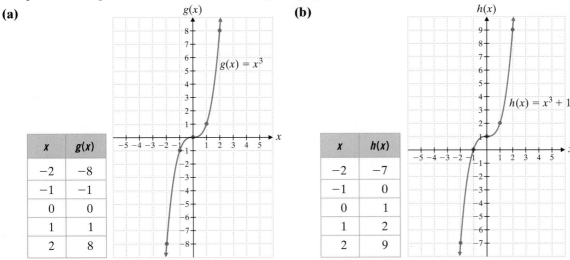

(a) $g(x)$

x	$g(x)$
-2	-8
-1	-1
0	0
1	1
2	8

(b) $h(x)$

x	$h(x)$
-2	-7
-1	0
0	1
1	2
2	9

Practice Problem 4   Consider the function $f(x) = x^3 - 2$. What do you think the graph of $f(x) = x^3 - 2$ will look like? Make a table of values for the function and draw the graph.

PRACTICE PROBLEM 4

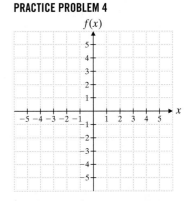

$f(x)$

TO THINK ABOUT: Example 4 Follow-up   Describe how the graph of $f(x) = x^3 - 2$ is related to the graph of $g(x) = x^3$.

In Examples 1–4 each function has had a domain of all real numbers. We were free to choose any real number for x. However, if a function is written in the form of a fraction and it contains a variable in the denominator, the domain of that function will not include any value for which the denominator becomes zero.

EXAMPLE 5 Graph $p(x) = \dfrac{4}{x}$.

Solution First we observe that we cannot choose x to be 0 because $\frac{4}{0}$ is not defined. (We can never divide by zero.) Therefore, the domain of this function is all real numbers except zero. To make a table of values, we will choose five values for x that are greater than 0 and five values for x that are less than 0.

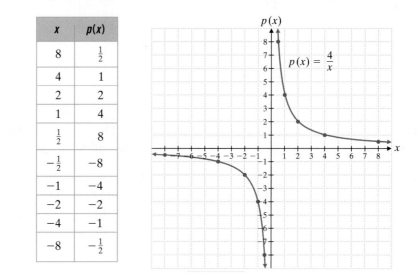

x	$p(x)$
8	$\frac{1}{2}$
4	1
2	2
1	4
$\frac{1}{2}$	8
$-\frac{1}{2}$	-8
-1	-4
-2	-2
-4	-1
-8	$-\frac{1}{2}$

The study of this type of function, called a **rational function,** will be continued in a more advanced course, such as college algebra or precalculus.

Practice Problem 5 Graph

$$q(x) = -\frac{4}{x}.$$

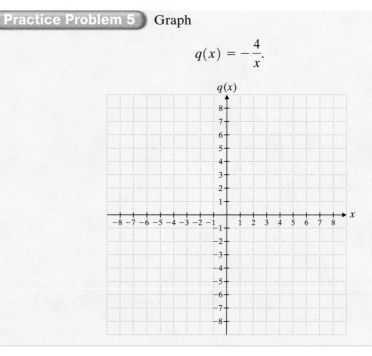

Graphing Calculator

Horizontal Shifts

Using a graphing calculator, graph

$y_1 = 4/x,$
$y_2 = 4/(x + 2)$, and
$y_3 = 4/(x - 3)$

on the same set of axes. Use a window of $[-8, 8]$ by $[-8, 8]$.

What pattern do you observe? What effect does c have on the graph of $y = 4/(x + c)$? What would the graph of $y = 4/(x + 4)$ look like? Verify using your calculator.

TO THINK ABOUT: Example 5 Follow-up Describe how the graph of $q(x) = -\dfrac{4}{x}$ is related to the graph of $p(x) = \dfrac{4}{x}$.

② Graphing a Function from a Table of Data

Sometimes when we study an event in daily life, we record data to better understand the event. In many cases we will not have an equation to work with, but rather a table of values that gives a general indication of some functional relationship. A graph of the values may help us to understand this underlying function.

In the following example, the number of items sold is the domain and the profit obtained from the sale is the range.

EXAMPLE 6   The marketing manager of a shoe company compiled the data in the table.

(a) Plot the data values and connect the points to see the graph of the underlying function.

(b) From the graph, determine the profit from selling 4000 pairs of shoes in one month.

(c) What kind of profit would you expect to make from selling 0 pairs of shoes in a month? What value do you obtain on the graph for $x = 0$? What does this mean?

x Pairs of Shoes Sold in a Month (In Thousands of Pairs)	$p(x)$ Monthly Profit from the Sales of Shoes (In Thousands of Dollars)
3	5
5	9
7	13

Solution

(a) We plot the three ordered pairs and connect the points by a straight line. This means the function is a linear function.

(b) We find that the value $x = 4$ on the graph corresponds to the value $y = 7$. Thus, we would expect that if we sold 4000 pairs of shoes, we would have a profit of $7000 for the month.

(c) In any business we would not expect to make a profit at all if we sold no items. Looking at our graph, we find that when $x = 0$, $y = -1$. Thus, we would predict a loss of $1000 in a month of zero sales of shoes. The loss would most likely be due to fixed expenses such as rent and supplies.

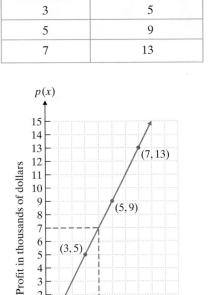

TO THINK ABOUT: Example 6 Follow-up   Can x in Example 6 be any real number? Well, not really. Obviously the number of pairs of shoes sold monthly must be a whole number. You cannot sell 345.859 pairs of shoes. However, when we connect the points on the graph we use a *continuous* line (i.e., a line with no breaks in it). When we study the function for all real numbers greater than or equal to zero, it helps us better understand the relationship between sales and profit. Using a continuous line helps us make better predictions for the future. Now, of course, we know as well that the graphed line does not really extend forever. There are physical limitations on the number of shoes that can be made and sold by any one shoe manufacturer.

Practice Problem 6   An accountant reviewed the profitability of a doctor's practice by comparing the number of patients the doctor saw with the profit. The data collected are in the table.

x Average Number of Patients Per Hour	$g(x)$ Weekly Profit
6	$2000
9	$4000
12	$6000

PRACTICE PROBLEM 6

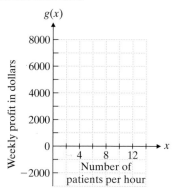

g(x)

(a) Plot the data values and connect the points to see the graph of the underlying function.
(b) Describe the function in as much detail as possible.
(c) What would the profit be if the doctor saw three patients per hour?
(d) What would the weekly loss be if the doctor saw zero patients per hour?
(e) Can we expect the function to increase indefinitely? Why or why not?

Given a set of data, we do not always know what the graph of the underlying function will look like. Not all functions are linear. When points do not appear to lie on a line, we connect them with a smooth curve.

Let's look at windchill, the effect of wind on the skin's reaction to the outside temperature. For example, if the temperature outside were $10°$ Fahrenheit (°F) and the wind were blowing at 20 miles per hour (mph), the effect would be the same as if the temperature were $24°$ below zero! A windchill table that describes this relationship is used in the next example.

EXAMPLE 7 The windchill values for various wind speeds when the temperature is $10°$ Fahrenheit are given in the table.

(a) Plot the points and connect them to see the graph of the underlying function.
(b) Estimate the windchill when the wind is blowing at 23 mph.
(c) Estimate the windchill when the wind is blowing at 40 mph.
(d) If the windchill is $-15°F$, at what speed is the wind blowing?
(e) Based on your analysis of the graph, does an increase in wind speed result in a greater change in the windchill at lower wind speeds or at higher wind speeds?

Wind Speed (mph)	Windchill (°F)
0	10
5	7
10	−9
15	−18
20	−24
25	−29
30	−33
35	−35

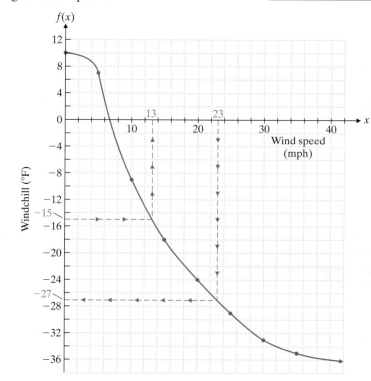

Solution

(a) First we need to determine which value is the independent variable and which value is the dependent variable. Since windchill depends on wind speed, wind speed is the independent variable and windchill is the dependent variable. Thus, the windchill values are the function values. Label the vertical axis "Windchill" and the horizontal axis "Wind Speed." Plot the points. Use a smooth curve to connect the points.

(b) Find 23 along the horizontal axis. Move down to the curve. Then move left until you intersect the vertical axis. The function value on the vertical axis for 23 mph is about $-27°F$.

(c) Since the last windchill number in the table is for a wind speed of 35 mph, we need to extend the curve. This is called *extrapolation*, and the function values obtained by this technique may be less accurate. At 40 mph, the windchill is about $-36°F$.

(d) Find -15 along the vertical scale. Move to the right until you intersect the curve. Then move up until you intersect the horizontal axis. This is the wind speed when the windchill is $-15°F$. This is about 13 mph.

(e) The curve goes downward more quickly at lower wind speeds. Thus, lower wind speeds have a greater effect on windchill. For example, when the wind speed goes from 0 to 15 mph, the windchill goes from $10°F$ to $-18°F$. This is a change of $28°F$. A change in wind speed from 20 to 35 mph produces a change in windchill of only $11°F$.

PRACTICE PROBLEM 7

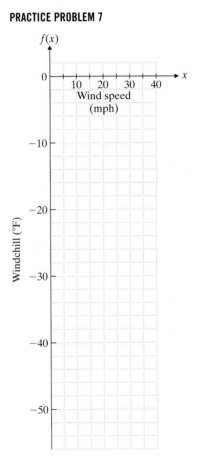

Practice Problem 7 A table of windchill values at $0°F$ is given below.

Wind Speed (mph)	0	5	10	15	20	25	30	35
Windchill (°F)	0	−5	−22	−31	−39	−44	−48	−52

(a) Plot the points and connect them to graph the underlying function.

(b) Estimate the windchill when the wind is blowing at 32 mph.

(c) Estimate the windchill when the wind is blowing at 40 mph.

(d) If the windchill is $-24°F$, at what speed is the wind blowing?

(e) What is the domain of the function? What is the range?

Scientists continue to study the present windchill table to see if it is truly accurate. Many scientists feel the figures need to be revised. Further research is needed to determine how the revision should be carried out. In addition to windchill tables, calculations can also be made using the internet. There are several sites that feature a "windchill calculator."

An interesting example of a place where windchill is critical is on top of Mount Washington in New Hampshire. In 1934 the observatory recorded a wind gust of 231 miles per hour. You can learn more at www.mountwashington.org.

EXAMPLE 8 Graph the function $f(x)$ suggested by the data given in the following table. The domain is $x \geq 2$, and the range is $f(x) \geq 0$.

(a) Determine an approximate value for $f(x)$ when $x = 8$.

(b) Determine an approximate value for x when $f(x) = 1.5$.

(c) What do you notice about the graph as the values of x get larger?

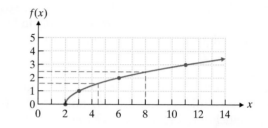

x	2	3	6	11
f(x)	0	1	2	3

Solution Assign the independent variable x to the horizontal axis. Assign the function values, the dependent variable, to the vertical axis. Plot the points and connect them with a smooth curve.

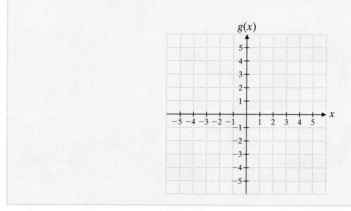

(a) Find $x = 8$ along the horizontal axis. Move up until you intersect the graph. Then move to the left until you intersect the vertical axis. This value is $f(x)$ when x is 8. Read the scale. Thus, $f(x)$ is about 2.5.

(b) Find $f(x) = 1.5$ along the vertical axis. Move to the right until you intersect the graph. Then move down until you intersect the horizontal axis. This value is x when $f(x)$ is 1.5. Read the scale. Thus, x is about 4.5.

(c) As the values of x get larger, the function values are increasing more slowly. In other words the *rate of change* decreases for larger values of x. Since the rate of change is the slope, we say that the slope is decreasing as x increases.

Practice Problem 8 Graph the function $g(x)$ suggested by the data given in the following table. The domain is $x \leq 4$, and the range is $g(x) \geq 0$.

(a) Determine an approximate value for $g(x)$ when $x = -3$.

(b) Determine an approximate value for x when $g(x) = 1.5$.

(c) As x goes from -5 to 0 to 3, what do you observe about the curve?

x	4	3	0	-5
g(x)	0	1	2	3

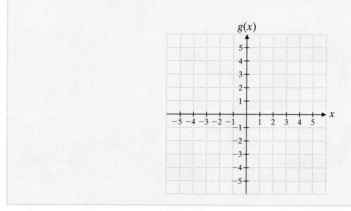

Graph each function.

1. $f(x) = \frac{3}{4}x + 2$

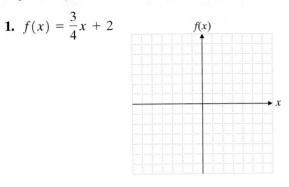

2. $f(x) = \frac{3}{2}x - 5$

3. $f(x) = -3x - 1$

4. $f(x) = -2x + 3$

Applications

5. *Cost of a Newsletter* The cost of producing a newsletter is an initial charge of $25 for formatting and editing plus $0.15 for printing each copy. Express the cost $C(x)$ of printing the newsletter as a function of the number of copies printed x. Obtain values for the function when $x = 0$, $x = 100$, $x = 200$, and $x = 300$. Graph the cost function.

6. *Population Growth* In 1996, the population of Paynesville was 18,000. The population has increased by 300 people each year since then and is expected to continue to do so. Express the population as a function $P(x)$, where x is the number of years since 1996. Obtain values for the function when $x = 0$, $x = 6$, $x = 10$, and $x = 18$. Graph the population function.

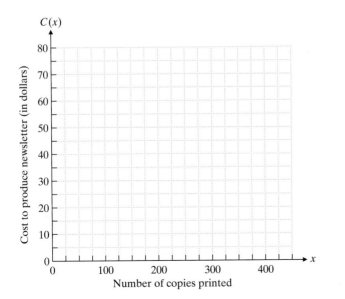

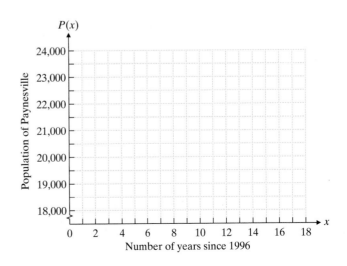

7. ***Water Pollution*** A wildlife biologist has determined that a certain stream in New Hampshire can support 45,000 fish if it is free of pollution. She has estimated that for every ton of pollutants in the stream, 1500 fewer fish can be supported. Express the fish population as a function $P(x)$, where x is the number of tons of pollutants found in the stream. Obtain values of the function when $x = 0$, $x = 10$, and $x = 30$. Graph the population function. What is the significance of $P(x)$ when $x = 30$?

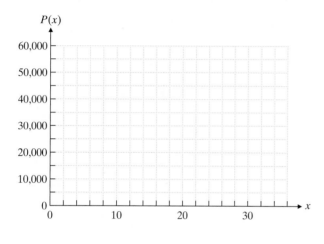

8. ***Insulation in a House*** The R value of insulation in a house is a measure of its ability to resist the loss of heat from the house. The R value of fiberglass insulation is a linear function of its thickness in inches. One type of fiberglass insulation that is 6 inches thick has an R value of 19. The R value in general of this type of insulation is obtained by multiplying 3.2 by the thickness x measured in inches and then adding the result to -0.2. Express the R value of this insulation as a function of x, the thickness in inches. Obtain values of the function when $x = 0$, $x = 1$, $x = 3.5$, and $x = 6$. Graph the function.

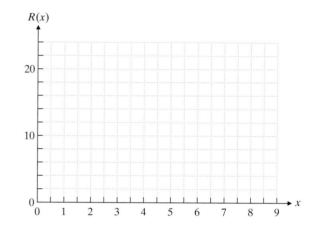

Graph each function.

9. $f(x) = |x - 1|$

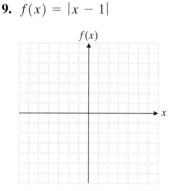

10. $g(x) = |x - 3|$

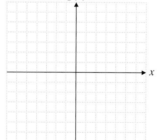

11. $g(x) = |x| - 5$

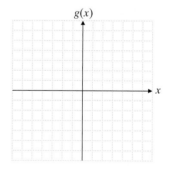

12. $f(x) = |x| + 2$

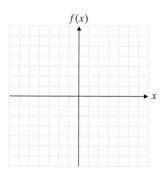

13. $g(x) = x^2 - 4$

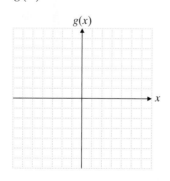

14. $f(x) = x^2 + 1$

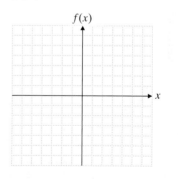

15. $g(x) = (x + 1)^2$

16. $f(x) = (x - 3)^2$

17. $g(x) = x^3 - 3$

18. $f(x) = x^3 + 2$

19. $p(x) = -x^3$

20. $s(x) = -x^3 + 2$

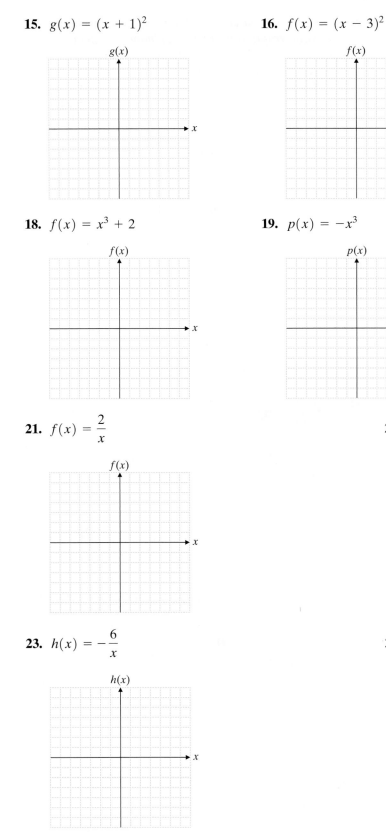

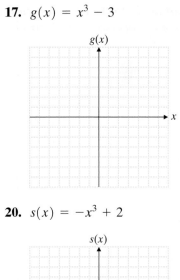

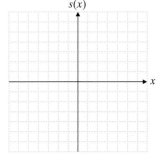

21. $f(x) = \dfrac{2}{x}$

22. $g(x) = -\dfrac{3}{x}$

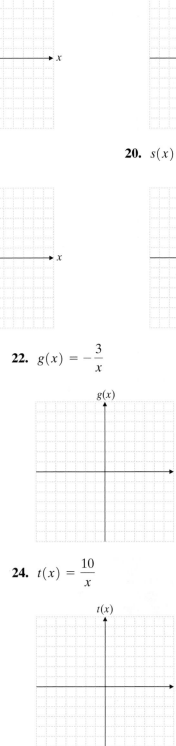

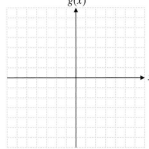

23. $h(x) = -\dfrac{6}{x}$

24. $t(x) = \dfrac{10}{x}$

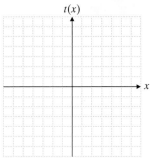

Optional Graphing Calculator Problem

25. On your graphing calculator, graph on the same set of axes: $y_1 = x^2$, $y_2 = 0.4x^2$, and $y_3 = 2.6x^2$. What effect does the coefficient have on the graph?

In each of the following problems, a table of values is given. Graph the function based on the table of values. Assume that both domain and range are all real numbers. After you have finished your graph in each case, estimate the value of $f(x)$ when $x = 2$.

26.

x	f(x)
−1	2
1	−2
3	−6

27.

x	f(x)
−5	−2
3	2
5	3

28.

x	f(x)
−2	0.25
−1	0.5
0	1
2.5	5.7

29.

x	f(x)
0	3
1	2.5
3	2.2
−1	4
−2	6

30.

x	f(x)
−3	−1
−2	2
−1	3
0	2
1	−1

31.

x	f(x)
−1	−3
0	0
1	1
3	−3

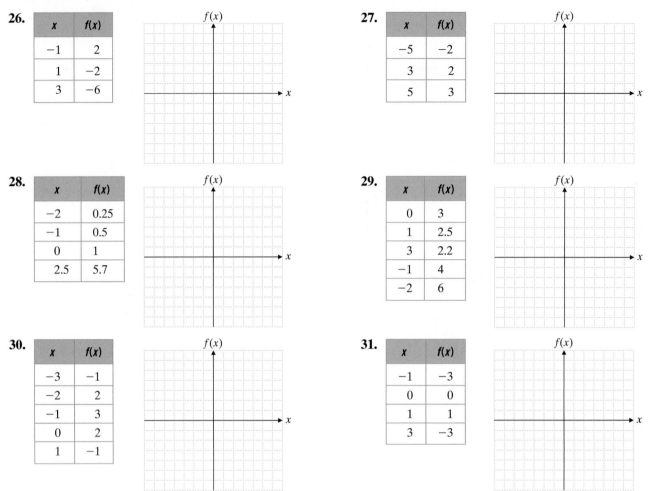

Applications

32. *Living with HIV/AIDS* The following table records the number of people worldwide living with HIV/AIDS for selected years.

Year, x	1991	1993	1995	1997	1999	2001	2003	2005	2007
Number of People Living with HIV/AIDS (in millions), n(x)	9.5	13.5	18	22	25.5	28.5	31	32	33

Source: www.avert.org

(a) Plot the points and connect them to graph the underlying function.

(b) Estimate $n(x)$ when x is 1998.

(c) What do you estimate $n(x)$ will be in 2009?

(d) How would you compare the rate of increase in $n(x)$ between 1991 and 1999 to the rate of increase in $n(x)$ between 1999 and 2007?

(e) During what year did the number of people living with HIV/AIDS reach 20 million?

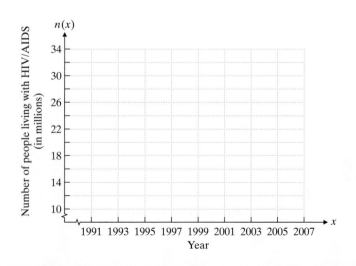

33. *Computer Manufacturing* The following table shows the relationship between the number of computers manufactured by a company each day and the company's profit for that day.

Number of Computers Made	20	30	40	50	60	70	80
Profit for the Company ($)	0	5000	8000	9000	8000	5000	0

(a) Plot the points and connect them to graph the underlying function.

(b) How many computers should be made each day to achieve the maximum profit?

(c) If the company wants to earn a profit of $8000 or more each day, how many computers should they manufacture each day?

(d) What will the profit picture be if the company manufactures eighty-two computers per day?

(e) Estimate the profit if the company manufactures forty-five computers per day.

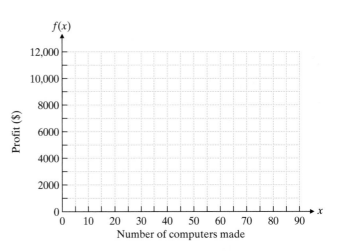

To Think About

34. *Heat Index* In hot weather, meteorologists use a heat index to indicate the relative comfort and safety of people subjected to high temperatures and humidity. The following table gives the heat index at 80°F for a relative humidity of 0% to 100%.

Relative Humidity	0%	20%	40%	60%	80%	100%
Heat Index at 80°F	73	77	79	82	86	91

Source: National Oceanic and Atmospheric Administration.

(a) Graph the heat index values at 80°F and connect the points. Is the function linear? Why or why not?

(b) Estimate the value of the function when the relative humidity is 50%.

(c) Between what levels of humidity does the heat index increase the fastest?

(d) For what level of humidity does 80°F have a heat index of 80°F?

(e) What is the significance of the heat index being only 77°F when the relative humidity is 20%?

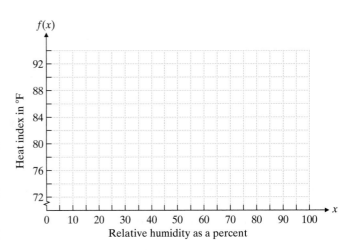

Cumulative Review *Solve for x.*

35. **[2.2.1]** $2(3ax - 4y) = 5(ax + 3)$

36. **[2.1.1]** $0.12(x - 4) = 1.16x - 8.02$

37. **[2.1.1]** $\frac{1}{2}(x + 2) - 5 = x - \frac{3}{4}(x + 4)$

38. **[1.2.2]** *Woodchuck Hibernation* A woodchuck breathes only ten times per hour while hibernating. How many times will it breathe during its hibernation months of December, January, and February (during a leap year)?

Quick Quiz 3.6

1. Graph $f(x) = |x - 4|$.

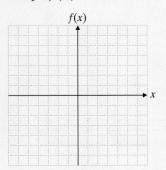

2. Graph $g(x) = 6 - x^2$.

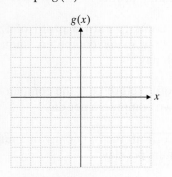

3. Graph $h(x) = x^3 - 2$.

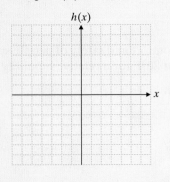

4. **Concept Check** Explain how you would find $f\left(\frac{1}{4}\right)$ for the following function.

$$f(x) = \frac{5}{8x - 5}$$

Putting Your Skills to Work: Use Math to Save Money

AUTOMOBILE LEASING VS. PURCHASE

Louvy has his eye on a brand new car. He thinks he should lease the car because his best friend Tranh has a car lease and says he can get the same deal for Louvy. On the other hand, Louvy's girlfriend Allie says it is always better to buy the car and finance it by taking out a loan.

Louvy does some research and finds that it is not at all simple. While a lease offers lower monthly payments, at the end of the lease period you are left with nothing, except expenses.

Look at the following comparison for lease vs. buy for the car Louvy is considering.

	Lease	Purchase
Automobile price	$23,000	$23,000
Interest rate	6%	6%
Length of loan	36 months	36 months
Down payment	$1000.00	$1000.00
Residual (value of car you are turning in, amount you pay if you wish to purchase it)	$11,000.00	Not applicable
Monthly payment	$388.06	$669.28

1. How much would Louvy pay over the entire length of the loan?

2. How much would Louvy pay over the entire length of the lease?

3. If Louvy decided to buy the car at the end of the lease, he would have to pay $11,000.00 in addition to his lease cost. What would that bring the total cost of that car up to?

Making It Personal for You

4. How much can you afford per month for payments for a car? How much would you pay in insurance, taxes, and gas?

5. Would you prefer to lease or buy? why?

Facts You Should Know

You may wish to lease a car if:

- You want a new vehicle every 2–3 years
- You don't drive an excessive number of miles each year

- You don't want major repairs risk
- You want a lower monthly payment

IF you lease a car:

- You pay only that portion of the vehicle you use
- There may be mileage restrictions, hidden fees, or security deposits.
- You can sometimes find a lease with no down payment
- You pay sales tax only on monthly fees
- You usually must keep the car for the entire lease period, or pay a heavy penalty

You may wish to buy a car if:

- You intend to keep it a long time
- You want to be debt-free after a time
- You qualify for a very low interest rate
- The long term cost is more important than the lower monthly payment

IF you buy a car:

- You pay for the entire vehicle
- You pay the sales tax on the entire price of the car
- There are no hidden mileage costs, except in wear and tear on the vehicle
- You can sell or trade the car during the period of the loan

Topic	Procedure	Examples
Graphing straight lines, p. 127.	An equation of the form $$Ax + By = C$$ where A, B, and C are real numbers and A and B are not both zero, is a linear equation and has a graph that is a straight line. To graph such an equation, plot any three points—two give the line and the third checks it. (Where possible, use the x- and y-intercepts.)	$2x + 3y = 12$ Form is $Ax + By = C$, so it is a straight line. Let $x = 0$; then $y = 4$. Let $y = 0$; then $x = 6$. Let $x = 3$; then $y = 2$. | x	y | |---|---| | 0 | 4 | | 6 | 0 | | 3 | 2 |
Intercepts, p. 127.	A line crosses the x-axis at its x-intercept $(a, 0)$. A line crosses the y-axis at its y-intercept $(0, b)$.	Find the intercepts of $7x - 2y = -14$. If $x = 0$, $-2y = -14$ and $y = 7$. The y-intercept is $(0, 7)$. If $y = 0$, $7x = -14$ and $x = -2$. The x-intercept is $(-2, 0)$.
Slope, p. 135.	The *slope* m of the straight line that contains the points (x_1, y_1) and (x_2, y_2) is defined by $$m = \frac{y_2 - y_1}{x_2 - x_1}, \quad \text{where } x_2 \neq x_1.$$	Find the slope of the line passing through $(6, -1)$ and $(-4, -5)$. $$m = \frac{-5 - (-1)}{-4 - 6} = \frac{-5 + 1}{-4 - 6} = \frac{-4}{-10} = \frac{2}{5}$$
Zero slope, p. 136.	All horizontal lines have *zero slope*. They can be described by equations of the form $y = b$, where b is a real number.	What is the slope of $2y = 8$? This equation can be simplified to $y = 4$. It is a horizontal line; the slope is zero.
Undefined slope, p. 136.	The slopes of all vertical lines are undefined. The lines can be described by equations of the form $x = a$, where a is a real number.	What is the slope of $5x = -15$? This equation can be simplified to $x = -3$. It is a vertical line. This line has an undefined slope.
Parallel and perpendicular lines, pp. 137, 138.	Two distinct lines with nonzero slopes m_1 and m_2 are **1.** parallel if $m_1 = m_2$. **2.** perpendicular if $m_1 = -\dfrac{1}{m_2}$.	Find the slope of a line parallel to $y = -\dfrac{3}{2}x + 6$. $$m = -\frac{3}{2}$$ Find the slope of a line perpendicular to $$y = -4x + 7, \qquad m = \frac{1}{4}$$
Standard form, p. 126.	The equation of a line is in standard form when it is written as $Ax + By = C$, where A, B, and C are real numbers and A and B are not both zero.	Place this equation in standard form: $y = -5(x + 6)$. $$y = -5x - 30$$ $$5x + y = -30$$
Slope–intercept form, p. 143.	The slope–intercept form of the equation of a line is $y = mx + b$, where the slope is m and the y-intercept is $(0, b)$.	Find the slope and y-intercept of $y = -\dfrac{7}{3}x + \dfrac{1}{4}$. The slope is $-\dfrac{7}{3}$; the y-intercept is $\left(0, \dfrac{1}{4}\right)$.
Point–slope form, p. 145.	The point–slope form of the equation of a line is $y - y_1 = m(x - x_1)$, where m is the slope and (x_1, y_1) are the coordinates of a point on the line.	Find an equation of the line passing through the points $(6, 0)$ and $(3, 4)$. $$m = \frac{4 - 0}{3 - 6} = -\frac{4}{3}$$ Then use the point–slope form. $$y - 0 = -\frac{4}{3}(x - 6)$$ $$y = -\frac{4}{3}x + 8$$

Topic	Procedure	Examples
Graphing the solution of a linear inequality in two variables, p. 154.	1. Replace the inequality symbol by an equals sign. This equation will be the boundary for the desired region. 2. Graph the line obtained in step 1, using a solid line if the original inequality contains a $\leq$ or $\geq$ symbol and a dashed line if the original inequality contains a $<$ or $>$ symbol. 3. Pick any point that does not lie on the boundary line. Substitute the coordinates into the original inequality. If you obtain a true inequality, shade the region on the side of the line containing the point. If you obtain a false inequality, shade the region on the side of the line opposite the test point.	Graph the solution of $3y - 4x \geq 6$. 1. The boundary line is $3y - 4x = 6$. 2. The line passes through $(0, 2)$ and $(-1.5, 0)$. The inequality includes the boundary; draw a solid line. 3. Pick $(0, 0)$ as a test point. $$3y - 4x \geq 6$$ $$3(0) - 4(0) \overset{?}{\geq} 6$$ $$0 - 0 \overset{?}{\geq} 6$$ $$0 \geq 6$$ Our test point fails. We shade the side that does not contain $(0, 0)$.
Finding the domain and the range of a relation, p. 160.	The set of all the first items of the ordered pairs in a relation is called the domain. The set of all the second items of the ordered pairs in a relation is called the range.	Find the domain and range of the relation $$A = \{(5, 6), (1, 6), (3, 4), (2, 3)\},$$ $$\text{Domain} = \{5, 1, 3, 2\}$$ $$\text{Range} = \{6, 4, 3\}$$
Determining whether a relation is a function, p. 162.	A function is a relation in which no two different ordered pairs have the same first coordinate.	Determine whether each of the following is a function. $$B = \{(6, 7), (3, 0), (6, 4)\}$$ $$C = \{(-9, 3), (16, 4), (9, 3)\}$$ B is not a function. Two different pairs have the same first coordinate. They are $(6, 7)$ and $(6, 4)$. C is a function. There are no different pairs with the same first coordinate.
Determining whether a graph represents the graph of a function, pp. 162, 164.	The graph of a function will have no two different ordered pairs with the same first coordinate. If a vertical line can intersect a graph more than once, the graph does not represent a function. If no such line can be drawn, the graph represents a function.	 This is a function. This is not a function. There are at least two different ordered pairs with the same first coordinate.

(Continued on next page)

Topic	Procedure	Examples
Function notation, p. 165.	Use function notation to evaluate the function at a given value. Replace the x by the quantity within the parentheses and then simplify.	$f(x) = 2x^2 - 3x + 4$ Find $f(-2)$. $f(-2) = 2(-2)^2 - 3(-2) + 4$ $\quad = 2(4) - 3(-2) + 4$ $\quad = 8 + 6 + 4 = 18$
Graphing functions, p. 170.	Prepare a table of ordered pairs (if one is not provided) that satisfy the function equation. Graph these ordered pairs. Connect the ordered pairs with a line or curve.	Graph $f(x) = \lvert x \rvert - 2$. First make a table. Then graph the ordered pairs and connect them. 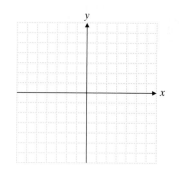 | x	$f(x)$
---	---
-2	0
-1	-1
0	-2
1	-1
2	0	

Chapter 3 Review Problems

Graph the straight line determined by each of the following equations.

1. $y = -\dfrac{1}{4}x - 1$

2. $y = -\dfrac{3}{2}x + 5$

3. $y - 2x + 4 = 0$

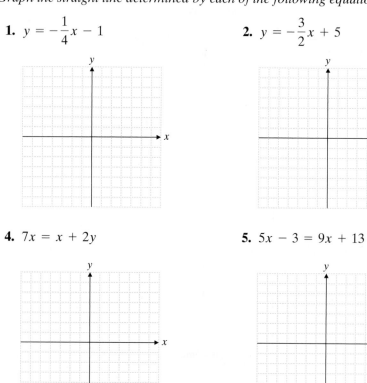

4. $7x = x + 2y$

5. $5x - 3 = 9x + 13$

6. $8y + 5 = 10y + 1$

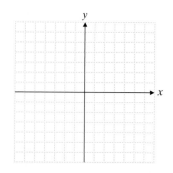

Find the slope, if possible, of the line connecting the given points.

7. $\left(\dfrac{1}{2}, 3\right)$ and $\left(\dfrac{3}{2}, 1\right)$

8. $(-8, -4)$ and $(2, -3)$

9. $(-3, 6)$ and $(-3, 1.8)$

10. $(7.5, -1)$ and $(0.3, -1)$

11. Find the slope of the line connecting $(9, -2)$ and $(1, -4)$.

12. Find the slope of a line perpendicular to the line passing through $\left(\dfrac{2}{3}, \dfrac{1}{3}\right)$ and $(4, 2)$.

13. A line has a slope of $\frac{1}{3}$ and a y-intercept of $(0, 5)$. Write its equation in standard form.

14. Find the standard form of the equation of the line that passes through $\left(\dfrac{1}{2}, -2\right)$ and has slope -4.

15. Find the standard form of the equation of the line that passes through $(-3, 1)$ and has slope 0.

16. **Computer Company Profit** A microcomputer company's profit in dollars is given by the equation $P = 140x - 2000$, where x is the number of microcomputers sold each day. **(a)** What is the slope of the equation $P = 140x - 2000$? **(b)** How many microcomputers must be sold each day for the company to make a profit?

In exercises 17–20, find the equation of the line satisfying the conditions given. Write your answer in standard form.

17. A line passing through $(5, 6)$ and $\left(-1, -\dfrac{1}{2}\right)$

18. A line that has an undefined slope and passes through $(-6, 5)$

19. A line perpendicular to $7x + 8y - 12 = 0$ and passing through $(-2, 5)$

20. A line parallel to $3x - 2y = 8$ and passing through $(5, 1)$

In exercises 21–23, find the slope and y-intercept of each of the following lines. Use these to write the equation of the line.

21.   **22.**   **23.**

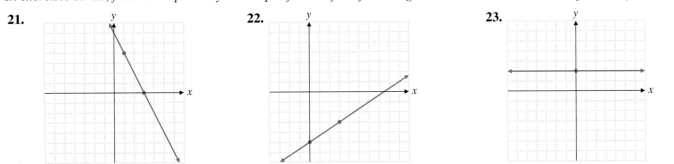

Graph the region described by the inequality.

24. $y < 2x + 4$

25. $y < 3x + 1$

26. $y > -\dfrac{1}{2}x + 3$

27. $y > -\dfrac{2}{3}x + 1$

28. $3x + 4y \leq -12$

29. $x \leq 3y$

30. $5x + 3y \leq -15$

31. $3x - 5 < 7$

32. $5y - 2 > 3y - 10$

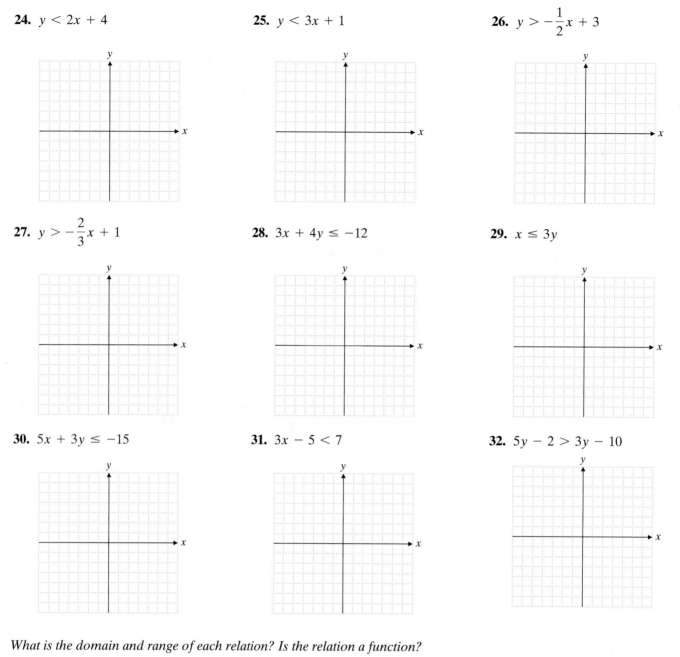

What is the domain and range of each relation? Is the relation a function?

33. $B = \{(-20, 18), (-18, 16), (-16, 14), (-12, 18)\}$

34. $A = \{(0, 0), (1, 1), (2, 4), (3, 9), (1, 16)\}$

Determine whether the graph represents a function.

35.

36.

37.

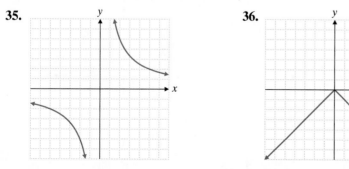

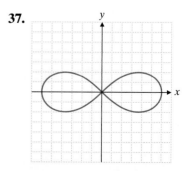

38. If $f(x) = -2x + 10$, find $f(-1)$ and $f(-5)$.

39. If $g(x) = 2x^2 + x - 4$, find $g(3)$ and $g(-1)$.

40. If $h(x) = x^3 + 2x^2 - 5x + 8$, find $h(-1)$.

41. If $p(x) = |-6x - 3|$, find $p(3)$.

Graph each function.

42. $f(x) = 2|x - 1|$

43. $g(x) = x^2 - 5$

44. $h(x) = x^3 + 3$

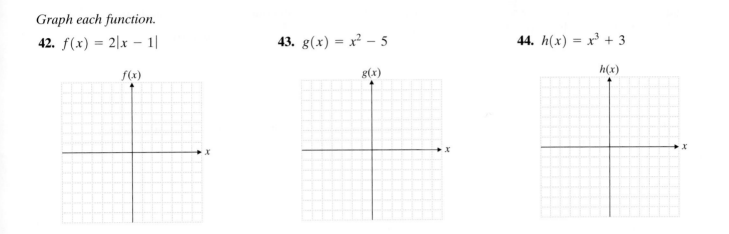

Plot the points and connect them to graph the underlying function. Estimate the value of $f(x)$ when $x = -2$.

45.

x	f(x)
−1	5
−3	−3
−4	−4
−5	−3
−6	0
−7	5

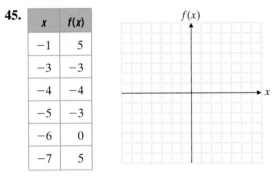

46.

x	f(x)
−3	4
−1	2
0	1
1	0
2	−1
3	0
4	1

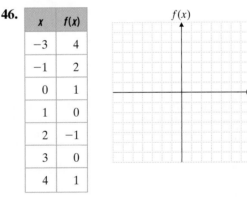

Complete the table of values for each function.

47. $f(x) = -\dfrac{4}{5}x + 3$

x	f(x)
−5	
0	
10	

48. $f(x) = 2x^2 - 3x + 4$

x	f(x)
−3	
0	
4	

49. $f(x) = 3x^3 - 4$

x	f(x)
−1	
0	
2	

50. $f(x) = \dfrac{7}{2x + 3}$

x	f(x)
−2	
0	
2	

Mixed Practice

51. Find the slope of the line connecting $(5, 0)$ and $(-4, 2)$.

52. A line n is perpendicular to the line passing through $(-3, 6)$ and $(1, 9)$. What is the slope of line n?

53. A line p is parallel to the line passing through $(4.5, 8)$ and $(2.5, -6)$. What is the slope of line p?

54. Find the standard form of the equation of the line that has slope $\frac{5}{6}$ and y-intercept $(0, -5)$.

55. Find the equation of the line parallel to $y = 5x - 2$ and passing through $(4, 10)$. Write the answer in slope–intercept form.

56. Find the slope–intercept form of the equation of the line that passes through $(5, 6)$ and $(-7, 3)$.

57. Find the equation of the line perpendicular to $3x - 6y = 9$ and passing through $(-2, -1)$. Write the answer in standard form.

58. Find the equation of a vertical line passing through $(5, 6)$.

Applications

59. *Car Rental Costs* The cost of renting a full-size sedan at Palm Beach Car Rental is $40 per day plus $0.20 per mile that the car is driven. Express the cost of renting a full-size sedan for one day at Palm Beach Car Rental as a function $f(x)$, where x is the number of miles the car is driven.

60. *Population Growth of a Town* The population of Garrison was 24,000 in 1995. Since then the town has grown by 200 people each year. Express the population as a function $f(x)$, where x is the number of years since 1995.

61. *Fish Population in a Stream* A biologist in Montana has determined that a certain mountain stream initially had 18,000 fish on June 1. The number of fish in the stream has decreased by 65 fish each day since June 1. Express the number of fish in this stream as a function $f(x)$, where x is the number of days after June 1.

Remember to use your Chapter Test Prep Video CD to see the worked-out solutions to the test problems you want to review.

Graph each line. Plot at least three points.

1. $y = \dfrac{1}{3}x - 2$

2. $2x - 3 = 1$

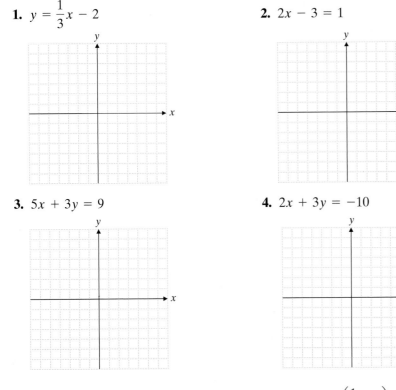

3. $5x + 3y = 9$

4. $2x + 3y = -10$

5. Find the slope of the line passing through $(2, -3)$ and $\left(\dfrac{1}{2}, -6\right)$.

6. Find the slope of a line passing through $(-7, 5)$ and $(6, 5)$.

7. Find the slope of the line $9x + 7y = 13$.

8. Write the standard form equation of the line that is perpendicular to $6x - 7y - 1 = 0$ and passes through $(0, -2)$.

9. Write the standard form of the equation of the line that passes through $(5, -2)$ and $(-3, -1)$.

10. Write the equation of the horizontal line passing through $\left(-\dfrac{1}{3}, 2\right)$.

11. Write the equation of the line with slope -5 and y-intercept $(0, -8)$. Write the answer in slope–intercept form.

Graph the regions.

12. $y \geq -4x$

13. $4x - 2y < -6$

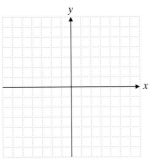

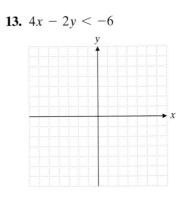

1. _____

2. _____

3. _____

4. _____

5. _____

6. _____

7. _____

8. _____

9. _____

10. _____

11. _____

12. _____

13. _____

14. _____

15. _____

16. _____

17. _____

18. _____

19. _____

20. _____

21. _____

14. What are the domain and range of the following relation?
$A = \{(0,0), (1,1), (1,-1), (2,4), (2,-4)\}$

15. If $f(x) = 2x - 3$, find $f\left(\dfrac{3}{4}\right)$.

16. If $g(x) = \dfrac{1}{2}x^2 + 3$, find $g(-4)$.

17. If $h(x) = \left|-\dfrac{2}{3}x + 4\right|$, find $h(-9)$.

18. If $p(x) = -2x^3 + 3x^2 + x - 4$, find $p(-2)$.

Graph each function.

19. $g(x) = 5 - x^2$

20. $h(x) = x^3 - 4$

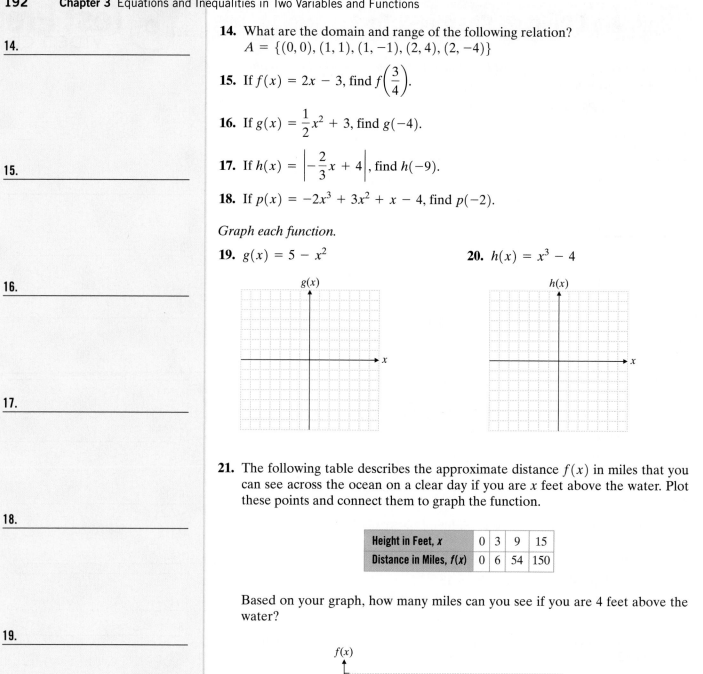

21. The following table describes the approximate distance $f(x)$ in miles that you can see across the ocean on a clear day if you are x feet above the water. Plot these points and connect them to graph the function.

Height in Feet, x	0	3	9	15
Distance in Miles, $f(x)$	0	6	54	150

Based on your graph, how many miles can you see if you are 4 feet above the water?

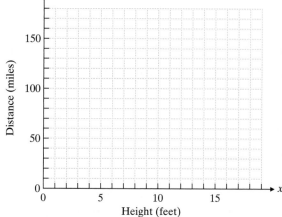

Cumulative Test for Chapters 1–3

Approximately one-half of this test covers the content of Chapters 1 and 2. The remainder covers the content of Chapter 3.

1. Name the property that justifies $(-6) + 6 = 0$.

2. Evaluate the expression. $3(4 - 6)^2 + \sqrt{16} + 12 \div (-3)$

3. Simplify. $(2a^{-3}b^4)^{-2}$

4. Simplify. $4(2x^2 - 1) - 2x(3x - 5y)$

5. Write using scientific notation. 0.000437

6. Solve and graph your solution. $3(x - 2) > 6$ or $5 - 3(x + 1) > 8$

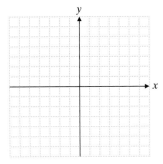

7. Solve for x. $2a - x = \frac{1}{3}(6x - y)$

▲ 8. A plastic insulator is made in a rectangular shape with a perimeter of 92 centimeters. The length is 1 centimeter longer than double the width. Find the dimensions of the insulator.

9. Marissa invested $7000 for 1 year. Part was invested at 4% simple interest and part at 7% simple interest. At the end of 1 year, she had earned $391 in interest. How much had she invested at each rate?

▲ 10. Find the area of the semicircle to the right. Round your answer to the nearest hundredth. (Use $\pi \approx 3.14$.)

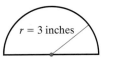

$r = 3$ inches

11. Graph the line $4x - 6y = 10$. Plot at least three points.

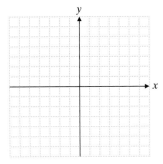

12. Find the slope of the line passing through $(6, 5)$ and $(-2, 1)$.

13. Write the standard form of the equation of the line that passes through $(5, 1)$ and $(4, 3)$.

14. Write the standard form of the equation of the line that passes through $(-1, 4)$ and is perpendicular to $y = -\frac{1}{3}x + 6$.

15. What are the domain and range of the relation $\left\{(3, 7), (5, 8), \left(\frac{1}{2}, -1\right), (2, 2)\right\}$? Is the relation a function?

16. Find $f(-3)$ for $f(x) = -2x^2 - 4x + 1$.

1. _____

2. _____

3. _____

4. _____

5. _____

6. _____

7. _____

8. _____

9. _____

10. _____

11. _____

12. _____

13. _____

14. _____

15. _____

16. _____

17. _____

18. _____

19. _____

20. _____

21. _____

22. _____

23. _____

24. _____

Graph the following functions.

17. $p(x) = -\dfrac{1}{3}x + 2$

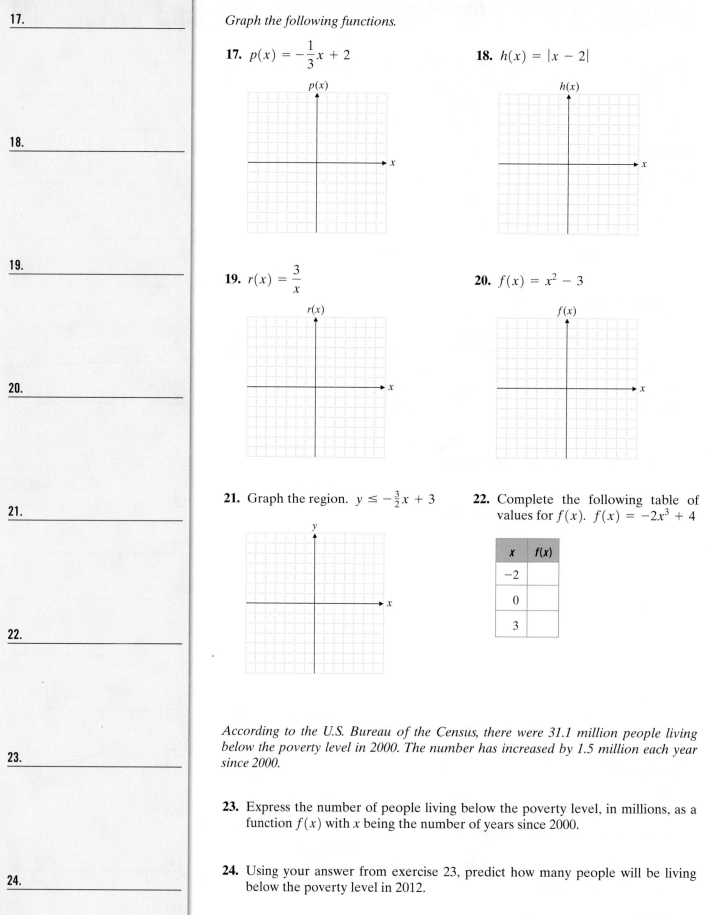

18. $h(x) = |x - 2|$

19. $r(x) = \dfrac{3}{x}$

20. $f(x) = x^2 - 3$

21. Graph the region. $y \leq -\frac{3}{2}x + 3$

22. Complete the following table of values for $f(x)$. $f(x) = -2x^3 + 4$

x	$f(x)$
-2	
0	
3	

According to the U.S. Bureau of the Census, there were 31.1 million people living below the poverty level in 2000. The number has increased by 1.5 million each year since 2000.

23. Express the number of people living below the poverty level, in millions, as a function $f(x)$ with x being the number of years since 2000.

24. Using your answer from exercise 23, predict how many people will be living below the poverty level in 2012.

Have you ever planned a canoe trip in a scenic area and discovered that you had to contend with a significant river current? Paddling downstream with the current is relaxing and pleasant. Paddling upstream against the current can be a real chore. The task may become extremely difficult if the current increases due to heavy rains and melting snow. The kind of mathematics you will learn in this chapter will allow you to solve a variety of problems regarding boating with and against the current.

Systems of Linear Equations and Inequalities

4.1   SYSTEMS OF LINEAR EQUATIONS IN TWO VARIABLES   196

4.2   SYSTEMS OF LINEAR EQUATIONS IN THREE VARIABLES   210

        HOW AM I DOING? SECTIONS 4.1–4.2   217

4.3   APPLICATIONS OF SYSTEMS OF LINEAR EQUATIONS   218

4.4   SYSTEMS OF LINEAR INEQUALITIES   229

        CHAPTER 4 ORGANIZER   238

        CHAPTER 4 REVIEW PROBLEMS   240

        HOW AM I DOING? CHAPTER 4 TEST   244

        CUMULATIVE TEST FOR CHAPTERS 1–4   245

Student Learning Objectives

After studying this section, you will be able to:

1 Determine whether an ordered pair is a solution to a system of two linear equations.

2 Solve a system of two linear equations by the graphing method.

3 Solve a system of two linear equations by the substitution method.

4 Solve a system of two linear equations by the addition (elimination) method.

5 Identify systems of linear equations that do not have a unique solution.

6 Choose an appropriate method to solve a system of linear equations algebraically.

1 Determining Whether an Ordered Pair Is a Solution to a System of Two Linear Equations

In Chapter 3 we found that a linear equation containing two variables, such as $4x + 3y = 12$, has an unlimited number of ordered pairs (x, y) that satisfy it. For example, $(3, 0), (0, 4),$ and $(-3, 8)$ all satisfy the equation $4x + 3y = 12$. We call *two* linear equations in two unknowns a **system of two linear equations in two variables.** Many such systems have exactly one solution. A **solution to a system** of two linear equations in two variables is an *ordered pair* that is a solution to *each* equation.

EXAMPLE 1 Determine whether $(3, -2)$ is a solution to the following system.

$$x + 3y = -3$$
$$4x + 3y = 6$$

Solution We will begin by substituting $(3, -2)$ into the first equation to see whether the ordered pair is a solution to the first equation.

$$3 + 3(-2) \overset{?}{=} -3$$
$$3 - 6 \overset{?}{=} -3$$
$$-3 = -3 \checkmark$$

Likewise, we will determine whether $(3, -2)$ is a solution to the second equation.

$$4(3) + 3(-2) \overset{?}{=} 6$$
$$12 - 6 \overset{?}{=} 6$$
$$6 = 6 \checkmark$$

Since $(3, -2)$ is a solution to each equation in the system, it is a solution to the system itself.

It is important to remember that we cannot confirm that a particular ordered pair is in fact the solution to a system of two equations unless we have checked to see whether the solution satisfies both equations. Merely checking one equation is not sufficient. Determining whether an ordered pair is a solution to a system of equations requires that we verify that the solution satisfies *both* equations.

Practice Problem 1 Determine whether $(-3, 4)$ is a solution to the following system.

$$2x + 3y = 6$$
$$3x - 4y = 7$$

2 Solving a System of Two Linear Equations by the Graphing Method

We can verify the solution to a system of linear equations by graphing each equation. If the lines intersect, the system has a unique solution. The point of intersection lies on both lines. Thus, it is a solution to each equation and the solution to the system. We will illustrate this by graphing the equations in Example 1. Notice that the coordinates of the point of intersection are $(3, -2)$. The solution to the system is $(3, -2)$.

This example shows that we can find the solution to a system of linear equations by graphing each line and determining the point of intersection.

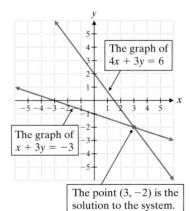

The graph of $4x + 3y = 6$

The graph of $x + 3y = -3$

The point $(3, -2)$ is the solution to the system.

EXAMPLE 2 Solve this system of equations by graphing.

$$2x + 3y = 12$$
$$x - y = 1$$

Solution   Using the methods that we developed in Chapter 3, we graph each line and determine the point at which the two lines intersect. The graph is to the right.

Finding the solution by the graphing method does not always lead to an accurate result, however, because it involves visual estimation of the point of intersection. Also, our plotting of one or more of the lines could be off slightly. Thus, we verify that our answer is correct by substituting $x = 3$ and $y = 2$ into the system of equations.

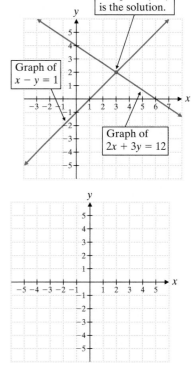

The point (3, 2) is the solution.

Graph of $x - y = 1$

Graph of $2x + 3y = 12$

$x - y = 1$	$2x + 3y = 12$
$3 - 2 \stackrel{?}{=} 1$	$2(3) + 3(2) \stackrel{?}{=} 12$
$1 = 1$ ✓	$12 = 12$ ✓

Thus, we have verified that the solution to the system is $(3, 2)$.

Practice Problem 2 Solve this system of equations by graphing. Check your solution.

$$3x + 2y = 10$$
$$x - y = 5$$

Many times when we graph a system, we find that the two lines intersect at one point. However, it is possible for a given system to have as its graph two parallel lines. In such a case there is no solution because there is no point that lies on both lines (i.e., no ordered pair that satisfies both equations). Such a system of equations is said to be **inconsistent.** Another possibility is that when we graph each equation in the system, we obtain one line. In such a case there are an infinite number of solutions. Any point (i.e., any ordered pair) that lies on the first line will also lie on the second line. A system of equations in two variables is said to have **dependent equations** if it has infinitely many solutions. We will discuss these situations in more detail after we have developed algebraic methods for solving systems of equations.

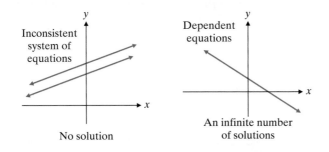

Inconsistent system of equations

No solution

Dependent equations

An infinite number of solutions

3 Solving a System of Two Linear Equations by the Substitution Method

One algebraic method of solving a system of linear equations in two variables is the **substitution method.** To use this method, we choose one equation and solve for one variable. It is usually best to solve for a variable that has a coefficient of $+1$ or -1, if possible. This will help us avoid introducing fractions. When we solve for one variable, we obtain an expression that contains the other variable. We *substitute* this expression into the second equation. Then we have one equation with one unknown, which we can easily solve. Once we know the value of this variable, we can substitute it into one of the original equations to find the value of the other variable.

Graphing Calculator

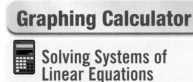

Solving Systems of Linear Equations

We can solve systems of equations graphically by using a graphing calculator. For example, to solve the system of equations in Example 2, first rewrite each equation in slope–intercept form.

$$y = -\frac{2}{3}x + 4$$
$$y = x - 1$$

Then graph $y_1 = -\frac{2}{3}x + 4$ and $y_2 = x - 1$ on the same screen.

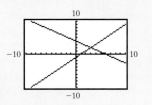

Next, use the Zoom and Trace features to find the intersection of the two lines.

Some graphing calculators have a command to find and calculate the intersection.

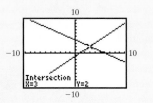

Try to find the solution to:

$$y_1 = -3x + 9$$
$$y_2 = 4x - 5$$

EXAMPLE 3 Find the solution to the following system of equations. Use the substitution method.

$$x + 3y = -7 \quad \textbf{(1)}$$
$$4x + 3y = -1 \quad \textbf{(2)}$$

Solution We can work with equation **(1)** or equation **(2)**. Let's choose equation **(1)** because x has a coefficient of 1. Now let us solve for x. This gives us equation **(3)**.

$$x = -7 - 3y \quad \textbf{(3)}$$

Now we substitute this expression for x into equation **(2)** and solve the equation for y.

$$4x + 3y = -1 \quad \textbf{(2)}$$
$$4(-7 - 3y) + 3y = -1$$
$$-28 - 12y + 3y = -1$$
$$-28 - 9y = -1$$
$$-9y = -1 + 28$$
$$-9y = 27$$
$$y = -3$$

Now we substitute $y = -3$ into equation **(1)** or **(2)** to find x. Let's use **(1)**:

$$x + 3(-3) = -7$$
$$x - 9 = -7$$
$$x = -7 + 9$$
$$x = 2$$

Therefore, our solution is the ordered pair $(2, -3)$.
Check. We must verify the solution in both of the *original* equations.

$$x + 3y = -7 \qquad\qquad 4x + 3y = -1$$
$$2 + 3(-3) \stackrel{?}{=} -7 \qquad 4(2) + 3(-3) \stackrel{?}{=} -1$$
$$2 - 9 \stackrel{?}{=} -7 \qquad\qquad 8 - 9 \stackrel{?}{=} -1$$
$$-7 = -7 \ \checkmark \qquad\qquad -1 = -1 \ \checkmark$$

Practice Problem 3 Use the substitution method to solve this system.

$$2x - y = 7$$
$$3x + 4y = -6$$

We summarize the substitution method here.

HOW TO SOLVE A SYSTEM OF TWO LINEAR EQUATIONS BY THE SUBSTITUTION METHOD

1. Choose one of the two equations and solve for one variable in terms of the other variable.

2. Substitute the expression from step 1 into the *other* equation.

3. You now have one equation with one variable. Solve this equation for that variable.

4. Substitute this value for the variable into one of the original equations to obtain a value for the second variable.

5. Check the solution in both original equations.

Optional Graphing Calculator Note.   Before solving the system in Example 3 with a graphing calculator, you will first need to solve each equation for y. Equation **(1)** can be written as $y_1 = -\dfrac{1}{3}x - \dfrac{7}{3}$ or as $y_1 = \dfrac{-x - 7}{3}$. Likewise, equation **(2)** can be written as $y_2 = -\dfrac{4}{3}x - \dfrac{1}{3}$ or as $y_2 = \dfrac{-4x - 1}{3}$.

EXAMPLE 4   Solve the following system of equations.

$$\frac{1}{2}x - \frac{1}{4}y = -\frac{3}{4} \quad \textbf{(1)}$$
$$3x - 2y = -6 \quad \textbf{(2)}$$

Solution   First clear equation **(1)** of fractions by multiplying each term by 4.

$$4\left(\frac{1}{2}x\right) - 4\left(\frac{1}{4}y\right) = 4\left(-\frac{3}{4}\right)$$
$$2x - y = -3 \quad \textbf{(3)}$$

The new system is as follows:

$$2x - y = -3 \quad \textbf{(3)}$$
$$3x - 2y = -6 \quad \textbf{(2)}$$

Step 1   Let's solve equation **(3)** for y. We select this because the y-variable has a coefficient of -1.

$$-y = -3 - 2x$$
$$y = 3 + 2x$$

Step 2   Substitute this expression for y into equation **(2).**

$$3x - 2(3 + 2x) = -6$$

Step 3   Solve this equation for x.

$$3x - 6 - 4x = -6$$
$$-6 - x = -6$$
$$-x = -6 + 6$$
$$-x = 0$$
$$x = 0$$

Step 4   Substitute $x = 0$ into equation **(2).**

$$3(0) - 2y = -6$$
$$-2y = -6$$
$$y = 3$$

So our solution is $(0, 3)$.

Step 5   We must verify the solution in both original equations.

$$\frac{1}{2}x - \frac{1}{4}y = -\frac{3}{4} \qquad\qquad 3x - 2y = -6$$
$$\frac{0}{2} - \frac{3}{4} \overset{?}{=} -\frac{3}{4} \qquad\qquad 3(0) - 2(3) \overset{?}{=} -6$$
$$-\frac{3}{4} = -\frac{3}{4} \;\checkmark \qquad\qquad\qquad -6 = -6 \;\checkmark$$

NOTE TO STUDENT: Fully worked-out solutions to all of the Practice Problems can be found at the back of the text starting at page SP-1

Practice Problem 4 Use the substitution method to solve this system.

$$\frac{1}{2}x + \frac{2}{3}y = 1$$
$$\frac{1}{3}x + y = -1$$

4 Solving a System of Two Linear Equations by the Addition (Elimination) Method

Another way to solve a system of two linear equations in two variables is to add the two equations so that a variable is eliminated. This technique is called the **addition method** or the **elimination method.** We usually have to multiply one or both of the equations by suitable factors so that we obtain opposite coefficients on one variable (either x or y) in the equations.

EXAMPLE 5 Solve the following system by the addition method.

$$5x + 8y = -1 \quad \textbf{(1)}$$
$$3x + y = 7 \quad \textbf{(2)}$$

Solution We can eliminate either the x- or the y-variable. Let's choose y. We multiply equation **(2)** by -8.

$$-8(3x) + (-8)(y) = -8(7)$$
$$-24x - 8y = -56 \quad \textbf{(3)}$$

We now add equations **(1)** and **(3)**.

$$\begin{array}{r} 5x + 8y = -1 \quad \textbf{(1)} \\ -24x - 8y = -56 \quad \textbf{(3)} \\ \hline -19x = -57 \end{array}$$

We solve for x.

$$x = \frac{-57}{-19} = 3$$

Now we substitute $x = 3$ into equation **(2)** (or equation **(1)**).

$$3(3) + y = 7$$
$$9 + y = 7$$
$$y = -2$$

Our solution is $(3, -2)$.

$$\begin{aligned} Check. \quad 5(3) + 8(-2) &\overset{?}{=} -1 \\ 15 + (-16) &\overset{?}{=} -1 \\ -1 &= -1 \checkmark \\ 3(3) + (-2) &\overset{?}{=} 7 \\ 9 + (-2) &\overset{?}{=} 7 \\ 7 &= 7 \checkmark \end{aligned}$$

Practice Problem 5 Use the addition method to solve this system.

$$-3x + y = 5$$
$$2x + 3y = 4$$

For convenience, we summarize the addition method here.

> **HOW TO SOLVE A SYSTEM OF TWO LINEAR EQUATIONS BY THE ADDITION (ELIMINATION) METHOD**
>
> 1. Arrange each equation in the form $ax + by = c$. (Remember that a, b, and c can be any real numbers.)
> 2. Multiply one or both equations by appropriate numbers so that the coefficients of one of the variables are opposites.
> 3. Add the two equations from step 2 so that one variable is eliminated.
> 4. Solve the resulting equation for the remaining variable.
> 5. Substitute this value into one of the *original* equations and solve to find the value of the other variable.
> 6. Check the solution in both of the original equations.

EXAMPLE 6   Solve the following system by the addition method.

$$3x + 2y = -8 \quad \textbf{(1)}$$
$$2x + 5y = 2 \quad \textbf{(2)}$$

Solution   To eliminate the variable x, we multiply equation **(1)** by 2 and equation **(2)** by -3. We now have the following equivalent system.

$$
\begin{aligned}
6x + 4y &= -16 \\
\underline{-6x - 15y} &= \underline{-6} \\
-11y &= -22 \quad \text{Add the equations.} \\
y &= 2 \quad\;\; \text{Solve for } y.
\end{aligned}
$$

Substitute $y = 2$ into equation **(1).**

$$
\begin{aligned}
3x + 2(2) &= -8 \\
3x + 4 &= -8 \\
3x &= -12 \\
x &= -4
\end{aligned}
$$

The solution to the system is $(-4, 2)$.

Check. Verify that this solution is correct.

Note. We could have easily eliminated the variable y in Example 6 by multiplying equation **(1)** by 5 and equation **(2)** by -2. Try it. Is the solution the same? Why?

Practice Problem 6   Use the addition (elimination) method to solve this system.

$$5x + 4y = 23$$
$$7x - 3y = 15$$

⑤ Identifying Systems of Linear Equations That Do Not Have a Unique Solution

So far we have examined only those systems that have one solution. But other systems must also be considered. These systems can best be illustrated with graphs. In general, the system of equations

$$ax + by = c$$
$$dx + ey = f$$

may have one solution, no solution, or an infinite number of solutions.

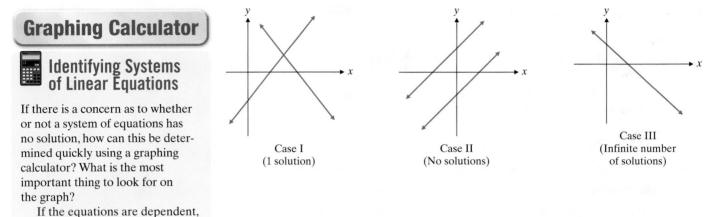

Case I
(1 solution)

Case II
(No solutions)

Case III
(Infinite number
of solutions)

Graphing Calculator

Identifying Systems of Linear Equations

If there is a concern as to whether or not a system of equations has no solution, how can this be determined quickly using a graphing calculator? What is the most important thing to look for on the graph?

If the equations are dependent, how can we be sure of this by looking at the display of a graphing calculator? Why? Determine whether there is one solution, no solution, or an infinite number of solutions for each of the following systems.

1. $y_1 = -2x + 1$
$y_2 = -6x + 3$

2. $y_1 = 3x + 6$
$y_2 = 3x + 2$

3. $y_1 = x - 3$
$y_2 = -2x + 12$

Case I: *One solution.* The two graphs intersect at one point, which is the solution. We say that the equations are **independent.** It is a **consistent system** of equations. There is a point (an ordered pair) *consistent* with both equations.

Case II: *No solution.* The two graphs are parallel and so do not intersect. We say that the system of equations is **inconsistent** because there is no point consistent with both equations.

Case III: *An infinite number of solutions.* The graphs of each equation yield the same line. Every ordered pair on this line is a solution to both of the equations. We say that the equations are **dependent.**

EXAMPLE 7 If possible, solve the system.

$$2x + 8y = 16 \quad (1)$$
$$4x + 16y = -8 \quad (2)$$

Solution To eliminate the variable *y*, we'll multiply equation **(1)** by -2.

$$-2(2x) + (-2)(8y) = (-2)(16)$$
$$-4x - 16y = -32 \quad (3)$$

We now have the following equivalent system.

$$-4x - 16y = -32 \quad (3)$$
$$4x + 16y = -8 \quad (2)$$

When we add equations **(3)** and **(2)**, we get

$$0 = -40,$$

which, of course, is false. Thus, we conclude that this system of equations is inconsistent, and **there is no solution.** Therefore, equations **(1)** and **(2)** do not intersect, as we can see on the graph to the left.

If we had used the substitution method to solve this system, we still would have obtained a false statement. When you try to solve an inconsistent system of linear equations by any method, you will always obtain a mathematical equation that is not true.

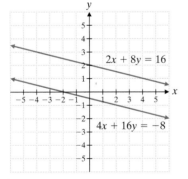

Practice Problem 7 If possible, solve the system.

$$4x - 2y = 6$$
$$-6x + 3y = 9$$

NOTE TO STUDENT: *Fully worked-out solutions to all of the Practice Problems can be found at the back of the text starting at page SP-1*

EXAMPLE 8   If possible, solve the system.

$$0.5x - 0.2y = 1.3 \quad \textbf{(1)}$$
$$-1.0x + 0.4y = -2.6 \quad \textbf{(2)}$$

Solution   Although we could work directly with the decimals, it is easier to multiply each equation by the appropriate power of 10 (10, 100, and so on) so that the coefficients of the new system are integers. Therefore, we will multiply equations **(1)** and **(2)** by 10 to obtain the following equivalent system.

$$5x - 2y = 13 \quad \textbf{(3)}$$
$$-10x + 4y = -26 \quad \textbf{(4)}$$

We can eliminate the variable y by multiplying each term of equation **(3)** by 2.

$$10x - 4y = 26 \quad \textbf{(5)}$$
$$\underline{-10x + 4y = -26 \quad \textbf{(4)}}$$
$$0 = 0 \qquad \text{Add the equations.}$$

This statement is always true; it is an **identity.** Hence, the two equations are dependent, and there is an infinite number of solutions. Any solution satisfying equation **(1)** will also satisfy equation **(2).** For example, $(3, 1)$ is a solution to equation **(1).** (Prove this.) Hence, it must also be a solution to equation **(2).** (Prove it). Thus, the equations actually describe the same line, as you can see on the graph.

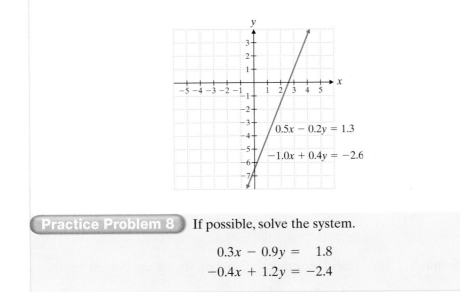

Practice Problem 8   If possible, solve the system.

$$0.3x - 0.9y = 1.8$$
$$-0.4x + 1.2y = -2.4$$

⑥ Choosing an Appropriate Method to Solve a System of Linear Equations Algebraically

At this point we will review the algebraic methods for solving systems of linear equations and discuss the advantages and disadvantages of each method.

Method	Advantage	Disadvantage
Substitution	Works well if one or more variables has a coefficient of 1 or −1.	Often becomes difficult to use if no variable has a coefficient of 1 or −1.
Addition	Works well if equations have fractional or decimal coefficients. Works well if no variable has a coefficient of 1 or −1.	None

EXAMPLE 9   Select a method and solve each system of equations.

(a) $x + y = 3080$
     $2x + 3y = 8740$

(b) $5x - 2y = 19$
       $-3x + 7y = 35$

Solution

(a) Since there are x- and y-values that have coefficients of 1, we will select the substitution method.

$$y = 3080 - x \quad \text{Solve the first equation for } y.$$
$$2x + 3(3080 - x) = 8740 \quad \text{Substitute the expression into the second equation.}$$
$$2x + 9240 - 3x = 8740 \quad \text{Remove parentheses.}$$
$$-1x = -500 \quad \text{Simplify.}$$
$$x = 500 \quad \text{Divide each side by } -1.$$

Substitute $x = 500$ into the first equation.

$$x + y = 3080$$
$$500 + y = 3080$$
$$y = 3080 - 500$$
$$y = 2580 \qquad \text{Simplify.}$$

The solution is $(500, 2580)$.

Check. Verify that this solution is correct.

(b) Because none of the x- and y-variables has a coefficient of 1 or -1, we select the addition method. We choose to eliminate the y-variable. Thus, we would like the coefficients of y to be -14 and 14.

$$7(5x) - 7(2y) = 7(19) \quad \text{Multiply each term of the first equation by 7.}$$
$$2(-3x) + 2(7y) = 2(35) \quad \text{Multiply each term of the second equation by 2.}$$
$$35x - 14y = 133 \quad \text{We now have an equivalent system of equations.}$$
$$\underline{-6x + 14y = 70}$$
$$29x = 203 \quad \text{Add the two equations.}$$
$$x = 7 \quad \text{Divide each side by 29.}$$

Substitute $x = 7$ into one of the original equations. We will use the first equation.

$$5(7) - 2y = 19$$
$$35 - 2y = 19 \quad \text{Solve for } y.$$
$$-2y = -16$$
$$y = 8$$

The solution is $(7, 8)$.

Check. Verify that this solution is correct.

NOTE TO STUDENT: Fully worked-out solutions to all of the Practice Problems can be found at the back of the text starting at page SP-1

Practice Problem 9   Select a method and solve each system of equations.

(a) $3x + 5y = 1485$
    $x + 2y = 564$

(b) $7x + 6y = 45$
    $6x - 5y = -2$

TO THINK ABOUT: Two Linear Equations with Two Variables

Now is a good time to look back over what we have learned. When you graph a system of two linear equations, what possible kinds of graphs will you obtain?

What will happen when you try to solve a system of two linear equations using algebraic methods? How many solutions are possible in each case? The following chart may help you to organize your answers to these questions.

Graph	Number of Solutions	Algebraic Interpretation
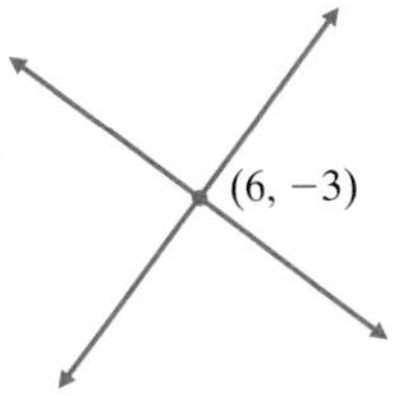 Two lines intersect at one point (6, −3)	**One unique solution**	You obtain one value for x and one value for y. For example, $$x = 6, \quad y = -3.$$
 Parallel lines	**No solution**	You obtain an equation that is inconsistent with known facts. For example, $$0 = 6.$$ The system of equations is inconsistent.
 Lines coincide	**Infinite number of solutions**	You obtain an equation that is always true. For example, $$8 = 8.$$ The equations are dependent.

Verbal and Writing Skills

1. Explain what happens when a system of two linear equations is inconsistent. What effect does it have in obtaining a solution? What would the graph of such a system look like?

2. Explain what happens when a system of two linear equations has dependent equations. What effect does it have in obtaining a solution? What would the graph of such a system look like?

3. How many possible solutions can a system of two linear equations in two unknowns have?

4. When you have graphed a system of two linear equations in two unknowns, how do you determine the solution of the system?

Determine whether the given ordered pair is a solution to the system of equations.

5. $\left(\dfrac{3}{2}, -1\right)$ $\begin{array}{l} 4x + 1 = 6 - y \\ 2x - 5y = 8 \end{array}$

6. $\left(-4, \dfrac{2}{3}\right)$ $\begin{array}{l} 2x - 3(y - 5) = 5 \\ 6y = x + 8 \end{array}$

Solve the system of equations by graphing. Check your solution.

7. $\begin{array}{l} 3x + y = 2 \\ 2x - y = 3 \end{array}$

8. $\begin{array}{l} 3x + y = 5 \\ 2x - y = 5 \end{array}$

9. $\begin{array}{l} 3x - 2y = 6 \\ 4x + y = -3 \end{array}$

10. $\begin{array}{l} 3x - y = 5 \\ 2x - 3y = -6 \end{array}$

11. $\begin{array}{l} y = -x + 3 \\ x + y = -\dfrac{2}{3} \end{array}$

12. $\begin{array}{l} y = \dfrac{1}{3}x - 2 \\ -x + 3y = 9 \end{array}$

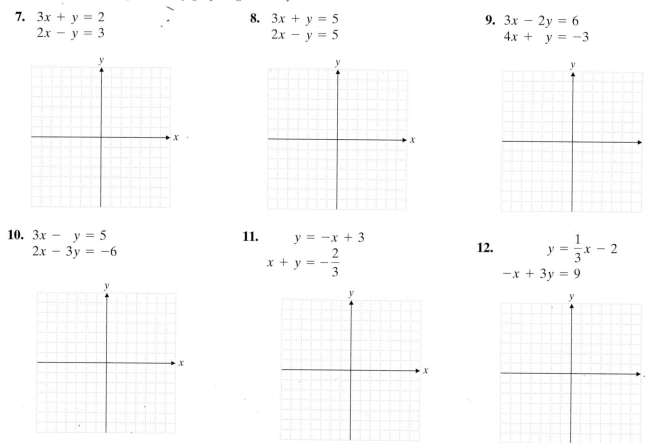

13. $y = -2x + 5$
$3y + 6x = 15$

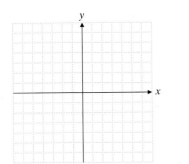

14. $x - 3 = 2y + 1$

$y - \dfrac{x}{2} = -2$

Find the solution to each system by the substitution method. Check your answers for exercises 15–18.

15. $3x + 4y = 14$
$x + 2y = -2$

16. $2x - 5y = -1$
$x - 3y = -3$

17. $-x + 3y = -8$
$2x - y = 6$

18. $10x + 3y = 8$
$2x + y = 2$

19. $2x - \dfrac{1}{2}y = -3$
$\dfrac{x}{5} + 2y = \dfrac{19}{5}$

20. $x - \dfrac{3}{2}y = 1$
$2x - 7y = 10$

21. $\dfrac{1}{2}x - \dfrac{1}{8}y = 3$
$\dfrac{2}{3}x + \dfrac{3}{4}y = 4$

22. $\dfrac{1}{5}x - y = 1$
$x + \dfrac{3}{10}y = 5$

Find the solution to each system by the addition (elimination) method. Check your answers for exercises 23–26.

23. $9x + 2y = 2$
$3x + 5y = 5$

24. $12x - 5y = -7$
$4x + 2y = 5$

25. $3s + 3t = 10$
$4s - 9t = -4$

26. $4s - 5t = 18$
$6s + 10t = -8$

27. $\dfrac{7}{2}x + \dfrac{5}{2}y = -4$
$3x + \dfrac{2}{3}y = 1$

28. $\dfrac{4}{3}x - y = 4$
$\dfrac{3}{4}x - y = \dfrac{1}{2}$

29. $1.6x + 1.5y = 1.8$
$0.4x + 0.3y = 0.6$

30. $2.5x + 0.6y = 0.2$
$0.5x - 1.2y = 0.7$

Mixed Practice

If possible, solve each system of equations. Use any method. If there is not a unique solution to a system, state a reason.

31. $7x - y = 6$
$3x + 2y = 22$

32. $8x - y = 17$
$4x + 3y = 33$

33. $3x + 4y = 8$
$5x + 6y = 10$

34. $7x + 5y = -25$
$3x + 7y = -1$

35. $2x + y = 4$
$\dfrac{2}{3}x + \dfrac{1}{4}y = 2$

36. $2x + 3y = 16$
$5x - \dfrac{3}{4}y = 7$

37. $0.2x = 0.1y - 1.2$
$2x - y = 6$

38. $0.1x - 0.6 = 0.3y$
$0.3x + 0.1y + 2.2 = 0$

39. $5x - 7y = 12$
$-10x + 14y = -24$

40. $3x - 11y = 9$
$-9x + 33y = 18$

41. $0.8x + 0.9y = 1.3$
$0.6x - 0.5y = 4.5$

42. $0.6y = 0.9x + 1$
$3x = 2y - 4$

43. $\dfrac{5}{3}b = \dfrac{1}{3} + a$
$9a - 15b = 2$

44. $a - 3b = \dfrac{3}{4}$
$\dfrac{a}{3} = \dfrac{1}{4} + b$

45. $\dfrac{2}{3}x - y = 4$
$2x - \dfrac{3}{4}y = 21$

46. $\dfrac{3}{7}x - y = -3$
$x - \dfrac{5}{3}y = 1$

47. $3.2x - 1.5y = -3$
$0.7x + y = 2$

48. $3x - 0.2y = 1$
$1.1x + 0.4y = -2$

49. $3 - (2x + 1) = y + 6$
$x + y + 5 = 1 - x$

50. $2(y - 3) = x + 3y$
$x + 2 = 3 - y$

To Think About

51. *Bathroom Tile* Wayne Burton is having some tile replaced in his bathroom. He has obtained an estimate from two tile companies. Old World Tile gave an estimate of $200 to remove the old tile and $50 per hour to place new tile on the wall. Modern Bathroom Headquarters gave an estimate of $300 to remove the old tile and $30 per hour to place new tile on the wall.

(a) Create a cost equation for each company where y is the total cost of the tile work and x is the number of hours of labor used to install new tile. Write a system of equations.

(b) Graph the two equations using the values $x = 0, 4,$ and 8.

(c) Determine from your graph how many hours of installing new tile will be required for the two companies to cost the same.

(d) Determine from your graph which company costs less to remove old tile and to install new tile if the time needed to install new tile is 6 hours.

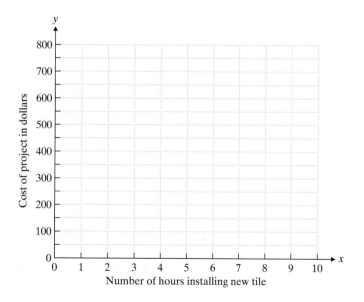

Optional Graphing Calculator Problems

On a graphing calculator, graph each system of equations on the same set of axes. Find the point of intersection to the nearest hundredth.

52. $y_1 = -0.81x + 2.3$
$y_2 = 1.6x + 0.8$

53. $y_1 = -1.7x + 3.8$
$y_2 = 0.7x - 2.1$

54. $5.86x + 6.22y = -8.89$
$-2.33x + 4.72y = -10.61$

55. $0.5x + 1.1y = 5.5$
$-3.1x + 0.9y = 13.1$

Cumulative Review

56. [1.2.2] *Winter Road Salt* Nine million tons of salt are applied to American highways for road deicing each year. The cost of buying and applying the salt totals $200 million. How much is this per pound? Round your answer to the nearest cent.

57. [1.2.2] *City Parking Space* Four-fifths of the automobiles that enter the city of Boston during rush hour will have to park in private or municipal parking lots. If there are 273,511 private or municipal lot spaces filled by cars entering the city during rush hour every morning, how many cars enter the city during rush hour? Round your answer to the nearest car.

Quick Quiz 4.1

1. Solve using the substitution method.

$5x - 3y = 14$
$2x - y = 6$

2. Solve using the addition method.

$6x + 7y = 26$
$5x - 2y = 6$

3. Solve by any method.

$\dfrac{2}{3}x + \dfrac{3}{5}y = -17$

$\dfrac{1}{2}x - \dfrac{1}{3}y = -1$

4. Concept Check Explain what happens when you go through the steps to solve the following system of equations. Why does this happen?

$6x - 4y = 8$
$-9x + 6y = -12$

Student Learning Objectives

After studying this section, you will be able to:

1 Determine whether an ordered triple is the solution to a system of three linear equations in three variables.

2 Find the solution to a system of three linear equations in three variables if none of the coefficients is zero.

3 Find the solution to a system of three linear equations in three variables if some of the coefficients are zero.

1 Determining Whether an Ordered Triple Is the Solution to a System of Three Linear Equations in Three Variables

We are now going to study **systems of three linear equations in three variables** (unknowns). A **solution** to a system of three linear equations in three unknowns is an **ordered triple** of real numbers (x, y, z) that satisfies each equation in the system.

EXAMPLE 1 Determine whether $(2, -5, 1)$ is a solution to the following system.

$$3x + y + 2z = 3$$
$$4x + 2y - z = -3$$
$$x + y + 5z = 2$$

Solution How can we determine whether $(2, -5, 1)$ is a solution to this system? We will substitute $x = 2$, $y = -5$, and $z = 1$ into each equation. If a true statement occurs each time, $(2, -5, 1)$ is a solution to each equation and hence, a solution to the system. For the first equation:

$$3(2) + (-5) + 2(1) \stackrel{?}{=} 3$$
$$6 - 5 + 2 \stackrel{?}{=} 3$$
$$3 = 3 \checkmark$$

For the second equation:

$$4(2) + 2(-5) - 1 \stackrel{?}{=} -3$$
$$8 - 10 - 1 \stackrel{?}{=} -3$$
$$-3 = -3 \checkmark$$

For the third equation:

$$2 + (-5) + 5(1) \stackrel{?}{=} 2$$
$$2 - 5 + 5 \stackrel{?}{=} 2$$
$$2 = 2 \checkmark$$

Since we obtained three true statements, the ordered triple $(2, -5, 1)$ is a solution to the system.

Practice Problem 1 Determine whether $(3, -2, 2)$ is a solution to this system.

$$2x + 4y + z = 0$$
$$x - 2y + 5z = 17$$
$$3x - 4y + z = 19$$

NOTE TO STUDENT: Fully worked-out solutions to all of the Practice Problems can be found at the back of the text starting at page SP-1

TO THINK ABOUT: Graphs in Three Variables

Can we graph an equation in three variables? How? What would the graph look like? What would the graph of the system in Example 1 look like? Describe the graph of the solution. At the end of Section 4.2 we will show you how the graphs might look.

2 Finding the Solution to a System of Three Linear Equations in Three Variables If None of the Coefficients Is Zero

One way to solve a system of three equations with three variables is to obtain from it a system of two equations in two variables; in other words, we eliminate one variable from both equations. We can then use the methods of Section 4.1 to solve the resulting system. You can find the third variable (the one that was eliminated) by substituting the two variables that you have found into one of the original equations.

EXAMPLE 2   Find the solution to (that is, solve) the following system of equations.

$$
\begin{aligned}
-2x + 5y + z &= 8 \quad \textbf{(1)} \\
-x + 2y + 3z &= 13 \quad \textbf{(2)} \\
x + 3y - z &= 5 \quad \textbf{(3)}
\end{aligned}
$$

Solution   Let's eliminate z because it can be done easily by adding equations **(1)** and **(3)**.

$$
\begin{aligned}
-2x + 5y + z &= 8 \quad \textbf{(1)} \\
\underline{x + 3y - z} &= \underline{5} \quad \textbf{(3)} \\
-x + 8y &= 13 \quad \textbf{(4)}
\end{aligned}
$$

Now we need to choose a *different pair* from the original system of equations and once again eliminate the same variable. In other words, we have to use equations **(1)** and **(2)** or equations **(2)** and **(3)** and eliminate z. Let's multiply each term of equation **(3)** by 3 (and call it equation **(6)**) and add the result to equation **(2)**.

$$
\begin{aligned}
-x + 2y + 3z &= 13 \quad \textbf{(2)} \\
\underline{3x + 9y - 3z} &= \underline{15} \quad \textbf{(6)} \\
2x + 11y &= 28 \quad \textbf{(5)}
\end{aligned}
$$

We now can solve the resulting system of two linear equations.

$$
\begin{aligned}
-x + 8y &= 13 \quad \textbf{(4)} \\
2x + 11y &= 28 \quad \textbf{(5)}
\end{aligned}
$$

Multiply each term of equation **(4)** by 2.

$$
\begin{aligned}
-2x + 16y &= 26 \\
\underline{2x + 11y} &= \underline{28} \\
27y &= 54 \quad \text{Add the equations.} \\
y &= 2 \quad \text{Solve for } y.
\end{aligned}
$$

Substituting $y = 2$ into equation **(4)**, we have the following:

$$
\begin{aligned}
-x + 8(2) &= 13 \\
-x &= -3 \\
x &= 3.
\end{aligned}
$$

Now substitute $x = 3$ and $y = 2$ into one of the original equations (any one will do) to solve for z. Let's use equation **(1)**.

$$
\begin{aligned}
-2x + 5y + z &= 8 \\
-2(3) + 5(2) + z &= 8 \\
-6 + 10 + z &= 8 \\
z &= 4
\end{aligned}
$$

The solution to the system is $(3, 2, 4)$.

Check.   Verify that $(3, 2, 4)$ satisfies *each* of the three *original* equations.

Practice Problem 2   Solve this system.

$$
\begin{aligned}
x + 2y + 3z &= 4 \\
2x + y - 2z &= 3 \\
3x + 3y + 4z &= 10
\end{aligned}
$$

Here's a summary of the procedure that we just used.

> ### HOW TO SOLVE A SYSTEM OF THREE LINEAR EQUATIONS IN THREE VARIABLES
>
> 1. Use the addition method to eliminate any variable from any pair of equations. (The choice of variable is arbitrary.)
> 2. Use appropriate steps to eliminate the *same variable* from a *different pair* of equations. (If you don't eliminate the same variable, you will still have three unknowns.)
> 3. Solve the resulting system of two equations in two variables.
> 4. Substitute the values obtained in step 3 into one of the three original equations. Solve for the remaining variable.
> 5. Check the solution in all of the original equations.

It is helpful to write all equations in the form $Ax + By + Cz = D$ before using this five-step method.

3   Finding the Solution to a System of Three Linear Equations in Three Variables If Some of the Coefficients Are Zero

If a system of three linear equations in three variables contains one or more equations of the form $Ax + By + Cz = D$, where one of the values of A, B, or C is zero, then we will slightly modify our approach to solving the system. We will select one equation that contains only two variables. Then we will take the remaining system of two equations and eliminate the variable that is missing in the equation that we selected.

EXAMPLE 3   Solve the system.

$$4x + 3y + 3z = 4 \quad \textbf{(1)}$$
$$3x + 2z = 2 \quad \textbf{(2)}$$
$$2x - 5y = -4 \quad \textbf{(3)}$$

Solution   Note that equation **(2)** has no y-term and equation **(3)** has no z-term. Obviously, that makes our work easier. Let's work with equations **(2)** and **(1)** to obtain an equation that contains only x and y.

Step 1   Multiply equation **(1)** by 2 and equation **(2)** by -3 to obtain the following system.

$$8x + 6y + 6z = 8 \quad \textbf{(4)}$$
$$\underline{-9x - 6z = -6} \quad \textbf{(5)}$$
$$-x + 6y = 2 \quad \textbf{(6)}$$

Step 2   This step is already done, since equation **(3)** has no z-term.

Step 3   Now we can solve the system formed by equations **(3)** and **(6)**.

$$2x - 5y = -4 \quad \textbf{(3)}$$
$$-x + 6y = 2 \quad \textbf{(6)}$$

If we multiply each term of equation **(6)** by 2, we obtain the system

$$2x - 5y = -4$$
$$\underline{-2x + 12y = 4}$$
$$7y = 0 \quad \text{Add.}$$
$$y = 0. \quad \text{Solve for } y.$$

Substituting $y = 0$ in equation **(6)**, we find the following:

$$-x + 6(0) = 2$$
$$-x = 2$$
$$x = -2.$$

Step 4 To find z, we substitute $x = -2$ and $y = 0$ into one of the original equations containing z. Since equation **(2)** has only two variables, let's use it.

$$3x + 2z = 2$$
$$3(-2) + 2z = 2$$
$$2z = 8$$
$$z = 4$$

The solution to the system is $(-2, 0, 4)$.

Check. Verify this solution by substituting these values into equations **(1)**, **(2)**, and **(3)**.

> **Practice Problem 3** Solve the system.
>
> $$2x + y + z = 11$$
> $$4y + 3z = -8$$
> $$x - 5y = 2$$

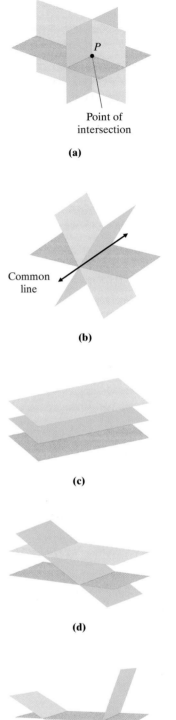

Point of intersection

(a)

Common line

(b)

(c)

(d)

(e)

A linear equation in three variables is a plane in three-dimensional space. A system of three linear equations in three variables is three planes. The solution to the system is the set of points at which all three planes intersect. There are three possible results. The three planes may intersect at one point. (See figure **(a)** in the margin.) This point is described by an ordered triple of the form (x, y, z) and lies in each plane.

The three planes may intersect at a line. (See figure **(b)** in the margin.) In this case the system has an infinite number of solutions; that is, all the points on the line are solutions to the system.

Finally, all three planes may not intersect at any points. It may mean that all three planes never share any point of intersection. (See figure **(c)** in the margin.) It may mean that two planes intersect. (See figures **(d)** and **(e)** in the margin.) In all such cases there is no solution to the system of equations.

1. Determine whether $(2, 1, -4)$ is a solution to the system.

$$2x - 3y + 2z = -7$$
$$x + 4y - z = 10$$
$$3x + 2y + z = 4$$

2. Determine whether $(-3, 0, 1)$ is a solution to the system.

$$2x + 5y - z = -7$$
$$x - 11y + 4z = 1$$
$$-5x + 8y - 12z = 3$$

3. Determine whether $(-1, 5, 1)$ is a solution to the system.

$$3x + 2y - z = 6$$
$$x - y - 2z = -8$$
$$4x + y + 2z = 5$$

4. Determine whether $(3, 2, 1)$ is a solution to the system.

$$x + y + 2z = 7$$
$$2x + y + z = 9$$
$$3x + 4y - 2z = 13$$

Solve each system.

5.
$$x + y + 2z = 0$$
$$2x - y - z = 1$$
$$x + 2y + 3z = 1$$

6.
$$2x + y + 3z = 2$$
$$x - y + 2z = -4$$
$$x + 3y - z = 1$$

7.
$$-x + 2y - z = -1$$
$$2x + y + z = 2$$
$$x - y - 2z = -13$$

8.
$$-4x + 2y + z = 1$$
$$x - y + 3z = -5$$
$$3x + y - 4z = 10$$

9.
$$8x - 5y + z = 15$$
$$3x + y - z = -7$$
$$x + 4y + z = -3$$

10.
$$-4x + y - 3z = 2$$
$$5x - 3y + 4z = 1$$
$$3x - 2y + 5z = 1$$

11.
$$x + 4y - z = -5$$
$$-2x - 3y + 2z = 5$$
$$x - \frac{2}{3}y + z = \frac{11}{3}$$

12.
$$x - 4y + 4z = -1$$
$$-x + \frac{y}{2} - \frac{5}{2}z = -3$$
$$-x + 3y - z = 5$$

13.
$$2x + 2z = -7 + 3y$$
$$\frac{3}{2}x + y + \frac{1}{2}z = 2$$
$$x + 4y = 10 + z$$

14.
$$8x - y = 2z$$
$$y = 3x + 4z + 2$$
$$5x + 2z = y - 1$$

15.
$$a = 8 + 3b - 2c$$
$$4a + 2b - 3c = 10$$
$$c = 10 + b - 2a$$

16.
$$a = c - b$$
$$3a - 2b + 6c = 1$$
$$c = 4 - 3b - 7a$$

17. $0.2a + 0.1b + 0.2c = 0.1$
$0.3a + 0.2b + 0.4c = -0.1$
$0.6a + 1.1b + 0.2c = 0.3$

18. $-0.1a + 0.2b + 0.3c = 0.1$
$0.2a - 0.6b + 0.3c = 0.5$
$0.3a - 1.2b - 0.4c = -0.4$

Find the solution for each system of equations. Round your answers to five decimal places.

19. $x - 4y + 4z = -3.72186$
$-x + 3y - z = 5.98115$
$2x - y + 5z = 7.93645$

20. $4x + 2y + 3z = 9$
$9x + 3y + 2z = 3$
$2.987x + 5.027y + 3.867z = 18.642$

Solve each system.

21. $x - y = 5$
$2y - z = 1$
$3x + 3y + z = 6$

22. $3x + 3z = -3$
$y - 2z = -4$
$-2x + y + z = 8$

23. $-y + 2z = 1$
$x + y + z = 2$
$-x + 3z = 2$

24. $-2x + y - 3z = 0$
$-2y - z = -1$
$x + 2y - z = 5$

25. $x - 2y + z = 0$
$-3x - y = -6$
$y - 2z = -7$

26. $x + 2z = 0$
$3x + 3y + z = 6$
$6y + 5z = -3$

27. $\dfrac{3a}{2} - b + 2c = 2$
$\dfrac{a}{2} + 2b - 2c = 4$
$-a + b = -6$

28. $\dfrac{2a}{3} - \dfrac{b}{3} = -1$
$2a + b + \dfrac{c}{3} = -3$
$-3b - c = 3$

Try to solve the system of equations. Explain your result in each case.

29.
$$2x + y = -3$$
$$2y + 16z = -18$$
$$-7x - 3y + 4z = 6$$

30.
$$6x - 2y + 2z = 2$$
$$4x + 8y - 2z = 5$$
$$-2x - 4y + z = -2$$

31.
$$3x + 3y - 3z = -1$$
$$4x + y - 2z = 1$$
$$-2x + 4y - 2z = -8$$

32.
$$-3x + 4y - z = -4$$
$$x + 2y + z = 4$$
$$-12x + 16y - 4z = -16$$

Cumulative Review

33. [2.3.1] Solve for x. $|2x - 1| = 7$

34. [1.4.5] Write using scientific notation. 76,300,000

35. [3.3.2] Explain the steps to find the standard form of the equation of the line that passes through $(1, 4)$ and $(-2, 3)$.

36. [3.3.3] Explain the steps to find the standard form of the equation of the line that is perpendicular to $y = -\dfrac{2}{3}x + 4$ and passes through $(-4, 2)$.

Quick Quiz 4.2

Solve each system.

1.
$$4x - y + 2z = 0$$
$$2x + y + z = 3$$
$$3x - y + z = -2$$

2.
$$x + 2y + 2z = -1$$
$$2x - y + z = 1$$
$$x + 3y - 6z = 7$$

3.
$$4x - 2y + 6z = 0$$
$$6y + 3z = 3$$
$$x + 2y - z = 5$$

4. Concept Check Explain how you would eliminate the variable z and obtain two equations with only the variables x and y in the following system.
$$2x + 4y - 2z = -22$$
$$4x + 3y + 5z = -10$$
$$5x - 2y + 3z = 13$$

How are you doing with your homework assignments in Sections 4.1 and 4.2? Do you feel you have mastered the material so far? Do you understand the concepts you have covered? Before you go further in the textbook, take some time to do each of the following problems.

4.1

1. Solve by the substitution method.

$4x - y = -1$
$3x + 2y = 13$

2. Solve by the addition method.

$5x + 2y = 0$
$-3x - 4y = 14$

Find the solution to each system of equations by any method. If there is no single solution to a system, state the reason.

3. $5x - 2y = 27$
$3x - 5y = -18$

4. $7x + 3y = 15$
$\dfrac{1}{3}x - \dfrac{1}{2}y = 2$

5. $2x = 3 + y$
$3y = 6x - 9$

6. $0.2x + 0.7y = -1$
$0.5x + 0.6y = -0.2$

7. $6x - 9y = 15$
$-4x + 6y = 8$

4.2

8. Determine whether $(2, 0, -3)$ is a solution to the system.

$4x - y + 3z = -1$
$-2x + 3y + 5z = -19$
$x - 2y + 4z = 10$

Find the solution to each system of equations.

9. $5x - 2y + z = -1$
$3x + y - 2z = 6$
$-2x + 3y - 5z = 7$

10. $2x - y + 3z = -1$
$5x + y + 6z = 0$
$2x - 2y + 3z = -2$

11. $x + y + 2z = 9$
$3x + 2y + 4z = 16$
$2y + z = 10$

12. $x - 2z = -5$
$y - 3z = -3$
$2x - z = -4$

Now turn to page SA-11 for the answers to each of these problems. Each answer also includes a reference to the objective in which the problem is first taught. If you missed any of these problems, you should stop and review the Examples and Practice Problems in the referenced objective. A little review now will help you master the material in the upcoming sections of the text.

1. _____

2. _____

3. _____

4. _____

5. _____

6. _____

7. _____

8. _____

9. _____

10. _____

11. _____

12. _____

Allosaurus

1 Solving Applications Requiring the Use of a System of Two Linear Equations in Two Unknowns

We will now examine how a system of linear equations can assist us in solving applied problems.

EXAMPLE 1   For the paleontology lecture on campus, advance tickets cost $5 and tickets at the door cost $6. The ticket sales this year came to $4540. The department chairman wants to raise prices next year to $7 for advance tickets and $9 for tickets at the door. He said that if exactly the same number of people attend next year, the ticket sales at these new prices will total $6560. If he is correct, how many tickets were sold in advance this year? How many tickets were sold at the door?

Solution

1. *Understand the problem.* Since we are looking for the number of tickets sold, we let

$$x = \text{number of tickets bought in advance and}$$
$$y = \text{number of tickets bought at the door}$$

2. *Write a system of two equations in two unknowns.* If advance tickets cost $5, then the total sales will be $5x$; similarly, total sales of door tickets will be $6y$. Since the total sales of both types of tickets was $4540, we have

$$5x + 6y = 4540.$$

By the same reasoning, we have

$$7x + 9y = 6560.$$

Thus, our system is as follows:

$$5x + 6y = 4540 \quad \textbf{(1)}$$
$$7x + 9y = 6560. \quad \textbf{(2)}$$

3. *Solve the system of equations and state the answer.* We multiply each term of equation **(1)** by -3 and each term of equation **(2)** by 2 to obtain the following equivalent system.

$$
\begin{array}{rcl}
-15x - 18y &=& -13{,}620 \\
14x + 18y &=& 13{,}120 \\
\hline
-x &=& -500
\end{array}
$$

Therefore, $x = 500$. Substituting $x = 500$ into equation **(1)**, we have the following:

$$5(500) + 6y = 4540$$
$$6y = 2040$$
$$y = 340.$$

Thus, 500 advance tickets and 340 door tickets were sold.

4. *Check.* We need to check our answers. Do they seem reasonable?

Would 500 advance tickets at $5 and 340 door tickets at $6 yield $4540?	Would 500 advance tickets at $7 and 340 door tickets at $9 yield $6560?
$5(500) + 6(340) \overset{?}{=} 4540$	$7(500) + 9(340) \overset{?}{=} 6560$
$2500 + 2040 \overset{?}{=} 4540$	$3500 + 3060 \overset{?}{=} 6560$
$4540 = 4540$ ✓	$6560 = 6560$ ✓

NOTE TO STUDENT: *Fully worked-out solutions to all of the Practice Problems can be found at the back of the text starting at page SP-1*

Practice Problem 1 Coach Perez purchased baseballs at $6 each and bats at $21 each last week for the college baseball team. The total cost of the purchase was $318. This week he noticed that the same items are on sale. Baseballs are now $5 each and bats are $17. He found that if he made the same purchase this week, it would cost only $259. How many baseballs and how many bats did he buy last week?

EXAMPLE 2 An electronics firm makes two types of switching devices. Type *A* takes 4 minutes to make and requires $3 worth of materials. Type *B* takes 5 minutes to make and requires $5 worth of materials. When the production manager reviewed the latest batch, he found that it took 35 hours to make these switches with a materials cost of $1900. How many switches of each type were produced for this latest batch?

Solution

1. *Understand the problem.* We are given a lot of information, but the major concern is to find out how many of the type *A* devices and the type *B* devices were produced. This becomes our starting point to define the variables we will use.

$$\text{Let } A = \text{the number of type } A \text{ devices produced and}$$
$$B = \text{the number of type } B \text{ devices produced}$$

2. *Write a system of two equations.* How should we construct the equations? What relationships exist between our variables (or unknowns)? According to the problem, the devices are related by time and by cost. So we set up one equation in terms of time (minutes in this case) and one in terms of cost (dollars). Each type *A* took 4 minutes to make, each type *B* took 5 minutes to make, and the total time was $60(35) = 2100$ minutes. Each type *A* used $3 worth of materials, each type *B* used $5 worth of materials, and the total materials cost was $1900. We can gather this information in a table. Making a table will help us form the equations.

	Type *A* Devices	Type *B* Devices	Total
Number of Minutes	4*A*	5*B*	2100
Cost of Materials	3*A*	5*B*	1900

$$4A + 5B = 2100$$
$$3A + 5B = 1900$$

Therefore, we have the following system.

$$4A + 5B = 2100 \quad \textbf{(1)}$$
$$3A + 5B = 1900 \quad \textbf{(2)}$$

3. *Solve the system of equations and state the answers.* Multiplying equation **(2)** by -1 and adding the equations, we find the following:

$$\begin{array}{r} 4A + 5B = 2100 \\ -3A - 5B = -1900 \\ \hline A = 200. \end{array}$$

Substituting $A = 200$ into equation **(1),** we have the following:

$$800 + 5B = 2100$$
$$5B = 1300$$
$$B = 260.$$

Thus, 200 type *A* devices and 260 type *B* devices were produced.

4. Check. If each type A requires 4 minutes and each type B requires 5 minutes, does this amount to a total time of 2100 minutes?

$$4A + 5B = 2100$$
$$4(200) + 5(260) \stackrel{?}{=} 2100$$
$$800 + 1300 \stackrel{?}{=} 2100$$
$$2100 = 2100 \checkmark$$

If each type A costs $3 and each type B costs $5, does this amount to a total cost of $1900?

$$3A + 5B = 1900$$
$$3(200) + 5(260) \stackrel{?}{=} 1900$$
$$600 + 1300 \stackrel{?}{=} 1900$$
$$1900 = 1900 \checkmark$$

NOTE TO STUDENT: Fully worked-out solutions to all of the Practice Problems can be found at the back of the text starting at page SP-1

Practice Problem 2 A furniture company makes both small and large chairs. It takes 30 minutes of machine time and 1 hour and 15 minutes of labor to build the small chair. The large chair requires 40 minutes of machine time and 1 hour and 20 minutes of labor. The company has 57 hours of labor time and 26 hours of machine time available each day. If all available time is used, how many chairs of each type can the company make?

When we encounter motion problems involving rate, time, or distance, it is useful to recall the formula $D = RT$ or distance = (rate)(time).

EXAMPLE 3 An airplane travels between two cities that are 1500 miles apart. The trip against the wind takes 3 hours. The return trip with the wind takes $2\frac{1}{2}$ hours. What is the speed of the plane in still air (in other words, how fast would the plane travel if there were no wind)? What is the speed of the wind?

Solution

1. Understand the problem. Our unknowns are the speed of the plane in still air and the speed of the wind.

Let

x = the speed of the plane in still air and

y = the speed of the wind

Let's make a sketch to help us see how these speeds are related to one another. When we travel against the wind, the wind is slowing us down. Since the wind speed opposes the plane's speed in still air, we must subtract: $x - y$.

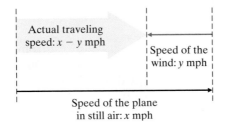

When we travel with the wind, the wind is helping us travel forward. Thus, the wind speed is added to the plane's speed in still air, and we add: $x + y$.

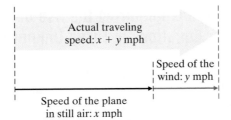

2. ***Write a system of two equations.*** To help us write our equations, we organize the information in a chart. The chart will be based on the formula $RT = D$, which is (rate)(time) = distance.

	R ·	T =	D
Flying against the wind	$x - y$	3	1500
Flying with the wind	$x + y$	2.5	1500

Using the rows of the chart, we obtain a system of equations.

$$(x - y)(3) = 1500$$
$$(x + y)(2.5) = 1500$$

If we remove the parentheses, we will obtain the following system.

$$3x - 3y = 1500 \quad \textbf{(1)}$$
$$2.5x + 2.5y = 1500 \quad \textbf{(2)}$$

3. ***Solve the system of equations and state the answer.*** It will be helpful to clear equation **(2)** of decimal coefficients. Although we could multiply each term by 10, doing so will result in large coefficients on x and y. For this equation, multiplying by 2 is a better choice.

$$3x - 3y = 1500 \quad \textbf{(1)}$$
$$5x + 5y = 3000 \quad \textbf{(3)}$$

If we multiply equation **(1)** by 5 and equation **(3)** by 3, we will obtain the following system.

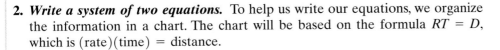

$$15x - 15y = 7500$$
$$\underline{15x + 15y = 9000}$$
$$30x \qquad\quad = 16{,}500$$
$$x = 550$$

Graphing Calculator

Exploration

· A visual interpretation of two equations in two unknowns is sometimes helpful. Study Example 3. Graph the two equations.

$$3x - 3y = 1500$$
$$2.5x + 2.5y = 1500$$

What is the significance of the point of intersection? If you were an air traffic controller, how would you interpret the linear equation $3x - 3y = 1500$? Why would this be useful? How would you interpret $2.5x + 2.5y = 1500$? Why would this be useful?

Substituting this result in equation **(1)**, we obtain the following:

$$3(550) - 3y = 1500$$
$$1650 - 3y = 1500$$
$$-3y = -150$$
$$y = 50.$$

Thus, the speed of the plane in still air is 550 miles per hour, and the speed of the wind is 50 miles per hour.

4. Check. The check is left to the student.

Practice Problem 3 An airplane travels west from city A to city B against the wind. It takes 3 hours to travel 1950 kilometers. On the return trip the plane travels east from city B to city C, a distance of 1600 kilometers in a time of 2 hours. On the return trip the plane travels with the wind. What is the speed of the plane in still air? What is the speed of the wind?

② Solving Applications Requiring the Use of a System of Three Linear Equations in Three Unknowns

EXAMPLE 4 A trucking firm has three sizes of trucks. The biggest truck holds 10 tons of gravel, the next size holds 6 tons, and the smallest holds 4 tons. The firm has a contract to provide fifteen trucks to haul 104 tons of gravel. To reduce fuel costs, the firm's manager wants to use two more of the fuel-efficient 10-ton trucks than the 6-ton trucks. How many trucks of each type should she use?

Solution

1. **Understand the problem.** Since we need to find three things (the numbers of 10-ton trucks, 6-ton trucks, and 4-ton trucks), it would be helpful to have three variables. Let

$$x = \text{the number of 10-ton trucks used,}$$
$$y = \text{the number of 6-ton trucks used, and}$$
$$z = \text{the number of 4-ton trucks used.}$$

2. **Write a system of three equations.** We know that fifteen trucks will be used; hence, we have the following:

$$x + y + z = 15. \quad \textbf{(1)}$$

How can we get our second equation? Well, we also know the *capacity* of each truck type, and we know the total tonnage to be hauled. The first type of truck hauls 10 tons, the second type 6 tons, and the third type 4 tons, and the total tonnage is 104 tons. Hence, we can write the following:

$$10x + 6y + 4z = 104. \quad \textbf{(2)}$$

We still need one more equation. What other given information can we use? The problem states that the manager wants to use two more 10-ton trucks than 6-ton trucks. Thus, we have the following:

$$x = 2 + y. \quad \textbf{(3)}$$

(We could also have written $x - y = 2$.) Hence, our system of equations is as follows:

$$x + y + z = 15 \quad \textbf{(1)}$$
$$10x + 6y + 4z = 104 \quad \textbf{(2)}$$
$$x - y \qquad = 2. \quad \textbf{(3)}$$

3. ***Solve the system of equations and state the answers.*** Equation **(3)** doesn't contain the variable z. Let's work with equations **(1)** and **(2)** to eliminate z. First, we multiply equation **(1)** by -4 and add it to equation **(2)**.

$$
\begin{array}{rcl}
-4x - 4y - 4z &=& -60 \quad \textbf{(4)} \\
10x + 6y + 4z &=& 104 \quad \textbf{(2)} \\
\hline
6x + 2y &=& 44 \quad \textbf{(5)}
\end{array}
$$

Make sure you understand how we got equation (5). Dividing each term of equation **(5)** by 2 and adding to equation **(3)** gives the following:

$$
\begin{array}{rcl}
3x + y &=& 22 \quad \textbf{(6)} \\
x - y &=& 2 \quad \textbf{(3)} \\
\hline
4x &=& 24 \\
x &=& 6.
\end{array}
$$

For $x = 6$, equation **(3)** yields the following:

$$
\begin{array}{rcl}
6 - y &=& 2 \\
4 &=& y.
\end{array}
$$

Now we substitute the known x- and y-values into equation **(1)**.

$$
\begin{array}{rcl}
6 + 4 + z &=& 15 \\
z &=& 5
\end{array}
$$

Thus, the manager needs six 10-ton trucks, four 6-ton trucks, and five 4-ton trucks.

4. ***Check.*** The check is left to the student.

Practice Problem 4   A factory uses three machines to wrap boxes for shipment. Machines A, B, and C working together can wrap 260 boxes in 1 hour. If machine A runs 3 hours and machine B runs 2 hours, they can wrap 390 boxes. If machine B runs 3 hours and machine C runs 4 hours, 655 boxes can be wrapped. How many boxes per hour can each machine wrap?

NOTE TO STUDENT: Fully worked-out solutions to all of the Practice Problems can be found at the back of the text starting at page SP-1

Developing Your Study Skills

Applications or Word Problems

Applications, or word problems, are the very life of mathematics! They are the reason for doing mathematics because they teach you how to use the mathematical skills you have developed. Learning mathematics without ever doing word problems is similar to learning all the skills of a sport without ever playing a game or learning all the notes on an instrument without ever playing a song.

The key to success is practice. Make yourself work through as many exercises as you can. You may not be able to do them all correctly at first, but keep trying. Do not give up when you reach a difficult exercise. If you cannot solve it, just try another one. Then go back and try the "difficult" one again later.

A misconception among students when they begin studying word problems is that each one is different. At first the exercises may seem this way, but as you practice more and more, you will begin to see the similarities, the different "types." You will see patterns, which will enable you to solve exercises of a given type more easily.

Applications

Use a system of two linear equations to solve each exercise.

1. The sum of 2 numbers is 87. If twice the smaller number is subtracted from the larger number, the result is 12. Find the two numbers.

2. The difference between two numbers is 10. If three times the larger is added to twice the smaller, the result is 70. Find the two numbers.

3. *Temporary Employment Agency* An employment agency specializing in temporary construction help pays heavy equipment operators $140 per day and general laborers $90 per day. If thirty-five people were hired and the payroll was $3950, how many heavy equipment operators were employed? How many laborers?

4. *Broadway Ticket Prices* A Broadway performance of *Les Misérables* had a paid attendance of 308 people. Balcony tickets cost $38 and orchestra tickets cost $60. Ticket sales receipts totaled $15,576. How many tickets of each type were sold?

5. *Amtrak Train Tickets* Ninety-eight passengers rode in an Amtrak train from Boston to Denver. Tickets for regular coach seats cost $120. Tickets for sleeper car seats cost $290. The receipts for the trip totaled $19,750. How many passengers purchased each type of ticket?

6. *Farm Operations* The Tupper Farm has 450 acres of land allotted for raising corn and wheat. The cost to cultivate corn is $42 per acre. The cost to cultivate wheat is $35 per acre. The Tuppers have $16,520 available to cultivate these crops. How many acres of each crop should the Tuppers plant?

7. *Computer Training* A large company wants to train its human resources department to use new spreadsheet software and a new payroll program. Experienced employees can learn the spreadsheet in 3 hours and the payroll in 4 hours. New employees need 4 hours to learn the spreadsheet and 7 hours to learn the payroll. The company can afford to pay for 115 hours of spreadsheet instruction and 170 hours of payroll instruction. How many of each type of employee can the company train?

8. *Radar Detector Manufacturing* Ventex makes auto radar detectors. Ventex has found that its basic model requires 3 hours of manufacturing for the inside components and 2 hours for the housing and controls. Its advanced model requires 5 hours to manufacture the inside components and 3 hours for the housing and controls. This week, the production division has available 1050 hours for producing inside components and 660 hours for housing and controls. How many detectors of each type can be made?

9. *Farm Management* A farmer has several packages of fertilizer for his new grain crop. The old packages contain 50 pounds of long-term-growth supplement and 60 pounds of weed killer. The new packages contain 65 pounds of long-term-growth supplement and 45 pounds of weed killer. Using past experience, the farmer estimates that he needs 3125 pounds of long-term-growth supplement and 2925 pounds of weed killer for the fields. How many old packages of fertilizer and how many new packages of fertilizer should he use?

10. *Hospital Dietician* A staff hospital dietician has two prepackaged mixtures of vitamin additives available for patients. Mixture 1 contains 5 grams of vitamin C and 3 grams of niacin; mixture 2 contains 6 grams of vitamin C and 5 grams of niacin. On an average day she needs 87 grams of niacin and 117 grams of vitamin C. How many packets of each mixture will she need?

11. *Coffee and Snack Expenses* On Monday, Harold picked up three doughnuts and four large coffees for the office staff. He paid $4.91. On Tuesday, Melinda picked up five doughnuts and six large coffees for the office staff. She paid $7.59. What is the cost of one doughnut? What is the cost of one large coffee?

12. *Advertising Costs* A local department store is preparing four-color sales brochures to insert into the *Salem Evening News*. The printer has a fixed charge to set up the printing of the brochure and a specific per-copy amount for each brochure printed. He quoted a price of $1350 for printing five thousand brochures and a price of $1750 for printing seven thousand brochures. What is the fixed charge to set up the printing? What is the per-copy cost for printing a brochure?

13. *Airspeed* Against the wind a small plane flew 210 miles in 1 hour and 10 minutes. The return trip took only 50 minutes. What was the speed of the wind? What was the speed of the plane in still air?

14. *Aircraft Operation* Against the wind, a commercial airline in South America flew 630 miles in 3 hours and 30 minutes. With a tailwind, the return trip took 3 hours. What was the speed of the wind? What was the speed of the plane in still air?

15. *Fishing and Boating* Don Williams uses his small motorboat to go 8 miles upstream to his favorite fishing spot. Against the current, the trip takes $\frac{2}{3}$ hour. With the current, the trip takes $\frac{1}{2}$ hour. How fast can the boat travel in still water? What is the speed of the current?

16. *Canoe Trip* It look Lance and Ivan 6 hours to travel 33 miles downstream by canoe on the Mississippi River. The next day they traveled for 8 hours upstream for 20 miles. What was the rate of the current? What was their average speed in still water?

17. *Basketball* In 1962, Wilt Chamberlain set an NBA single-game scoring record. He scored 64 times for a total of 100 points. He made no 3-point shots (these were not part of the professional basketball game until 1979), but made several free throws worth 1 point each and several regular shots worth 2 points each. How many free throws did he make? How many 2-point shots did he make?

18. *Basketball* Shaquille O'Neal scored 38 points in a recent basketball game. He scored no free throws, but he made a number of 2-point shots and 3-point shots. He scored 16 times. How many 2-point baskets did he make? How many 3-point baskets did he make?

19. *Telephone Charges* Nick's telephone company charges $0.05 per minute for weekend calls and $0.08 for calls made on weekdays. This month Nick was billed for 625 minutes. The charge for these minutes was $43.40. How many minutes did he talk on weekdays and how many minutes did he talk on the weekends?

20. *Office Supply Costs* A new catalog from an office supply company shows that some of its prices will increase next month. This month, copier paper costs $2.70 per ream and printer cartridges cost $15.50. If Chris submits his order this month, the cost will be $462. Next month, when paper will cost $3.00 per ream and the cartridges will cost $16, his order would cost $495. How many reams of paper and how many printer cartridges are in his order?

21. *Highway Department Purchasing* This year the state highway department in Montana purchased 256 identical cars and 183 identical trucks for official use. The purchase price was $5,791,948. Due to a budget shortfall, next year the department plans to purchase only 64 cars and 107 trucks. It will be charged the same price for each car and for each truck. Next year it plans to spend $2,507,612. How much does the department pay for each car and for each truck?

22. *Concert Ticket Prices* A recent concert at Gordon College had a paid audience of 987 people. Advance tickets were $9.95 and tickets at the door were $12.95. A total of $10,738.65 was collected in ticket sales. How many of each type of ticket were sold?

Use a system of three linear equations to solve exercises 23–30.

23. *Bookstore Supplies* Johanna bought 15 items at the college bookstore. The items cost a total of $23.00. The pens cost $0.50 each, the notebooks were $3.00 each, and the highlighters cost $1.50 each. She bought 2 more notebooks than highlighters. How many of each item did she buy?

24. *Basketball* In January 2006, Kobe Bryant of the Los Angeles Lakers had the highest-scoring game of his career. He made 46 shots for a total of 81 points. He made several free throws worth 1 point each, several regular shots worth 2 points each, and several 3-point shots. He made three times more regular shots than 3-point shots. How many of each type of shot did he make?

25. *High School Play* A total of three hundred people attended the high school play. The admission prices were $5 for adults, $3 for high school students, and $2 for any children not yet in high school. The ticket sales totaled $1010. The school principal suggested that next year they raise prices to $7 for adults, $4 for high school students, and $3 for children not yet in high school. He said that if exactly the same number of people attend next year, the ticket sales at the higher prices will total $1390. How many adults, high school students, and children not yet in high school attended this year?

26. *CPR Training* The college conducted a CPR training class for students, faculty, and staff. Faculty were charged $10, staff were charged $8, and students were charged $2 to attend the class. A total of four hundred people came. The receipts for all who attended totaled $2130. The college president remarked that if he had charged faculty $15 and staff $10 and let students come free, the receipts this year would have been $2425. How many students, faculty, and staff came to the CPR training class?

27. *City Subway Token Costs* A total of twelve thousand passengers normally ride the green line of the MBTA during the morning rush hour. The token prices for a ride are $0.25 for children under 12, $1 for adults, and $0.50 for senior citizens, and the revenue from these riders is $10,700. If the token prices were raised to $0.35 for children under 12 and $1.50 for adults, and the senior citizen price were unchanged, the expected revenue from these riders would be $15,820. How many riders in each category normally ride the green line during the morning rush hour?

28. *Commission Sales* The owner of Danvers Ford found that he sold a total of 520 cars, Escapes, and Explorers last year. He paid the sales staff a commission of $100 for every car, $200 for every Escape, and $300 for every Explorer sold. The total of these commissions last year was $87,000. In the coming year he is contemplating an increase so that the commission will be $150 for every car and $250 for every Escape, with no change in the commission for Explorer sales. If the sales are the same this year as they were last year, the commissions will total $106,500. How many vehicles in each category were sold last year?

29. *Pizza Costs* The Hamilton House of Pizza delivered twenty pepperoni pizzas to Gordon College on the first night of final exams. The cost of these pizzas totaled $181. A small pizza costs $5 and contains 3 ounces of pepperoni. A medium pizza costs $9 and contains 4 ounces of pepperoni. A large pizza costs $12 and contains 5 ounces of pepperoni. The owner of the pizza shop used 5 pounds 2 ounces of pepperoni in making these twenty pizzas. How many pizzas of each size were delivered to Gordon College?

30. *Roast Beef Sandwich Costs* One of the favorite meeting places for local college students is Nick's Roast Beef in Beverly, Massachusetts. Last night from 8 P.M. to 9 P.M. Nick served twenty-four roast beef sandwiches. He sliced 15 pounds 8 ounces of roast beef to make these sandwiches and collected $131 for them. The medium roast beef sandwich has 6 ounces of beef and costs $4. The large roast beef sandwich has 10 ounces of beef and costs $5. The extra large roast beef sandwich has 14 ounces of beef and costs $7. How many of each size of roast beef sandwich did Nick sell from 8 P.M. to 9 P.M.?

31. *Packing Fruit* Florida Fruits packs three types of gift boxes containing oranges, grapefruit, and tangerines. Box *A* contains 12 oranges, 5 grapefruit, and 3 tangerines. Box *B* contains 10 oranges, 6 grapefruit, and 4 tangerines. Box *C* contains 5 oranges, 8 grapefruit, and 5 tangerines. The shipping manager has available 91 oranges, 63 grapefruit, and 40 tangerines. How many gift boxes of each type can she prepare?

Cumulative Review *Solve for the variable indicated.*

32. **[2.1.1]** Solve for x. $\dfrac{1}{3}(4 - 2x) = \dfrac{1}{2}x - 3$

33. **[2.1.1]** Solve for x. $0.06x + 0.15(0.5 - x) = 0.04$

34. **[2.1.1]** Solve for y. $2(y - 3) - (2y + 4) = -6y$

35. **[2.2.1]** Solve for x. $6a(2x - 3y) = 7ax - 3$

Quick Quiz 4.3

1. A plane flew 1200 miles with a tailwind in 2.5 hours. The return trip against the wind took 3 hours. Find the speed of the wind. Find the speed of the plane in still air.

2. A man needed a car in Alaska to travel to fishing sites. He found a company that rented him a car but charged him a daily rental fee and a mileage charge. He rented the car for 8 days and drove 300 miles and was charged $355. Next week his friend rented the same car for 9 days and drove 260 miles and was charged $380. What was the daily rental charge? How much did the company charge per mile?

3. Three friends went on a trip to Uganda. They went to the same store and purchased the exact same items. Nancy purchased 3 drawings, 2 carved elephants, and 1 set of drums for $55. John purchased 2 drawings, 3 carved elephants, and 1 set of drums for $65. Steve purchased 4 drawings, 3 carved elephants, and 2 sets of drums for $85. What was the price of each drawing, carved elephants, and set of drums?

4. **Concept Check** Explain how you would set up two equations using the information in problem 1 above if the plane flew 1500 miles instead of 1200 miles.

4.4 SYSTEMS OF LINEAR INEQUALITIES

① Graphing a System of Linear Inequalities

We learned how to graph a linear inequality in two variables in Section 3.4. We call two linear inequalities in two variables a **system of linear inequalities in two variables.** We now consider how to graph such a system. The solution to a system of inequalities is the intersection of the solution sets of the individual inequalities of the system.

EXAMPLE 1 Graph the solution of the system.

$$y \leq -3x + 2$$
$$-2x + y \geq -1$$

Solution

In this example, we will first graph each inequality separately. The graph of $y \leq -3x + 2$ is the region on or below the line $y = -3x + 2$.

The graph of $-2x + y \geq -1$ consists of the region on or above the line $-2x + y = -1$.

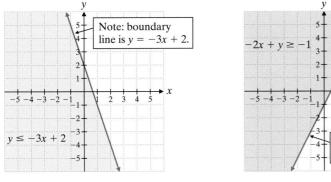

We will now place these graphs on one rectangular coordinate system. The darker shaded region is the intersection of the two graphs. Thus, the solution to the system of two inequalities is the darker shaded region and its boundary lines.

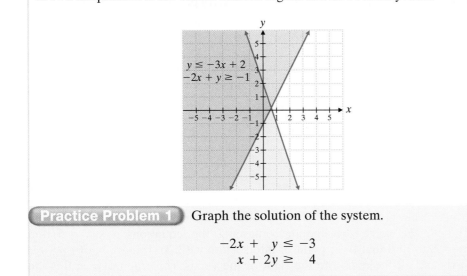

Practice Problem 1 Graph the solution of the system.

$$-2x + y \leq -3$$
$$x + 2y \geq 4$$

Usually we sketch the graphs of the individual inequalities on one set of axes. We will illustrate that concept with the following example.

Student Learning Objective

After studying this section, you will be able to:

① Graph a system of linear inequalities.

PRACTICE PROBLEM 1

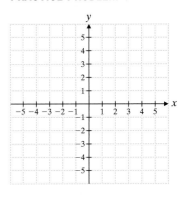

NOTE TO STUDENT: Fully worked-out solutions to all of the Practice Problems can be found at the back of the text starting at page SP-1

EXAMPLE 2 Graph the solution of the system.

$$y < 4$$
$$y > \frac{3}{2}x - 2$$

Solution The graph of $y < 4$ is the region below the line $y = 4$. It does not include the line since we have the $<$ symbol. Thus, we use a dashed line to indicate that the boundary line is not part of the answer. The graph of $y > \frac{3}{2}x - 2$ is the region above the line $y = \frac{3}{2}x - 2$. Again, we use the dashed line to indicate that the boundary line is not part of the answer. The final solution is the darker shaded region. The solution does *not* include the dashed boundary lines.

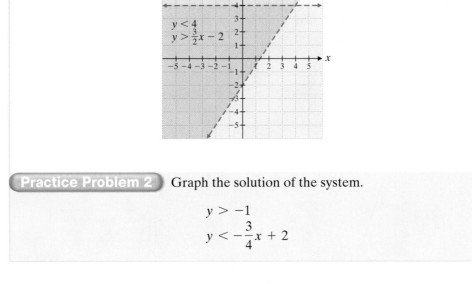

PRACTICE PROBLEM 2

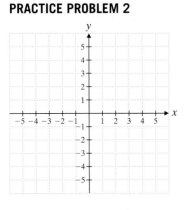

NOTE TO STUDENT: Fully worked-out solutions to all of the Practice Problems can be found at the back of the text starting at page SP-1

Practice Problem 2 Graph the solution of the system.

$$y > -1$$
$$y < -\frac{3}{4}x + 2$$

There are times when we require the exact location of the point where two boundary lines intersect. In these cases the boundary points are labeled on the final sketch of the solution.

EXAMPLE 3 Graph the solution of the following system of inequalities. Find the coordinates of any points where boundary lines intersect.

$$x + y \le 5$$
$$x + 2y \le 8$$
$$x \ge 0$$
$$y \ge 0$$

Solution The graph of $x + y \le 5$ is the region on and below the line $x + y = 5$. The graph of $x + 2y \le 8$ is the region on and below the line $x + 2y = 8$. We solve the system containing the equations $x + y = 5$ and $x + 2y = 8$ to find that their point of intersection is $(2, 3)$. The graph of $x \ge 0$ is the y-axis and all the region to the right of the y-axis. The graph of $y \ge 0$ is the x-axis and all the region above the x-axis. Thus, the solution to the system is the shaded region and its boundary lines. There are four points where boundary lines intersect.

These points are called the **vertices** of the solution. Thus, the vertices of the solution are $(0, 0)$, $(0, 4)$, $(2, 3)$, and $(5, 0)$.

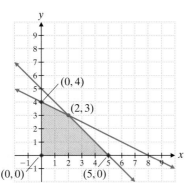

Practice Problem 3 Graph the solution to the system of inequalities. Find the vertices of the solution.

$$\begin{aligned} x + y &\le 6 \\ 3x + y &\le 12 \\ x &\ge 0 \\ y &\ge 0 \end{aligned}$$

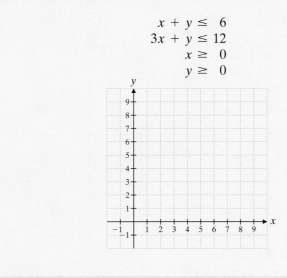

Graphing Calculator

Graphing Systems of Linear Inequalities

On some graphing calculators, you can graph a system of linear inequalities. To graph the system in Example 1, first rewrite it as:

$$\begin{aligned} y &\le -3x + 2 \\ y &\ge 2x - 1. \end{aligned}$$

Enter each expression into the Y = editor of your graphing calculator and then select the appropriate direction for shading. Display:

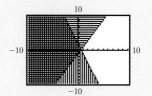

Notice that one inequality is shaded vertically and the other is shaded horizontally. The intersection is the solution. Note also that the graphing calculator will not indicate whether or not the boundary of the solution region is included in the solution.

Developing Your Study Skills

How to Review for an Exam

Reviewing adequately for an exam enables you to bring together the concepts you have learned over several sections. For your review, you will need to:

1. Reread your textbook. Make a list of any terms, rules, or formulas you need to know for the exam. Be sure you understand them all.

2. Reread your notes. Go over returned homework and quizzes. Redo the exercises you missed.

3. Practice some of each type of exercise covered in the chapter(s) you are to be tested on.

4. Use the end-of-chapter materials provided in your textbook. Read carefully through the Chapter Organizer. Take the Chapter Test. When you are finished, check your answers. Redo any exercises you missed.

5. Get help if any concepts give you difficulty.

Verbal and Writing Skills

1. In the graph of the system $y > 3x + 1$ and $y < -2x + 5$, would the boundary lines be solid or dashed? Why?

2. In the graph of the system $y \geq -6x + 3$ and $y \leq -4x - 2$, would the boundary lines be solid or dashed? Why?

3. Stephanie wanted to know if the point $(3, -4)$ lies in the region that is a solution for $y < -2x + 3$ and $y > 5x - 3$. How could she determine if this is true?

4. John wanted to know if the point $(-5, 2)$ lies in the region that is a solution for $x + 2y < 3$ and $-4x + y > 2$. How could he determine if this is true?

Graph the solution of each of the following systems.

5. $y \geq 2x - 1$
   $x + y \leq 6$

6. $y \geq x - 3$
   $x + y \geq 2$

7. $y \geq -2x$
   $y \geq 3x + 5$

8. $y \geq x$
   $y \leq -x + 2$

9. $y \geq 2x - 3$
   $y \leq \dfrac{2}{3}x$

10. $y \leq \dfrac{1}{2}x - 3$
    $y \geq -\dfrac{1}{2}x$

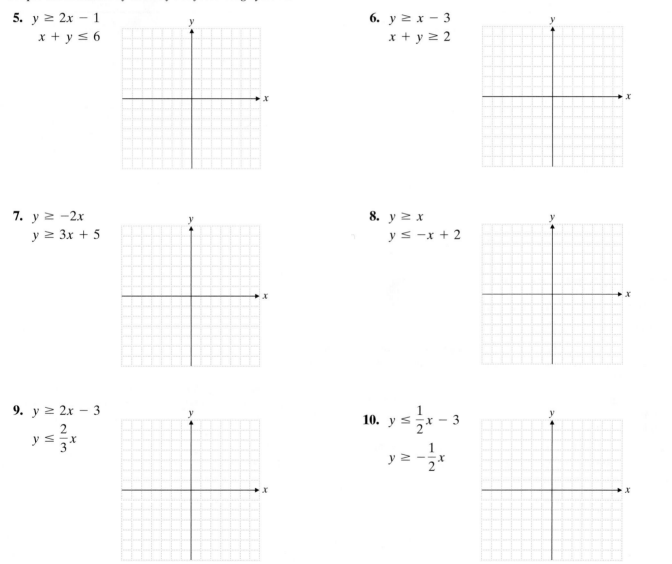

11. $x - y \geq -1$
     $-3x - y \leq 4$

12. $3x - y \leq 3$
     $-x + y \leq 4$

13. $x + 2y < 6$
        $y < 3$

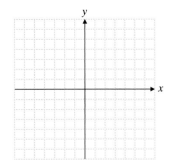

14. $-x + 2y < -6$
         $y > -3$

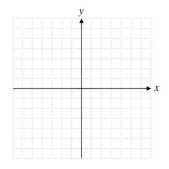

15. $y < 4$
       $x > -2$

16. $y > -3$
       $x < 2$

Mixed Practice

Graph the solution of each of the following systems.

17. $x - 4y \geq -4$
        $3x + y \leq 3$

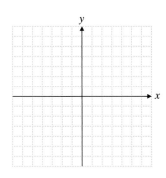

18. $5x - 2y \leq 10$
        $x - y \geq -1$

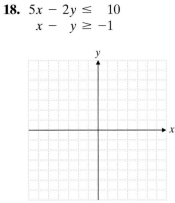

19. $3x + 2y < 6$
$3x + 2y > -6$

20. $2x - y < 2$
$2x - y > -2$

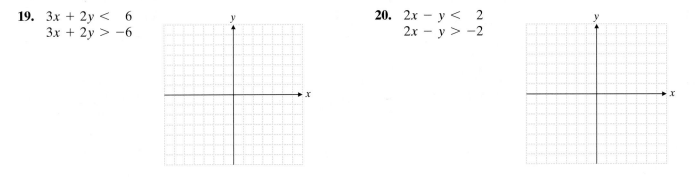

Graph the solution of the following systems of inequalities. Find the vertices of the solution.

21. $x + y \leq 5$
$2x - y \geq 1$

22. $x + y \geq 2$
$y + 4x \leq -1$

23. $x + 3y \leq 12$
$y < x$

24. $x + 2y \leq 4$
$y < -x$

25. $y \leq x$
$x + y \geq 1$
$x \leq 5$

26. $x + y \leq 3$
$x - y \leq 1$
$x \geq -1$

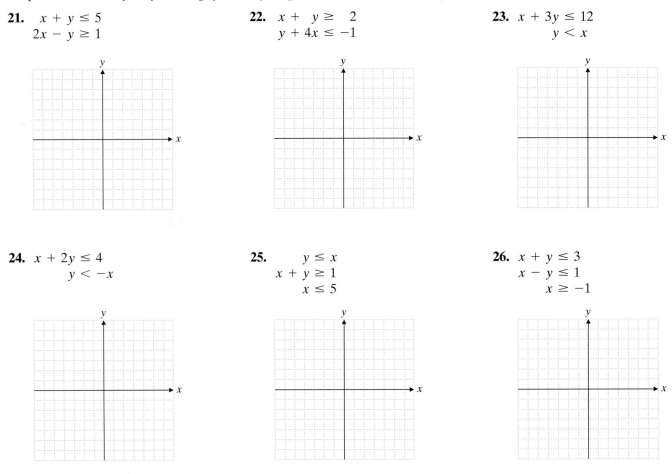

To Think About

Graph the region determined by each of the following systems.

27. $y \leq 3x + 6$
$4y + 3x \leq 3$
$x \geq -2$
$y \geq -3$

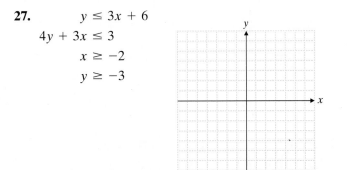

28. $-x + y \leq 100$
$x + 3y \leq 150$
$x \geq -80$
$y \geq 20$

(*Hint:* Use a
scale of each
square = 20 units
on both axes.)

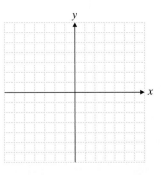

Applications

Hint: In exercises 29 and 30, if the coordinates of a boundary point contain fractions, it is wise to obtain the point of intersection algebraically rather than graphically.

29. ***Hospital Staffing Levels*** The equation that represents the proper level of medical staffing in the cardiac care unit of a local hospital is $N \le 2D$, where N is the number of nurses on duty and D is the number of doctors on duty. In order to control costs, the equation $4N + 3D \le 20$ is appropriate. The number of doctors and nurses on duty at any time cannot be negative, so $N \ge 0$ and $D \ge 0$.

    (a) Graph the region satisfying all of the medical requirements for the cardiac care unit. Use the following special graph grid, where D is measured on the horizontal axis and N is measured on the vertical axis.

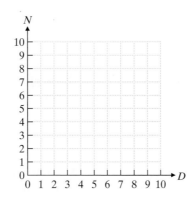

    (b) If there are three doctors and two nurses on duty in the cardiac care unit, are all of the medical requirements satisfied?

    (c) If there is one doctor and four nurses on duty in the cardiac care unit, are all of the medical requirements satisfied?

30. ***Traffic Control*** The equation that represents the proper traffic control and emergency vehicle response availability in the city of Salem is $P + 3F \le 18$, where P is the number of police cars on active duty and F is the number of fire trucks that have left the firehouse and are involved in a response to a call. In order to comply with staffing limitations, the equation $4P + F \le 28$ is appropriate. The number of police cars on active duty and the number of fire trucks that have left the firehouse cannot be negative, so $P \ge 0$ and $F \ge 0$.

    (a) Graph the regions satisfying all of the availability and staffing limitation requirements for the city of Salem. Use the following special graph grid, where P is measured on the horizontal axis and F is measured on the vertical axis.

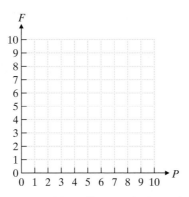

    (b) If four police cars are on active duty and four fire trucks have left the firehouse in response to a call, are all of the requirements satisfied?

    (c) If two police cars are on active duty and six fire trucks have left the firehouse in response to a call, are all of the requirements satisfied?

Cumulative Review

31. **[1.6.1]** Evaluate for $x = 2$, $y = -1$.
    $-3x^2y - x^2 + 5y^2$

32. **[1.5.3]** Simplify. $2x - 2[y + 3(x - y)]$

33. **[4.3.1]** *Driving Range Receipts* Golf Galaxy Indoor Driving Range took in $6050 on one rainy day and six sunny days. The next week the driving range took in $7400 on four rainy days and three sunny days. What is the average amount of money taken in at Golf Galaxy on a rainy day? On a sunny day?

34. **[4.3.2]** *Office Lunch Expenses* Two weeks ago, Raquel took the office lunch order to Kramer's Deli and bought 3 chicken sandwiches, 2 side salads, and 3 sodas for $17.30. One week ago it was Gerry's turn to take the trip to Kramer's. He bought 3 chicken sandwiches, 4 side salads, and 4 sodas for $23.15. This week Sheryl made the trip. She purchased 4 chicken sandwiches, 3 side salads, and 5 sodas for $24.75. What is the cost of one chicken sandwich? Of one side salad? Of one soda?

Quick Quiz 4.4

1. In graphing the inequality $3x + 5y < 15$, do you shade the region above the line or below the line?

2. In graphing the following region, should the boundary lines be dashed lines or solid lines?
$$4x + 5y \geq 20$$
$$2x - 3y \geq 6$$

3. Graph the solution of the following system of linear inequalities.
$$3x + 2y > 6$$
$$x - 2y < 2$$

4. **Concept Check** Explain how you would graph the region described by the following.
$$y > x + 2$$
$$x < 3$$

DETERMINING YOUR EARNINGS

Recorded on our paycheck stub are the **gross earnings** (earnings before deductions), and **net earnings** (earnings after deductions such as taxes, health benefits...). Does this net dollar amount represent the amount of money left to spend? No, because we must also consider all the expenses associated with working, and then deduct this amount from our net earnings. That is, we must find our **net employment income,** the actual amount we really earn and have left to spend. Consider the story of Sharon.

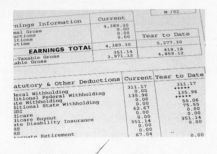

Sharon is currently not employed outside the home and has two children, one preschool age and one in second grade. Before deciding if she should take a job, Sharon estimated her job-related expenses as follows: **preschool daycare**—$175 per week; **after-school program**—$75 per week; **transportation**—$0.35 per mile; **clothing expenses**—$25 per week; and **extra expenses**—$50 per week. During vacations, Sharon will not incur any job-related expenses; she accounted for this fact in her budget. Sharon is considering the following two positions.

Job 1: Donivan Tech Donivan Tech offered Sharon a full-time management position with a contract that offers a yearly salary of $50,310 and 3 weeks paid vacation. Since a salaried position does not pay an hourly rate, Sharon will not be paid for hours worked above 40 hours per week. Sharon estimates that she will work an average of 45 hours per week when she considers travel time for the 45-mile round trip commute to work, and the extra work commitment required for managers. Federal and state taxes, as well as our taxable income brackets, often vary, but Sharon estimates her payroll deductions will be 17% of her gross salary.

Job 2: J&R Financial Group A part-time position as a Personal Assistant Manager was offered to Sharon by J&R Financial. This position requires a work schedule of 6 hours per day, 5 days a week, pays an hourly rate of $20.50, and includes 2 weeks paid vacation. Sharon determined that the round-trip commute to work is only 5 miles, and therefore will not increase her workday significantly. Since the workdays are shorter, after-school daycare for her child in second grade will be reduced to $25 per week, and daycare for her preschool child will decrease to $150 a week. Sharon estimates that the reduced work schedule will also decrease her extra expenses to $25 per week and payroll deductions to 12% of her gross salary, but it will not affect her clothing expenses.

Fill in the blanks with the missing information. If necessary, round your answer to the nearest hundredth.

1. Donivan Tech
   Gross earnings per week: _____
   Hourly wage for a 45-hour work week: _____

2. J&R Financial Group
   Gross earnings per week: _____
   Yearly gross earnings: _____

3. Based on Sharon's gross income, _____ pays the higher hourly rate. The higher rate is _____ per hour higher.

For exercises 4–6, place your answers in the table that follows. If necessary, round your answer to the nearest hundredth.

4. (a) Calculate yearly job-related expenses for the position at Donivan Tech. Be sure to include payroll deductions.
   (b) Job-related expenses would account for what percentage of Sharon's gross yearly earnings?

5. (a) Calculate the yearly job-related expenses for the position at J&R Financial Group. Be sure to include payroll deductions.
   (b) Job-related expenses would account for what percentage of Sharon's gross yearly earnings?

6. For each of the positions, what is Sharon's net employment income per year? Per week? Per hour?
   (a) Donivan Tech (b) J&R Financial Group

7. Which job do you think Sharon should take? Why?

	Donivan Tech Salaried Full-Time Position (3 wk vacation)	J&R Financial Hourly Part-Time Position (2 wk vacation)
Total Job-Related Expenses Per Year	4. (a)	5. (a)
Net Per Year Employment Income	6. (a)	6. (b)
Net Per Week Employment Income	6. (a)	6. (b)
Net Per Hour Employment Income	6. (a)	6. (b)
Percent: Work-Related Expenses	4. (b)	5. (b)

Chapter 4 Organizer

Topic	Procedure	Examples
Finding a solution to a system of equations by the graphing method, p. 196.	1. Graph the first equation. 2. Graph the second equation. 3. Approximate from your graph the coordinate of the point where the two lines intersect, if they intersect at one point. 4. If the lines are parallel, there is no solution. If the lines coincide, there is an infinite number of solutions.	Solve by graphing. $x + y = 6$ $2x - y = 6$ Graph each line. The solution is $(4, 2)$.
Solving a system of two linear equations by the substitution method, p. 198.	The substitution method is most appropriate when *at least one variable has a coefficient of* 1 *or* −1. 1. Solve for one variable in one of the equations. 2. In the other equation, replace that variable with the expression you obtained in step 1. 3. Solve the resulting equation. 4. Substitute the numerical value you obtain for a variable into one of the original equations and solve for the other variable. 5. Check the solution in both original equations.	Solve. $\quad 2x + y = 11 \quad \textbf{(1)}$ $\qquad\qquad x + 3y = 18 \quad \textbf{(2)}$ $y = 11 - 2x$ from equation **(1)**. Substitute into **(2)**. $$x + 3(11 - 2x) = 18$$ $$x + 33 - 6x = 18$$ $$-5x = -15$$ $$x = 3$$ Substitute $x = 3$ into $2x + y = 11$. $$2(3) + y = 11$$ $$y = 5$$ The solution is $(3, 5)$.
Solving a system of two linear equations by the addition (or elimination) method, p. 200.	The addition method is most appropriate when the variables *all have coefficients other than* 1 *or* −1. 1. Arrange each equation in the form $ax + by = c$. 2. Multiply one or both equations by appropriate numerical values so that when the two resulting equations are added, one variable is eliminated. 3. Solve the resulting equation. 4. Substitute the numerical value you obtain for the variable into one of the original equations. 5. Solve this equation to find the other variable.	Solve. $\quad 2x + 3y = 5 \quad \textbf{(1)}$ $\qquad\qquad -3x - 4y = -2 \quad \textbf{(2)}$ Multiply equation **(1)** by 3 and **(2)** by 2. $$6x + 9y = 15$$ $$\underline{-6x - 8y = -4}$$ $$y = 11$$ Substitute $y = 11$ into equation **(1)**. $$2x + 3(11) = 5$$ $$2x + 33 = 5$$ $$2x = -28$$ $$x = -14$$ The solution is $(-14, 11)$.
Inconsistent system of equations, p. 202.	If there is *no solution* to a system of linear equations, the system of equations is inconsistent. When you try to solve an inconsistent system algebraically, you obtain an equation that is not true, such as $0 = 5$.	Solve. $\quad 4x + 3y = 10 \quad \textbf{(1)}$ $\qquad\qquad -8x - 6y = 5 \quad \textbf{(2)}$ Multiply equation **(1)** by 2 and add to **(2)**. $$8x + 6y = 20$$ $$\underline{-8x - 6y = 5}$$ $$0 = 25$$ But $0 \neq 25$. Thus, there is no solution. The system of equations is inconsistent.

Topic	Procedure	Examples
Dependent equations, *p. 202.*	If there is an *infinite number of solutions* to a system of linear equations, at least one pair of equations is dependent. When you try to solve a system that contains dependent equations, you will obtain an equation that is always true (such as $0 = 0$ or $3 = 3$). These equations are called *identities*.	Attempt to solve the system. $$x - 2y = -5 \quad \textbf{(1)}$$ $$-3x + 6y = 15 \quad \textbf{(2)}$$ Multiply equation **(1)** by 3 and add to **(2)**. $$\begin{array}{r} 3x - 6y = -15 \\ -3x + 6y = 15 \\ \hline 0 = 0 \end{array}$$ There is an infinite number of solutions. The equations are dependent.
Solving a system of three linear equations by algebraic methods, *p. 211.*	If there is one solution to a system of three linear equations in three unknowns, it may be obtained in the following manner. 1. Choose two equations from the system. 2. Multiply one or both of the equations by the appropriate constants so that by adding the two equations together, one variable can be eliminated. 3. Choose a *different* pair of the three original equations and eliminate the *same* variable using the procedure of step 2. 4. Solve the system formed by the two equations resulting from steps 2 and 3 to find values for both variables. 5. Substitute the values obtained in step 4 into one of the original three equations to find the value of the third variable.	Solve. $$2x - y - 2z = -1 \quad \textbf{(1)}$$ $$x - 2y - z = 1 \quad \textbf{(2)}$$ $$x + y + z = 4 \quad \textbf{(3)}$$ Add equations **(2)** and **(3)** to eliminate z. $$2x - y = 5 \quad \textbf{(4)}$$ Multiply equation **(3)** by 2 and add to **(1)**. $$4x + y = 7 \quad \textbf{(5)}$$ Add equations **(4)** and **(5)**. $$\begin{array}{r} 2x - y = 5 \\ 4x + y = 7 \\ \hline 6x = 12 \\ x = 2 \end{array}$$ Substitute $x = 2$ into equation **(5)**. $$\begin{aligned} 4(2) + y &= 7 \\ y &= -1 \end{aligned}$$ Substitute $x = 2$, $y = -1$ into equation **(3)**. $$\begin{aligned} 2 + (-1) + z &= 4 \\ z &= 3 \end{aligned}$$ The solution is $(2, -1, 3)$.
Graphing the solution to a system of inequalities in two variables, *p. 229.*	1. Determine the region that satisfies each individual inequality. 2. Shade the common region that satisfies all the inequalities.	Graph. $$3x + 2y \le 10$$ $$-x + 2y \ge 2$$ **1.** $3x + 2y \le 10$ can be graphed more easily as $y \le -\dfrac{3}{2}x + 5$. We draw a solid line and shade the region below it. $-x + 2y \ge 2$ can be graphed more easily as $y \ge \dfrac{1}{2}x + 1$. We draw a solid line and shade the region above it. **2.** The common region is shaded.

Chapter 4 Review Problems

Solve the following systems by graphing.

1. $x + 2y = 8$
$x - y = 2$

2. $x + y = 2$
$3x - y = 6$

3. $2x + y = 6$
$3x + 4y = 4$

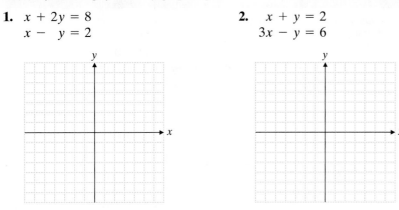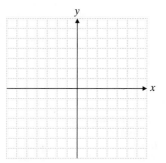

Solve the following systems by substitution.

4. $3x - 2y = -9$
$2x + y = 1$

5. $-6x - y = 1$
$3x - 4y = 31$

6. $4x + 5y = 2$
$3x - y = 11$

7. $3x - 4y = -12$
$x + 2y = -4$

Solve the following systems by addition.

8. $-2x + 5y = -12$
$3x + y = 1$

9. $-3x + 4y = 9$
$5x + 3y = -15$

10. $7x - 4y = 2$
$6x - 5y = -3$

11. $5x + 2y = 3$
$7x + 5y = -20$

Solve by any appropriate method. If there is no unique solution, state why.

12. $x = 3 - 2y$
$3x + 6y = 8$

13. $x + 5y = 10$
$y = 2 - \dfrac{1}{5}x$

14. $4x - 6y = 3$
$8x + 9y = -1$

15. $5x - 2y = 15$
$3x + y = -2$

16. $x + \dfrac{1}{3}y = 1$
$\dfrac{1}{4}x - \dfrac{3}{4}y = -\dfrac{9}{4}$

17. $\dfrac{2}{3}x + y = 1$
$\dfrac{1}{3}x + y = \dfrac{5}{6}$

18. $3a + 8b = 0$
$9a + 2b = 11$

19. $3a + 5b = -2$
$10b = -6a - 4$

20. $x + 3 = 3y + 1$
$1 - 2(x - 2) = 6y + 1$

21. $10(x + 1) - 13 = -8y$
$4(2 - y) = 5(x + 1)$

22. $0.3x - 0.2y = 0.7$
$-0.6x + 0.4y = 0.3$

23. $0.2x - 0.1y = 0.8$
$0.1x + 0.3y = 1.1$

Solve by an appropriate method.

24. $3x - 2y - z = 3$
$2x + y + z = 1$
$-x - y + z = -4$

25. $-2x + y - z = -7$
$x - 2y - z = 2$
$6x + 4y + 2z = 4$

26. $2x + 5y + z = 3$
$x + y + 5z = 42$
$2x + y = 7$

27. $x + 2y + z = 5$
$3x - 8y = 17$
$2y + z = -2$

28. $3x + 2y - z = -6$
$x - 3y + 4z = 11$
$2x + 5y - 3z = -11$

29. $-3x - 4y + z = -4$
$x + 6y + 3z = -8$
$5x + 3y - z = 14$

30. $3x + 2y = 7$
$2x + 7z = -26$
$5y + z = 6$

31. $x - y = 2$
$5x + 7y - 5z = 2$
$3x - 5y + 2z = -2$

Use a system of linear equations to solve each of the following exercises.

32. *Commercial Airline* A plane flies 720 miles against the wind in 3 hours. The return trip with the wind takes only $2\frac{1}{2}$ hours. Find the speed of the wind. Find the speed of the plane in still air.

33. *Football* During the 2008 Pro Bowl between the American Football Conference and the National Football Conference, 63 points were scored in touchdowns and field goals. The total number of touchdowns (6 points each) and field goals (3 points each) was 12. How many touchdowns and how many field goals were scored?

34. *Temporary Help Expenses* When the circus came to town last year, they hired general laborers at $70 per day and mechanics at $90 per day. They paid $1950 for this temporary help for one day. This year they hired exactly the same number of people of each type, but they paid $80 for general laborers and $100 for mechanics for the one day. This year they paid $2200 for temporary help. How many general laborers did they hire? How many mechanics did they hire?

35. *Children's Theater Prices* A total of 330 tickets were sold for the opening performance of *Frog and Toad*. Children's admission tickets were $8, and adults' tickets were $13. The ticket receipts for the performance totaled $3215. How many children's tickets were sold? How many adults' tickets were sold?

36. *Baseball Equipment* A baseball coach bought two hats, five shirts, and four pairs of pants for $129. His assistant purchased one hat, one shirt, and two pairs of pants for $42. The next week the coach bought two hats, three shirts, and one pair of pants for $63. What was the cost of each item?

37. *Math Exam Scores* Jess, Chris, and Nick scored a combined total of 249 points on their last math exam. Jess's score was 20 points higher than Chris's score. Twice Nick's score was 6 more than the sum of Chris's and Jess's scores. What did each of them score on the exam?

38. *Food Costs* Four jars of jelly, three jars of peanut butter, and five jars of honey cost $9.80. Two jars of jelly, two jars of peanut butter, and one jar of honey cost $4.20. Three jars of jelly, four jars of peanut butter, and two jars of honey cost $7.70. Find the cost for one jar of jelly, one jar of peanut butter, and one jar of honey.

39. *Transportation Logistics* The church youth group is planning a trip to Mount Washington. A total of 127 people need rides. The church has available buses that hold forty passengers, and several parents have volunteered station wagons that hold eight passengers or sedans that hold five passengers. The youth leader is planning to use nine vehicles to transport the people. One parent said that if they didn't use any buses, tripled the number of station wagons, and doubled the number of sedans, they would be able to transport 126 people. How many buses, station wagons, and sedans are they planning to use if they use nine vehicles?

Mixed Practice

Solve by any appropriate method.

40. $-x - 5z = -5$
$13x + 2z = 2$

41. $x - y = 1$
$5x + y = 7$

42. $2x + 5y = 4$
$5x - 7y = -29$

43. $\dfrac{x}{2} - 3y = -6$
$\dfrac{4}{3}x + 2y = 4$

44. $\dfrac{x-2}{4} = y - 2$

$\dfrac{-3y+1}{2} = x + y$

45. $\dfrac{x}{2} - y = -12$

$x + \dfrac{3}{4}y = 9$

46. $3(2+x) = y + 1$

$5(x - y) = -7 - 3y$

47. $7(x+3) = 2y + 25$

$3(x-6) = -2(y+1)$

48. $0.3x - 0.4y = 0.9$

$0.2x - 0.3y = 0.4$

49. $1.2x - y = 1.6$

$x + 1.5y = 6$

50. $x - \dfrac{y}{2} + \dfrac{1}{2}z = -1$

$2x + \dfrac{5}{2}z = -1$

$\dfrac{3}{2}y + 2z = 1$

51. $2x - 3y + 2z = 0$

$x + 2y - z = 2$

$2x + y + 3z = -1$

52. $x - 4y + 4z = -1$

$2x - y + 5z = -3$

$x - 3y + z = 4$

53. $x - 2y + z = -5$

$2x + z = -10$

$y - z = 15$

Graph the solution of each of the following systems of linear inequalities.

54. $y \geq -\dfrac{1}{2}x - 1$

$-x + y \leq 5$

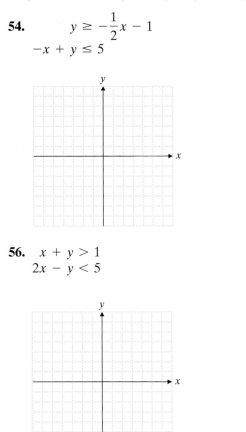

55. $-2x + 3y < 6$

$y > -2$

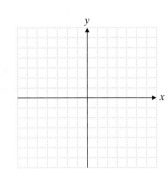

56. $x + y > 1$

$2x - y < 5$

57. $x + y \geq 4$

$y \leq x$

$x \leq 6$

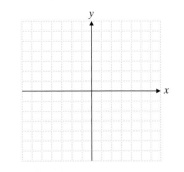

How Am I Doing? Chapter 4 Test

Remember to use your Chapter Test Prep Video CD to see the worked-out solutions to the test problems you want to review.

Solve each system of equations. If there is no solution to the system, give a reason. In exercises 3–9, you may use any method.

1. _____

2. _____

3. _____

4. _____

5. _____

6. _____

7. _____

8. _____

9. _____

10. _____

11. _____

12. _____

13. _____

14. _____

1. Solve using the substitution method.

$$x - y = 3$$
$$2x - 3y = -1$$

2. Solve using the addition method.

$$3x + 2y = 1$$
$$5x + 3y = 3$$

3. $$5x - 3y = 3$$
$$7x + y = 25$$

4. $$\frac{1}{4}a - \frac{3}{4}b = -1$$
$$\frac{1}{3}a + b = \frac{5}{3}$$

5. $$\frac{1}{3}x + \frac{5}{6}y = 2$$
$$\frac{3}{5}x - y = -\frac{7}{5}$$

6. $$8x - 3y = 5$$
$$-16x + 6y = 8$$

7. $$3x + 5y - 2z = -5$$
$$2x + 3y - z = -2$$
$$2x + 4y + 6z = 18$$

8. $$3x + 2y = 0$$
$$2x - y + 3z = 8$$
$$5x + 3y + z = 4$$

9. $$x + 5y + 4z = -3$$
$$x - y - 2z = -3$$
$$x + 2y + 3z = -5$$

Use a system of linear equations to solve the following exercises.

10. A plane flew 1000 miles with a tailwind in 2 hours. The return trip against the wind took $2\frac{1}{2}$ hours. Find the speed of the wind and the speed of the plane in still air.

11. The math club is selling items with the college logo to raise money. Sam bought 4 pens, a mug, and a T-shirt for $20.00. Alicia bought 2 pens and 2 mugs for $11.00. Ramon bought 6 pens, a mug, and 2 T-shirts for $33.00. What was the price of each pen, mug, and T-shirt?

12. Sue Miller had to move some supplies to Camp Cherith for the summer camp program. She rented a Portland Rent-A-Truck in April for 5 days and drove 150 miles. She paid $180 for the rental in April. Then in May she rented the same truck again for 7 days and drove 320 miles. She paid $274 for the rental in May. How much does Portland Rent-A-Truck charge for a daily rental of the truck? How much do they charge per mile?

Solve the following systems of linear inequalities by graphing.

13. $$x + 2y \leq 6$$
$$-2x + y \geq -2$$

14. $$3x + y > 8$$
$$x - 2y > 5$$

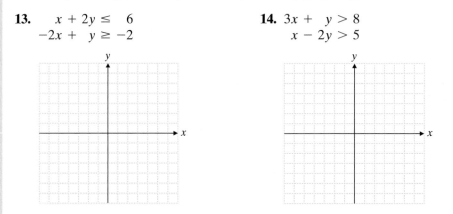

Cumulative Test for Chapters 1–4

Approximately one half of this test covers the content of Chapters 1–3. The remainder covers the content of Chapter 4.

1. State what property is illustrated. $7 + 0 = 7$

2. Evaluate. $\sqrt{64} - (3 - 5)^3 + (-50) \div 5$

3. Simplify. $(5x^{-2})(3x^{-4}y^2)$

4. Simplify. $2x - 4[x - 3(2x + 1)]$

5. Solve for P. $A = P(3 + 4rt)$

6. Solve for x. $\dfrac{1}{4}x + 5 = \dfrac{1}{3}(x - 2)$

7. Graph the line $4x - 8y = 10$. Plot at least three points.

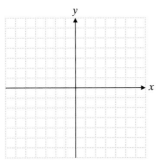

8. Find the slope of the line passing through $(5, 2)$ and $(-1, 10)$.

Solve the following linear inequalities and graph the solution on a number line.

9. $4x + 3 - 13x - 7 < 2(3 - 4x)$

10. $\dfrac{2x - 1}{3} \le 7$ and $2(x + 1) \ge 12$

11. Find the standard form of the equation of the line passing through $(2, -3)$ and perpendicular to $5x + 6y = -2$.

▲ **12.** A triangle has a perimeter of 69 meters. The second side is 7 meters longer than the first side. The third side is 6 meters shorter than double the length of the first side. Find the length of each side.

13. Victor invests $6000 in a bank. Part is invested at 7% simple interest and part at 9% simple interest. In 1 year Victor earns $510 in interest. How much did he invest at each amount?

1. _____

2. _____

3. _____

4. _____

5. _____

6. _____

7. _____

8. _____

9. _____

10. _____

11. _____

12. _____

13. _____

14. _____

14. Solve the system.

$$5x + 2y = 2$$
$$4x + 3y = -4$$

15. Solve the system.

$$x - \frac{1}{2}y = 4$$
$$2x + \frac{1}{3}y = 0$$

15. _____

16. Jane went shopping at Old Navy. She bought six T-shirts and nine sweatshirts for her children. She spent a total of $156. Her sister, Lisa, spent a total of $124 on her children. She bought eight T-shirts and five sweatshirts. How much did a T-shirt cost? How much did a sweatshirt cost?

16. _____

17. Solve the system.

$$7x - 6y = 17$$
$$3x + y = 18$$

17. _____

18. Solve the system.

$$x + 3y + z = 5$$
$$2x - 3y - 2z = 0$$
$$x - 2y + 3z = -9$$

19. What happens when you attempt to solve the following system? Why is this?

$$-5x + 6y = 2$$
$$10x - 12y = -4$$

18. _____

20. Graph the solution of the following system of inequalities.

$$x - y \geq -4$$
$$x + 2y \geq 2$$

19. _____

20. _____

CHAPTER
5

Coral reefs around the world serve as the habitat for a variety of fish and other sea creatures. However, scientists are now studying coral reef diseases that are threatening the health of coral reefs worldwide. Scientists use polynomials to describe certain measures of health of a coral reef. In this chapter you will learn to perform a variety of mathematical operations using polynomials.

Polynomials

5.1   **INTRODUCTION TO POLYNOMIALS AND POLYNOMIAL FUNCTIONS: ADDING, SUBTRACTING, AND MULTIPLYING**   248

5.2   **DIVIDING POLYNOMIALS**   255

5.3   **SYNTHETIC DIVISION**   260

5.4   **REMOVING COMMON FACTORS; FACTORING BY GROUPING**   264

5.5   **FACTORING TRINOMIALS**   270

        HOW AM I DOING? SECTIONS 5.1– 5.5   277

5.6   **SPECIAL CASES OF FACTORING**   278

5.7   **FACTORING A POLYNOMIAL COMPLETELY**   285

5.8   **SOLVING EQUATIONS AND APPLICATIONS USING POLYNOMIALS**   289

        CHAPTER 5 ORGANIZER   298

        CHAPTER 5 REVIEW PROBLEMS   300

        HOW AM I DOING? CHAPTER 5 TEST   303

        CUMULATIVE TEST FOR CHAPTERS 1–5   305

Student Learning Objectives

After studying this section, you will be able to:

1. Identify types and degrees of polynomials.

2. Evaluate polynomial functions.

3. Add and subtract polynomials.

4. Multiply two binomials by FOIL.

5. Multiply two binomials $(a + b)(a - b)$.

6. Multiply two binomials $(a - b)^2$ or $(a + b)^2$.

7. Multiply polynomials with more than two terms.

 Identifying Types and Degrees of Polynomials

A **polynomial** is an algebraic expression of one or more terms. A **term** is a number, a variable raised to a nonnegative integer power, or a product of numbers and variables raised to nonnegative integer powers. There must be no division by a variable. Three types of polynomials that you will see often are **monomials, binomials,** and **trinomials.**

1. A **monomial** has *one* term.
2. A **binomial** has *two* terms.
3. A **trinomial** has *three* terms.

Number of Variables	Monomials	Binomials	Trinomials	Other Polynomials
One Variable	$8x^3$	$2y^2 + 3y$	$5x^2 + 2x - 6$	$x^4 + 2x^3 - x^2 + 9$
Two Variables	$6x^2y$	$3x^2 - 5y^3$	$8x^2 + 5xy - 3y^2$	$x^3y + 5xy^2 + 3xy - 7y^5$
Three Variables	$12uvw^3$	$11a^2b + 5c^2$	$4a^2b^4 + 7c^4 - 2a^5$	$3c^2 + 4c - 8d + 2e - e^2$

The following are *not* polynomials.

$$2x^{-3} + 5x^2 - 3 \qquad 4ab^{\frac{1}{2}} \qquad \frac{2}{x} + \frac{3}{y}$$

TO THINK ABOUT: Understanding Polynomials Give a reason each expression above is not a polynomial.

Polynomials are also classified by degree. The **degree of a term** is the sum of the exponents of its variables. The **degree of a polynomial** is the degree of the highest-degree term in the polynomial. If the polynomial has no variable, then it has degree zero.

EXAMPLE 1 Name the type of polynomial and give its degree.

(a) $5x^6 + 3x^2 + 2$

(b) $7x + 6$

(c) $5x^2y + 3xy^3 + 6xy$

(d) $7x^4y^5$

Solution

(a) This is a trinomial of degree 6.

(b) This is a binomial of degree 1. Remember that if a variable has no exponent,

(c) This is a trinomial of degree 4. the exponent is understood to be 1.

(d) This is a monomial of degree 9.

NOTE TO STUDENT: *Fully worked-out solutions to all of the Practice Problems can be found at the back of the text starting at page SP-1*

Practice Problem 1 State the type of polynomial and give its degree.

(a) $3x^5 - 6x^4 + x^2$ T 5

(b) $5x^2 + 2$ B 2

(c) $3ab + 5a^2b^2 - 6a^4b$ Tr 4

(d) $16x^4y^6$ M 10

Some polynomials contain only one variable. A **polynomial in x** is an expression of the form

$$a_n x^n + a_{n-1} x^{n-1} + a_{n-2} x^{n-2} + \cdots + a_0$$

where n is a nonnegative integer and the constants $a_n, a_{n-1}, a_{n-2}, \ldots, a_0$ are real numbers. We usually write polynomials in **descending order** of the variable. That is, the exponents on the variables decrease from left to right. For example, the polynomial $4x^5 - 2x^3 + 6x^2 + 5x - 8$ is written in descending order.

② Evaluating Polynomial Functions

A **polynomial function** is a function that is defined by a polynomial.
For example,

$$p(x) = 5x^2 - 3x + 6 \quad \text{and} \quad p(x) = 2x^5 - 3x^3 + 8x - 15$$

are both polynomial functions.

To evaluate a polynomial function, we use the skills developed in Section 3.5.

EXAMPLE 2 Evaluate the polynomial function $p(x) = -3x^3 + 2x^2 - 5x + 6$ to find **(a)** $p(-3)$ and **(b)** $p(6)$.

Solution

(a) $p(-3) = -3(-3)^3 + 2(-3)^2 - 5(-3) + 6$
$\qquad\quad = -3(-27) + 2(9) - 5(-3) + 6$
$\qquad\quad = 81 + 18 + 15 + 6$
$\qquad\quad = 120$

(b) $p(6) = -3(6)^3 + 2(6)^2 - 5(6) + 6$
$\qquad\quad = -3(216) + 2(36) - 5(6) + 6$
$\qquad\quad = -648 + 72 - 30 + 6$
$\qquad\quad = -600$

Practice Problem 2 Evaluate the polynomial function
$p(x) = 2x^4 - 3x^3 + 6x - 8$ to find

(a) $p(-2)$ **(b)** $p(5)$

③ Adding and Subtracting Polynomials

We can add and subtract polynomials by combining like terms as we learned in Section 1.5.

EXAMPLE 3 Add. $(5x^2 - 3x - 8) + (-3x^2 - 7x + 9)$

Solution

$5x^2 - 3x - 8 - 3x^2 - 7x + 9$ We remove the parentheses and combine like terms.
$= 2x^2 - 10x + 1$

Practice Problem 3 Add. $(-7x^2 + 5x - 9) + (2x^2 - 3x + 5)$

To subtract real numbers, we add the opposite of the second number to the first. Thus, for real numbers a and b, we have $a - (b) = a + (-b)$. Similarly for polynomials, to subtract polynomials we add the opposite of the second polynomial to the first.

EXAMPLE 4   Subtract. $(-5x^2 - 19x + 15) - (3x^2 - 4x + 13)$

Solution

$(-5x^2 - 19x + 15) + (-3x^2 + 4x - 13)$ We add the opposite of the second
                                              polynomial to the first polynomial.

$= -8x^2 - 15x + 2$

Practice Problem 4   Subtract. $(2x^2 - 14x + 9) - (-3x^2 + 10x + 7)$

4 Multiplying Two Binomials by FOIL

The FOIL method for multiplying two binomials has been developed to help you keep track of the order of the terms to be multiplied. The acronym FOIL means the following:

F	**First**
O	**Outer**
I	**Inner**
L	**Last**

That is, we multiply the first terms, then the outer terms, then the inner terms, and finally, the last terms.

EXAMPLE 5   Multiply. $(5x + 2)(7x - 3)$

Solution

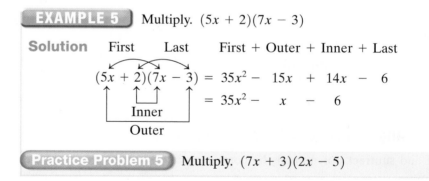

First   Last   First + Outer + Inner + Last

$(5x + 2)(7x - 3) = 35x^2 - 15x + 14x - 6$

$= 35x^2 - x - 6$

Practice Problem 5   Multiply. $(7x + 3)(2x - 5)$

EXAMPLE 6   Multiply. $(7x^2 - 8)(2x - 3)$

Solution

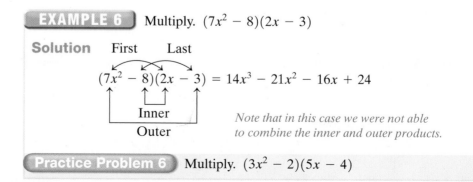

First   Last

$(7x^2 - 8)(2x - 3) = 14x^3 - 21x^2 - 16x + 24$

Note that in this case we were not able to combine the inner and outer products.

Practice Problem 6   Multiply. $(3x^2 - 2)(5x - 4)$

5 Multiplying $(a + b)(a - b)$

Products of the form $(a + b)(a - b)$ deserve special attention.

$$(a + b)(a - b) = a^2 - ab + ab - b^2 = a^2 - b^2$$

Notice that the middle terms, $-ab$ and $+ab$, equal zero when combined. The product is the difference of two squares, $a^2 - b^2$. This is always true when you multiply binomials of the form $(a + b)(a - b)$. You should memorize the following formula.

$$(a + b)(a - b) = a^2 - b^2$$

EXAMPLE 7 Multiply. $(2a - 9b)(2a + 9b)$

Solution $(2a - 9b)(2a + 9b) = (2a)^2 - (9b)^2 = 4a^2 - 81b^2$

Of course, we could have used the FOIL method, but recognizing the special product allowed us to save time.

Practice Problem 7 Multiply. $(7x - 2y)(7x + 2y)$

6 Multiplying $(a - b)^2$ or $(a + b)^2$

Another special product is the square of a binomial.

$$(a - b)^2 = (a - b)(a - b) = a^2 - ab - ab + b^2 = a^2 - 2ab + b^2$$

Once you understand the pattern, you should memorize these two formulas.

$$(a - b)^2 = a^2 - 2ab + b^2 \qquad (a + b)^2 = a^2 + 2ab + b^2$$

This procedure is also called **expanding a binomial.** *Note:* $(a - b)^2 \neq a^2 - b^2$ and $(a + b)^2 \neq a^2 + b^2$.

EXAMPLE 8 Multiply.

(a) $(5a - 8b)^2$ **(b)** $(3u + 11v^2)^2$

Solution

(a) $(5a - 8b)^2 = (5a)^2 - 2(5a)(8b) + (8b)^2 = 25a^2 - 80ab + 64b^2$

(b) Here $a = 3u$ and $b = 11v^2$.

$$(3u + 11v^2)^2 = (3u)^2 + 2(3u)(11v^2) + (11v^2)^2$$
$$= 9u^2 + 66uv^2 + 121v^4$$

Practice Problem 8 Multiply.

(a) $(4u + 5v)^2$ **(b)** $(7x^2 - 3y^2)^2$

7 Multiplying Polynomials with More Than Two Terms

The distributive property is the basis for multiplying polynomials. Recall that

$$a(b + c) = ab + ac.$$

We can use this property to multiply a polynomial by a monomial.

$$3xy(5x^3 + 2x^2 - 4x + 1) = 3xy(5x^3) + 3xy(2x^2) - 3xy(4x) + 3xy(1)$$
$$= 15x^4y + 6x^3y - 12x^2y + 3xy$$

A similar procedure can be used to multiply two binomials.

$$(3x + 5)(6x + 7) = (3x + 5)6x + (3x + 5)7 \quad \text{We use the distributive}$$
$$\text{property again.}$$
$$= (3x)(6x) + (5)(6x) + (3x)(7) + (5)(7)$$
$$= 18x^2 + 30x + 21x + 35$$
$$= 18x^2 + 51x + 35$$

The multiplication of a binomial and a trinomial is more involved. One way to multiply two polynomials is to write them vertically, as we do when multiplying two- and three-digit numbers. We then multiply them in the usual way.

EXAMPLE 9   Multiply. $(4x^2 - 2x + 3)(-3x + 4)$

Solution
$$
\begin{array}{r}
4x^2 - 2x + 3 \\
-3x + 4 \\
\hline
16x^2 - 8x + 12 \\
-12x^3 + 6x^2 - 9x \\
\hline
-12x^3 + 22x^2 - 17x + 12
\end{array}
$$
Multiply $(4x^2 - 2x + 3)(+4)$.
Multiply $(4x^2 - 2x + 3)(-3x)$.
Add the two products.

Practice Problem 9   Multiply. $(2x^2 - 3x + 1)(x^2 - 5x)$

Another way to multiply polynomials is to multiply horizontally. We redo Example 9 in the following example.

EXAMPLE 10   Multiply horizontally. $(4x^2 - 2x + 3)(-3x + 4)$

Solution   By the distributive law, we have the following:

$$(4x^2 - 2x + 3)(-3x + 4) = (4x^2 - 2x + 3)(-3x) + (4x^2 - 2x + 3)(4)$$
$$= -12x^3 + 6x^2 - 9x + 16x^2 - 8x + 12$$
$$= -12x^3 + 22x^2 - 17x + 12.$$

In actual practice you will find that you can do some of these steps mentally.

Practice Problem 10   Multiply horizontally. $(2x^2 - 3x + 1)(x^2 - 5x)$

Developing Your Study Skills

Taking an Exam

Allow yourself plenty of time to get to your exam. You may even find it helpful to arrive a little early in order to collect your thoughts and ready yourself. This will help you feel more relaxed.

After you receive your exam, you will find it helpful to do the following.

1. Take two or three moderately deep breaths. Inhale; then exhale slowly. You will feel your entire body begin to relax.

2. Write down on the back of the exam any formulas or ideas that you need to remember.

3. Look over the entire test quickly in order to pace yourself and use your time wisely. Notice how many points each exercise is worth. Spend more time on items of greater worth.

4. Read directions carefully and be sure to answer all questions clearly. Keep your work neat and easy to read.

5. Ask your instructor about anything that is not clear to you.

6. Work the exercises and answer the questions that are easiest for you first. Then go back to the more difficult ones.

7. Do not get bogged down on one exercise for too long because it may jeopardize your chances of finishing other exercises. Leave the tough exercise and go back to it when you have time later.

8. Check your work. This will help you catch minor errors.

9. Stay calm if others leave before you do. You are entitled to use the full amount of allotted time.

Name the type of polynomial and give its degree.

1. $2x^2 - 5x + 3$

2. $7x^3 + 6x^2 - 2$
trinomials

3. $-3.2a^4bc^3$

4. $26.8a^3bc^2$

5. $\frac{3}{5}m^3n - \frac{2}{5}mn$

6. $\frac{2}{7}m^2n^2 + \frac{1}{2}mn^2$

For the polynomial function $p(x) = 5x^2 - 9x - 12$, evaluate the following:

7. $p(3)$

8. $p(-4)$

For the polynomial function $g(x) = -4x^3 - x^2 + 5x - 1$, evaluate the following:

9. $g(-2)$

10. $g(1)$

For the polynomial function $h(x) = 2x^4 - x^3 + 2x^2 - 4x - 3$, evaluate the following:

11. $h(-1)$

12. $h(3)$

Add or subtract the following polynomials as indicated.

13. $(x^2 + 3x - 2) + (-2x^2 - 5x + 1) + (x^2 - x - 5)$

14. $(2x^2 - 5x - 1) + (3x^2 - 7x + 3) + (-5x^2 + x + 1)$

15. $(7m^3 + 4m^2 - m + 2.5) - (-3m^3 + 5m + 3.8)$

16. $(3x^3 + 2x^2 - 8x - 9.2) - (-5x^3 + x^2 - x - 12.7)$

17. $(5a^3 - 2a^2 - 6a + 8) + (5a + 6) - (-a^2 - a + 2)$

18. $(a^5 + 3a^2) + (2a^4 - a^3 - 3a^2 + 2) - (a^4 + 3a^3 - 5)$

19. $\left(\frac{2}{3}x^2 + 5x\right) + \left(\frac{1}{2}x^2 - \frac{1}{3}x\right)$

20. $\left(\frac{5}{6}x^2 - \frac{1}{4}x\right) + \left(\frac{1}{6}x^2 + 3x\right)$

21. $(2.3x^3 - 5.6x^2 - 2) - (5.5x^3 - 7.4x^2 + 2)$

22. $(5.9x^3 + 3.4x^2 - 7) - (2.9x^3 - 9.6x^2 + 3)$

Multiply.

23. $(5x + 8)(2x + 9)$

24. $(6x + 7)(3x + 2)$

25. $(5w + 2d)(3a - 4b)$

26. $(7a + 8b)(5d - 8w)$

27. $(-6x + y)(2x - 5y)$

28. $(4a + 3y)(-8x - 3y)$

29. $(7r - s^2)(-4a - 11s^2)$

30. $(-3r - 2s^2)(5r - 6s^2)$

Multiply mentally. See Examples 7 and 8.

31. $(5x - 8y)(5x + 8y)$

32. $(2a - 7b)(2a + 7b)$

33. $(5a - 2b)^2$

34. $(6a + 5b)^2$

35. $(7m - 1)^2$

36. $(5r + 3)^2$

37. $(6 + 5x^2)(6 - 5x^2)$

38. $(8 - 3x^3)(8 + 3x^3)$

39. $(3m^3 + 1)^2$

40. $(4r^3 - 5)^2$

Multiply.

41. $2x(3x^2 - 5x + 1)$

42. $-5x(x^2 - 6x - 2)$

43. $-\dfrac{1}{2}ab(4a - 5b - 10)$

44. $\dfrac{3}{4}ab^2(a - 8b + 6)$

45. $(2x - 3)(x^2 - x + 1)$

46. $(4x + 1)(2x^2 + x + 1)$

47. $(3x^2 - 2xy - 6y^2)(2x - y)$

48. $(5x^2 + 3xy - 7y^2)(3x - 2y)$

49. $(3a^3 + 4a^2 - a - 1)(a - 5)$

50. $(4b^3 - b^2 + 2b - 6)(3b + 1)$

First multiply any two binomials in the exercise; then multiply the result by the third binomial.

51. $(x + 2)(x - 3)(2x - 5)$

52. $(x - 6)(x + 2)(3x + 2)$

53. $(a - 2)(5 + a)(3 - 2a)$

54. $(4 + 3a)(1 - 2a)(3 - a)$

Applications

▲ **55.** *Geometry* The area of the base of a rectangular box measures $2x^2 + 5x + 8$ cm². The height of the box measures $3x + 5$ cm. Find the volume of the box.

▲ **56.** *Geometry* A rectangular garden has $3n^2 + 4n + 7$ flowers planted in each row. The garden has $2n + 5$ rows. Find the number of flowers in the garden.

Cumulative Review

57. **[2.6.3]** Solve. $\dfrac{1}{2}x + 4 \le \dfrac{2}{3}(x - 3) + 1$

58. **[2.1.1]** Solve. $2(x + 1) - 3 = 4 - (x + 5)$

Quick Quiz 5.1

1. Combine. $(7x - 4) + (5x - 6) - (3x^2 - 9x)$

2. Multiply. $(5x - 2y^2)^2$

3. Multiply. $(2x - 3)(4x^2 + 2x - 1)$

4. **Concept Check** Explain how you would multiply the following. $(3x + 1)(x - 4)(2x - 3)$

① Dividing a Polynomial by a Monomial

The easiest type of polynomial division occurs when the divisor is a monomial. We perform this type of division just as if we were dividing numbers. First we write the indicated division as the sum of separate fractions, and then we reduce each fraction (if possible).

EXAMPLE 1 Divide. $(15x^3 - 10x^2 + 40x) \div 5x$

Solution
$$\frac{15x^3 - 10x^2 + 40x}{5x} = \frac{15x^3}{5x} - \frac{10x^2}{5x} + \frac{40x}{5x}$$
$$= 3x^2 - 2x + 8$$

Practice Problem 1 Divide. $(-16x^4 + 16x^3 + 8x^2 + 64x) \div 8x$

② Dividing a Polynomial by a Polynomial

When we divide polynomials by binomials or trinomials, we perform long division. This is much like the long division method for dividing numbers. Both polynomials must be written in descending order.

First we write the problem in the form of long division.

$$2x + 3\overline{)6x^2 + 17x + 12}$$

The divisor is $2x + 3$; the dividend is $6x^2 + 17x + 12$. Now we divide the first term of the dividend ($6x^2$) by the first term of the divisor ($2x$).

$$\boxed{3x} \qquad \boxed{6x^2 \div 2x = 3x}$$
$$2x + 3\overline{)6x^2 + 17x + 12}$$

Now we multiply $3x$ (the first term of the quotient) by the divisor $2x + 3$.

$$\begin{array}{r} 3x \\ 2x + 3\overline{)6x^2 + 17x + 12} \\ 6x^2 + 9x \end{array} \qquad \longleftarrow \boxed{\text{The product } 3x(2x + 3).}$$

Next, just as in long division with numbers, we subtract and bring down the next term.

$$\begin{array}{r} 3x \\ 2x + 3\overline{)6x^2 + 17x + 12} \\ 6x^2 + 9x \\ \hline 8x + 12 \end{array}$$
$$\boxed{\text{Subtract } 6x^2 + 9x \text{ from } 6x^2 + 17x.}$$
$$\boxed{\text{Bring down the next term.}}$$

Now we divide the first term of this new dividend ($8x$) by the first term of the divisor ($2x$).

$$\begin{array}{r} 3x + \boxed{4} \\ 2x + 3\overline{)6x^2 + 17x + 12} \\ 6x^2 + 9x \\ \hline 8x + 12 \\ 8x + 12 \\ \hline 0 \end{array} \qquad \boxed{8x \div 2x = 4}$$
$$\boxed{\text{The product } 4(2x + 3).}$$

Note that we then multiplied $(2x + 3)(4)$ and subtracted, just as we did before. We continue this process until the remainder is zero. Thus, we find that

$$\frac{6x^2 + 17x + 12}{2x + 3} = 3x + 4.$$

The remainder is not always zero, as we shall see in Example 2.

Graphing Calculator

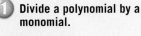 **Verifying Answers When Dividing Polynomials**

One way to verify that the division was performed correctly is to graph

$$y_1 = 3x + 4$$
and $y_2 = \dfrac{6x^2 + 17x + 12}{2x + 3}.$

If the graphs appear to coincide, then we have an independent verification that

$$\frac{6x^2 + 17x + 12}{2x + 3} = 3x + 4.$$

DIVIDING A POLYNOMIAL BY A BINOMIAL OR TRINOMIAL

1. Write the division problem in long division form. Write both polynomials in descending order; write missing terms with a coefficient of zero.

2. Divide the *first* term of the divisor into the first term of the dividend. The result is the first term of the quotient.

3. Multiply the first term of the quotient by *every* term in the divisor.

4. Write the product under the dividend (align like terms) and subtract.

5. Treat this difference as a new dividend after bringing down the next term of the original dividend. Repeat steps 2 through 4. Continue until the remainder is zero or a polynomial of lower degree than the *first term* of the divisor.

6. If there is a remainder, write it as the numerator of a fraction with the divisor as the denominator. Add this fraction to the quotient.

EXAMPLE 2   Divide. $(6x^3 + 7x^2 + 3) \div (3x - 1)$

Solution   There is no x-term in the dividend, so we write $0x$.

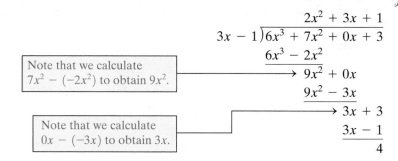

$$
\begin{array}{r}
2x^2 + 3x + 1 \\
3x - 1 \overline{)6x^3 + 7x^2 + 0x + 3} \\
\underline{6x^3 - 2x^2} \\
9x^2 + 0x \\
\underline{9x^2 - 3x} \\
3x + 3 \\
\underline{3x - 1} \\
4
\end{array}
$$

Note that we calculate $7x^2 - (-2x^2)$ to obtain $9x^2$.

Note that we calculate $0x - (-3x)$ to obtain $3x$.

The quotient is $2x^2 + 3x + 1$ with a remainder of 4. We may write this as

$$2x^2 + 3x + 1 + \frac{4}{3x - 1}$$

Check:

$$(3x - 1)(2x^2 + 3x + 1) + 4 \stackrel{?}{=} 6x^3 + 7x^2 + 3$$
$$6x^3 + 7x^2 - 0x - 1 + 4 \stackrel{?}{=} 6x^3 + 7x^2 + 3$$
$$6x^3 + 7x^2 + 3 = 6x^3 + 7x^2 + 3 \;\checkmark$$

Practice Problem 2   Divide. $(8x^3 - 10x^2 - 9x + 14) \div (4x - 3)$

EXAMPLE 3   Divide. $\dfrac{64x^3 - 125}{4x - 5}$

Solution   This fraction is equivalent to the problem $(64x^3 - 125) \div (4x - 5)$.

Note that two terms are missing in the dividend. We write them with zero coefficients.

$$
\begin{array}{r}
16x^2 + 20x + 25 \\
4x - 5 \overline{)64x^3 + 0x^2 + 0x - 125} \\
\underline{64x^3 - 80x^2} \\
80x^2 + 0x \\
\underline{80x^2 - 100x} \\
100x - 125 \\
\underline{100x - 125} \\
0
\end{array}
$$

Note that $0x^2 - (-80x^2) = 80x^2$.

Note that $0x - (-100x) = 100x$.

The quotient is $16x^2 + 20x + 25$.

  Check: Verify that $(4x - 5)(16x^2 + 20x + 25) = 64x^3 - 125$.

Practice Problem 3   Divide. $(8x^3 + 27) \div (2x + 3)$

NOTE TO STUDENT: Fully worked-out solutions to all of the Practice Problems can be found at the back of the text starting at page SP-1

EXAMPLE 4   Divide. $(7x^3 - 10x - 7x^2 + 2x^4 + 8) \div (2x^2 + x - 2)$

Solution   Arrange the dividend in descending order before dividing.

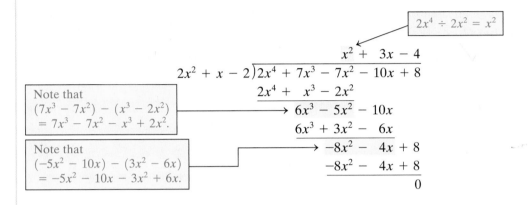

$$\boxed{2x^4 \div 2x^2 = x^2}$$

$$
\begin{array}{r}
x^2 + 3x - 4 \\
2x^2 + x - 2\,\overline{)\,2x^4 + 7x^3 - 7x^2 - 10x + 8} \\
\underline{2x^4 + x^3 - 2x^2} \\
6x^3 - 5x^2 - 10x \\
\underline{6x^3 + 3x^2 - 6x} \\
-8x^2 - 4x + 8 \\
\underline{-8x^2 - 4x + 8} \\
0
\end{array}
$$

Note that
$(7x^3 - 7x^2) - (x^3 - 2x^2)$
$= 7x^3 - 7x^2 - x^3 + 2x^2$.

Note that
$(-5x^2 - 10x) - (3x^2 - 6x)$
$= -5x^2 - 10x - 3x^2 + 6x$.

The quotient is $x^2 + 3x - 4$.

Check: Verify that

$$(2x^2 + x - 2)(x^2 + 3x - 4) = 2x^4 + 7x^3 - 7x^2 - 10x + 8.$$

Practice Problem 4   Divide. $(2x^4 + 3x^3 + 7x - 3 - x^2) \div (2x^2 - x + 3)$

Developing Your Study Skills

Taking Notes in a Lecture Class

An important part of studying mathematics is taking notes. In order to take meaningful notes, you must be an active listener. Keep your mind on what the instructor is saying, and be ready with questions whenever you do not understand something.

If you have previewed the lesson material, you will be prepared to take good notes. The important concepts will seem somewhat familiar. You will have a better idea of what needs to be written down. If you frantically try to write all that the instructor says or copy all the examples done in class, you may find your notes to be nearly worthless when you are home alone. You may find that you are unable to make sense of what you have written.

Write down *important* ideas and examples as the instructor lectures, making sure that you are listening and following the logic. Include any helpful hints or suggestions that your instructor gives you or refers to in your text. You will be amazed at how easily these are forgotten if they are not written down.

Successful note taking requires active listening and processing. Stay alert in class. You will realize the advantages of taking your own notes over copying those of someone else.

Taking Notes in an Online Class

In an online class, you will need notes to help you focus on three things.

1. What are the important types of problems I need to be able to identify?

2. How do I solve each of these types of problems?

3. Is there anything in this section that gives me difficulty?

The best thing you can do to learn the material is to very carefully study each example in a section and make a few notes (one or two sentences) about each example. If you have difficulty with a particular example, place a * next to it and spend some extra time going over that example. You may need to talk to the instructor, a tutor, or a classmate to help you master that particular kind of problem.

Divide.

1. $(24x^2 - 8x - 44) \div 4$

2. $(18x^2 - 63x + 81) \div 9$

3. $(27x^4 - 9x^3 + 63x^2) \div (-9x)$

4. $(22x^4 + 33x^3 - 121x^2) \div (-11x)$

5. $\dfrac{8b^4 - 6b^3 - b^2}{2b^2}$

6. $\dfrac{12w^4 - 6w^3 + w^2}{3w^2}$

7. $\dfrac{9a^3b^3 - 15a^3b^2 + 3a^2b}{3a^2b}$

8. $\dfrac{5mn^5 - 15mn^4 + 30mn^3}{5mn^3}$

Divide. Check your answers for exercises 11–16.

9. $(5x^2 - 17x + 6) \div (x - 3)$

10. $(6x^2 - 31x + 5) \div (x - 5)$

11. $(15x^2 + 23x + 4) \div (5x + 1)$

12. $(12x^2 + 11x + 2) \div (4x + 1)$

13. $(20x^2 - 17x + 3) \div (4x - 1)$

14. $(21x^2 - 2x - 8) \div (3x - 2)$

15. $(x^3 - x^2 + 11x - 1) \div (x + 1)$

16. $(x^3 + 2x^2 - 3x + 2) \div (x + 1)$

17. $(2x^3 - x^2 - 7) \div (x - 2)$

18. $(4x^3 - 6x - 11) \div (2x - 4)$

19. $\dfrac{4x^3 - 6x^2 - 3}{2x + 1}$

20. $\dfrac{9x^3 + 5x - 4}{3x - 1}$

21. $\dfrac{2x^4 - 3x^3 + 8x^2 - 18}{2x - 3}$

22. $\dfrac{18x^4 - 5x^2 - 11x + 8}{3x - 1}$

23. $\dfrac{6t^4 - 5t^3 - 8t^2 + 16t - 8}{3t^2 + 2t - 4}$

24. $\dfrac{2t^4 + 5t^3 - 11t^2 - 20t + 12}{t^2 + t - 6}$

Applications

▲ **25.** *Space Station* For the space station an engineer has designed a new rectangular solar panel that has an area of $18x^3 - 21x^2 + 11x - 2$ square meters. The length of the solar panel is $6x^2 - 5x + 2$ meters. What is the width of the solar panel?

▲ **26.** *Space Station* For the space station an engineer has designed a new rectangular solar panel that has an area of $8x^3 + 22x^2 - 29x + 6$ square meters. The length of the solar panel is $2x^2 + 7x - 2$ meters. What is the width of the solar panel?

Cumulative Review

27. **[3.2.1]** Find the slope of the line passing through $\left(-\dfrac{1}{3}, 0\right)$ and $\left(\dfrac{1}{2}, -1\right)$.

28. **[3.3.3]** Find the slope of a line that is perpendicular to $3y - 2x = 7$.

Solve for x.

29. **[2.2.1]** $5(x - 2) + 3y = 3x - (y - 1)$

30. **[2.2.1]** $\dfrac{2x + 4}{3} - \dfrac{y}{2} = x - 2y + 1$

Quick Quiz 5.2 Divide.

1. $(16x^3 - 20x^2 - 8x) \div (-4x)$

2. $(x^3 - 6x^2 + 7x - 2) \div (x - 1)$

3. $(2x^4 - x^3 - 12x^2 - 17x - 12) \div (2x + 3)$

4. **Concept Check** Explain how you would check your answer if you divided $(x^2 - 9x - 5) \div (x - 3)$ and obtained a result of $x - 6 + \dfrac{-23}{x - 3}$.

Student Learning Objective

After studying this section, you will be able to:

1 Use synthetic division to divide polynomials.

1 Using Synthetic Division to Divide Polynomials

When dividing a polynomial by a binomial of the form $x + b$ you may find a procedure known as **synthetic division** quite efficient. Notice the following division exercises. The right-hand problem is the same as the left, but without the variables.

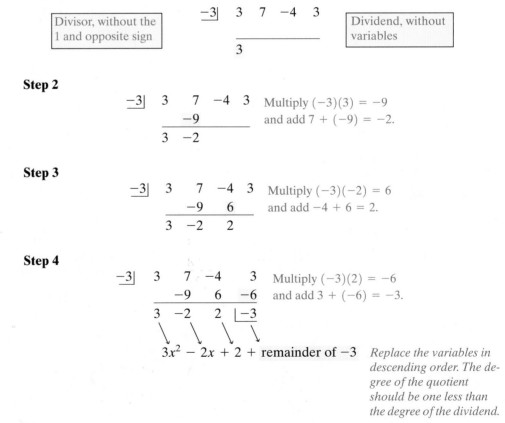

$$
\begin{array}{r}
3x^2 - 2x + 2 \\
x + 3\overline{)3x^3 + 7x^2 - 4x + 3} \\
\underline{3x^3 + 9x^2} \\
-2x^2 - 4x \\
\underline{-2x^2 - 6x} \\
2x + 3 \\
\underline{2x + 6} \\
-3
\end{array}
\qquad
\begin{array}{r}
3 - 2 \ \ 2 \\
1 + 3\overline{)3 \ \ 7 - 4 \ \ 3} \\
\underline{3 \ \ 9} \\
-2 - 4 \\
\underline{-2 - 6} \\
2 \ \ 3 \\
\underline{2 \ \ 6} \\
-3
\end{array}
$$

Eliminating the variables makes synthetic division efficient, and we can make the procedure simpler yet. Note that the colored numbers (3, -2, and 2) appear twice in the previous example, once in the quotient and again in the subtraction. Synthetic division makes it possible to write each number only once. Also, in synthetic division we change the subtraction that division otherwise requires to addition. We do this by dropping the 1, which is the coefficient of x in the divisor, and taking the opposite of the second number in the divisor. In our first example, this means dropping the 1 and changing 3 to -3. The following steps detail synthetic division.

Step 1

Divisor, without the 1 and opposite sign

$$
\begin{array}{r|rrrr}
-3 & 3 & 7 & -4 & 3 \\
\hline
& 3
\end{array}
$$

Dividend, without variables

Step 2

$$
\begin{array}{r|rrrr}
-3 & 3 & 7 & -4 & 3 \\
& & -9 & & \\
\hline
& 3 & -2
\end{array}
$$

Multiply $(-3)(3) = -9$ and add $7 + (-9) = -2$.

Step 3

$$
\begin{array}{r|rrrr}
-3 & 3 & 7 & -4 & 3 \\
& & -9 & 6 & \\
\hline
& 3 & -2 & 2
\end{array}
$$

Multiply $(-3)(-2) = 6$ and add $-4 + 6 = 2$.

Step 4

$$
\begin{array}{r|rrrr}
-3 & 3 & 7 & -4 & 3 \\
& & -9 & 6 & -6 \\
\hline
& 3 & -2 & 2 & \boxed{-3}
\end{array}
$$

Multiply $(-3)(2) = -6$ and add $3 + (-6) = -3$.

$$3x^2 - 2x + 2 + \text{remainder of } -3$$

Replace the variables in descending order. The degree of the quotient should be one less than the degree of the dividend.

The result is read from the bottom row. Our answer is $3x^2 - 2x + 2 + \dfrac{-3}{x + 3}$.

EXAMPLE 1 Divide using synthetic division. $(3x^3 - x^2 + 4x + 8) \div (x + 2)$

Solution

$$
\begin{array}{r|rrrr}
-2 & 3 & -1 & 4 & 8 \\
   &   & -6 & 14 & -36 \\
\hline
   & 3 & -7 & 18 & \underline{|-28} \\
\end{array}
$$

The quotient is $3x^2 - 7x + 18 + \dfrac{-28}{x + 2}$.

Practice Problem 1 Divide using synthetic division.

$$(x^3 - 3x^2 + 4x - 5) \div (x + 3)$$

When a term is missing in the sequence of descending powers of x, we use a zero to indicate the coefficient of that term.

EXAMPLE 2 Divide using synthetic division.

$$(3x^4 - 21x^3 + 31x^2 - 25) \div (x - 5)$$

Solution Since $b = -5$, we use 5 as the divisor for synthetic division.

$$
\begin{array}{r|rrrrr}
5 & 3 & -21 & 31 & 0 & -25 \\
  &   & 15 & -30 & 5 & 25 \\
\hline
  & 3 & -6 & 1 & 5 & \underline{|0} \\
\end{array}
$$
Note that the remainder is zero.

The quotient is $3x^3 - 6x^2 + x + 5$.

Practice Problem 2 Divide using synthetic division.

$$(2x^4 - x^2 - 39x - 36) \div (x - 3)$$

NOTE TO STUDENT: Fully worked-out solutions to all of the Practice Problems can be found at the back of the text starting at page SP-1

EXAMPLE 3 Divide using synthetic division.

$$(3x^4 - 4x^3 + 8x^2 - 5x - 5) \div (x - 2)$$

Solution

$$
\begin{array}{r|rrrrr}
2 & 3 & -4 & 8 & -5 & -5 \\
  &   & 6 & 4 & 24 & 38 \\
\hline
  & 3 & 2 & 12 & 19 & \underline{|33} \\
\end{array}
$$

The quotient is $3x^3 + 2x^2 + 12x + 19 + \dfrac{33}{x - 2}$.

Practice Problem 3 Divide using synthetic division.

$$(2x^4 - 9x^3 + 5x^2 + 13x - 5) \div (x - 3)$$

Divide using synthetic division.

1. $(2x^2 - 11x - 8) \div (x - 6)$

2. $(2x^2 - 15x - 23) \div (x - 9)$

3. $(3x^3 + x^2 - x + 4) \div (x + 1)$

4. $(3x^3 + 10x^2 + 6x - 4) \div (x + 2)$

5. $(x^3 + 7x^2 + 17x + 15) \div (x + 3)$

6. $(3x^3 - x^2 + 4x + 8) \div (x + 2)$

7. $(4x^3 - 11x^2 - 20x + 5) \div (x - 4)$

8. $(5x^3 + 13x^2 - 18x + 9) \div (x - 1)$

9. $(x^3 - 2x^2 + 8) \div (x + 2)$

10. $(2x^3 + 7x^2 - 5) \div (x + 3)$

11. $(6x^4 + 13x^3 + 35x - 24) \div (x + 3)$

12. $(x^4 - 2x^3 - 11x^2 + 34) \div (x + 2)$

13. $(2x^4 + 3x^3 + x^2 + 2x + 5) \div (x + 1)$

14. $(x^4 - 4x^3 + 5x^2 - 6x + 1) \div (x - 3)$

15. $(2x^4 - 5x - 3) \div (x - 1)$

16. $(3x^4 + 8x - 10) \div (x + 2)$

17. $(2x^5 - 13x^3 + 10x^2 + 6) \div (x + 3)$

18. $(x^5 + 5x^4 - 20x^2 + 30) \div (x + 4)$

19. $(x^6 - 5x^3 + x^2 + 12) \div (x + 1)$

20. $(x^6 - 4) \div (x + 1)$

How do we use synthetic division when the divisor is in the form $ax + b$? We divide the divisor by a to get $x + \dfrac{b}{a}$. Then, after performing the synthetic division, we divide each term of the result by a. The number that is the remainder does not change.

To divide $(2x^3 + 7x^2 - 5x - 4) \div (2x + 1)$, we would use $-\dfrac{1}{2}\bigg|\ 2\quad 7\quad -5\quad -4$ and then divide each term of the result by 2.

In exercises 21 and 22, divide using synthetic division.

21. $(4x^3 - 6x^2 + 6) \div (2x + 3)$

22. $(2x^3 - 3x^2 + 6x + 4) \div (2x + 1)$

Cumulative Review *A total of 21 people were killed and 150 people injured in the Great Boston Molasses Flood in January 1919. A molasses storage tank burst and spilled 2 million gallons of molasses through the streets of Boston.*

▲ **23.** **[1.2.2]** *Boston Molasses Flood* How many cubic feet of molasses were contained in the 2-million-gallon molasses tank? (Use 1 gallon ≈ 0.134 cubic feet.)

▲ **24.** **[1.6.2]** *Boston Molasses Flood* At one point the moving flood of molasses appeared as a huge cylindrically shaped object with a radius of 200 feet. At that point how deep was the molasses? (Round to the nearest tenth.)

25. **[3.5.3]** If $p(x) = 2x^4 - 3x^2 + 6x - 1$, find $p(-3)$.

Quick Quiz 5.3 Divide using synthetic division.

1. $(x^3 - 2x^2 - 5x - 2) \div (x + 1)$

2. $(x^3 - 8x^2 + 24) \div (x - 2)$

3. $(x^4 - 7x^2 + 2x - 12) \div (x + 3)$

4. **Concept Check** When doing a problem such as $(2x^4 - x + 3) \div (x - 2)$, why is it necessary to use zeros to represent $0x^3 + 0x^2$ when performing synthetic division?

Student Learning Objectives

After studying this section, you will be able to:

1 Factor out the greatest common factor from a polynomial.

2 Factor a polynomial by the grouping method.

We learned to multiply polynomials in Section 5.1. When two or more algebraic expressions (monomials, binomials, and so on) are multiplied, each expression is called a **factor.**

In the rest of this chapter, we will learn how to find the factors of a polynomial. **Factoring** is the opposite of multiplication and is an extremely important mathematical technique.

1 Factoring Out the Greatest Common Factor

To factor out a common factor, we make use of the distributive property.

$$ab + ac = a(b + c)$$

The **greatest common factor** is simply the largest factor that is common to all terms of the expression. It must contain

1. The largest possible common factor of the numerical coefficients and
2. The largest possible common variable factor

EXAMPLE 1 Factor out the greatest common factor.

(a) $7x^2 - 14x$ (b) $40a^3 - 20a^2$

Solution

(a) $7x^2 - 14x = 7 \cdot x \cdot x - 7 \cdot 2 \cdot x = 7x(x - 2)$

Be careful. The greatest common factor is $7x$, not 7.

(b) $40a^3 - 20a^2 = 20a^2(2a - 1)$

The greatest common factor is $20a^2$.

Suppose we had written $10a(4a^2 - 2a)$ or $10a(2a)(2a - 1)$ as our answer. Although we have factored the expression, we have not found the *greatest* common factor.

Practice Problem 1 Factor out the greatest common factor.

(a) $19x^3 - 38x^2$ (b) $100a^4 - 50a^2$

EXAMPLE 2 Factor out the greatest common factor.

(a) $9x^2 - 18xy - 15y^2$ (b) $4a^3 - 12a^2b^2 - 8ab^3 + 6ab$

Solution

(a) $9x^2 - 18xy - 15y^2 = 3(3x^2 - 6xy - 5y^2)$

The greatest common factor is 3.

(b) $4a^3 - 12a^2b^2 - 8ab^3 + 6ab = 2a(2a^2 - 6ab^2 - 4b^3 + 3b)$

The greatest common factor is $2a$.

Practice Problem 2 Factor out the greatest common factor.

(a) $21x^3 - 18x^2y + 24xy^2$ (b) $12xy^2 - 14x^2y + 20x^2y^2 + 36x^3y$

NOTE TO STUDENT: Fully worked-out solutions to all of the Practice Problems can be found at the back of the text starting at page SP-1

How do you know whether you have factored correctly? You should do two things to verify your answer.

1. Examine the polynomial in the parentheses. Its terms should not have any remaining common factors.
2. Multiply the two factors. You should obtain the original expression.

In the next two examples, you will be asked to **factor** a polynomial (i.e., to find the factors that, when multiplied, give the polynomial as a product). For each of these examples, this will require you to factor out the greatest common factor.

EXAMPLE 3 Factor $6x^3 - 9x^2y - 6x^2y^2$. Check your answer.

Solution $6x^3 - 9x^2y - 6x^2y^2 = 3x^2(2x - 3y - 2y^2)$

$3 \cdot 2 \cdot x \cdot x \cdot x - 3 \cdot 3 \cdot x \cdot x \cdot y - 3 \cdot 2 \cdot x \cdot x \cdot y \cdot y$

Check:

1. $(2x - 3y - 2y^2)$ has no common factors. If it did, we would know that we had not factored out the *greatest* common factor.
2. Multiply the two factors.

$$3x^2(2x - 3y - 2y^2) = 6x^3 - 9x^2y - 6x^2y^2$$

Observe that we do obtain the original polynomial.

Practice Problem 3 Factor $9a^3 - 12a^2b^2 - 15a^4$. Check your answer.

The greatest common factor need not be a monomial. It may be a binomial or even a trinomial. For example, note the following:

$$5a(x + 3) + 2(x + 3) = (x + 3)(5a + 2)$$
$$5a(x + 4) + 2(x + 4) = (x + 4)(5a + 2)$$

The common factors are binomials.

EXAMPLE 4 Factor.

(a) $2x(x + 5) - 3(x + 5)$ **(b)** $5a(a + b) - 2b(a + b) - (a + b)$

Solution

(a) $2x(x + 5) - 3(x + 5) = (x + 5)(2x - 3)$ The common factor is $x + 5$.

(b) $5a(a + b) - 2b(a + b) - (a + b) = 5a(a + b) - 2b(a + b) - 1(a + b)$
$$= (a + b)(5a - 2b - 1)$$

The common factor is $a + b$.

Practice Problem 4 Factor $7x(x + 2y) - 8y(x + 2y) - (x + 2y)$.

2 Factoring by Grouping

Because the common factors in Example 4 were grouped inside parentheses, it was easy to pick them out. However, this rarely happens, so we have to learn how to manipulate expressions to find the greatest common factor.

Polynomials with four terms can often be factored by the method of Example 4(a). However, the parentheses are not always present in the original problem. When they are not present, we look for a way to remove a common factor from the first two terms. We then factor out a common factor from the first two terms and a common factor from the second two terms. Often, we can then find the greatest common factor of the original expression.

EXAMPLE 5 Factor $ax + 2ay + 2bx + 4by$.

Solution

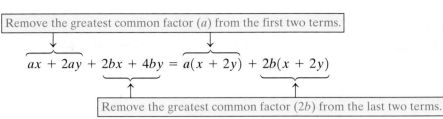

Remove the greatest common factor (a) from the first two terms.

$$ax + 2ay + 2bx + 4by = a(x + 2y) + 2b(x + 2y)$$

Remove the greatest common factor $(2b)$ from the last two terms.

Now we can see that $(x + 2y)$ is a common factor.

$$a(x + 2y) + 2b(x + 2y) = (x + 2y)(a + 2b)$$

NOTE TO STUDENT: Fully worked-out solutions to all of the Practice Problems can be found at the back of the text starting at page SP-1

Practice Problem 5 Factor $bx + 5by + 2wx + 10wy$.

EXAMPLE 6 Factor $2x^2 - 18y - 12x + 3xy$.

Solution First write the polynomial in this order: $2x^2 - 12x + 3xy - 18y$

Remove the greatest common factor $(2x)$ from the first two terms.

$$2x^2 - 12x + 3xy - 18y = 2x(x - 6) + 3y(x - 6) = (x - 6)(2x + 3)$$

Remove the greatest common factor $(3y)$ from the last two terms.

Practice Problem 6 Factor $5x^2 - 12y - 15x + 4xy$.

To factor an expression by this method, it may be necessary to rearrange the order of the four terms so that the first two terms do have a common factor.

EXAMPLE 7 Factor $xy - 6 + 3x - 2y$.

Solution

$xy + 3x - 2y - 6$ Rearrange the terms so that the first two terms have a common factor and the last two terms have a common factor.

$= x(y + 3) - 2(y + 3)$ Factor out a common factor of x from the first two terms and -2 from the second two terms.

$= (y + 3)(x - 2)$ Factor out the common binomial factor $y + 3$.

Practice Problem 7 Factor $xy - 12 - 4x + 3y$.

EXAMPLE 8 Factor $2x^3 + 21 - 7x^2 - 6x$. Check your answer by multiplication.

Solution

$2x^3 - 7x^2 - 6x + 21$ Rearrange the terms.

$= x^2(2x - 7) - 3(2x - 7)$ Factor out a common factor from each group of two terms.

$= (2x - 7)(x^2 - 3)$ Factor out the common binomial factor $2x - 7$.

Check:

$(2x - 7)(x^2 - 3) = 2x^3 - 6x - 7x^2 + 21$ Multiply the two binomials.

$= 2x^3 + 21 - 7x^2 - 6x$ Rearrange the terms.

The product is identical to the original expression.

Practice Problem 8 Factor $2x^3 - 15 - 10x + 3x^2$.

Developing Your Study Skills

Reading the Textbook

Your homework time each day should begin with a careful reading of the section(s) assigned in your textbook. Usually, much time and effort have gone into the selection of a particular text, and your instructor has decided that this is the book that will help you be successful in this mathematics class. Textbooks are expensive, but they can be a good investment if you take advantage of them by reading them.

Reading a mathematics textbook is unlike reading the types of books that you may find in your literature, history, psychology, or sociology courses. Mathematics texts are technical books that provide you with exercises to practice. Learning from a mathematics text requires slow and careful reading of each word, which takes time and effort.

Begin reading your textbook with a paper and pencil in hand. As you come across a new definition or concept, underline it in the text and/or write it down in your notebook. Whenever you encounter an unfamiliar term, look it up and make a note of it. When you come to an example, work through it step-by-step. Be sure to read each word and to follow directions carefully.

Notice the helpful hints the author provides. They guide you to correct solutions and prevent you from making errors. Take advantage of these pieces of expert advice.

Be sure that you understand what you are reading. Make a note of any of the things that you do not understand, and ask your instructor about them. Do not hurry through the material. Learning mathematics takes time.

Factor out the greatest common factor.

1. $80 - 10y$

2. $16x - 16$

3. $5a^2 - 25a$

4. $7a^2 - 14a$

5. $4a^2b^3 - 8ab + 32a$

6. $15c^2d^2 + 10c^2 - 60c$

7. $30y^4 + 24y^3 + 18y^2$

8. $16y^5 - 24y^4 - 40y^3$

9. $15ab^2 + 5ab - 10a^3b$

10. $-12x^2y - 18xy + 6x$

11. $10a^2b^3 - 30a^3b^3 + 10a^3b^2 - 40a^4b^2$

12. $28x^3y^2 - 12x^2y^4 + 4x^3y^4 - 32x^2y^2$

13. $3x(x + y) - 2(x + y)$

14. $5a(a + 3b) + 4(a + 3b)$

15. $5b(a - 3b) + 8(-3b + a)$

16. $4y(x - 5y) - 3(-5y + x)$

17. $3x(a + 5b) + (a + 5b)$

Hint: Is the expression in the first parentheses equal to the expression in the second parentheses in exercises 15 and 16?

18. $2w(s - 3t) - (s - 3t)$

19. $2a^2(3x - y) - 5b^3(3x - y)$

20. $7a^3(5a + 4) - 2(5a + 4)$

21. $3x(5x + y) - 8y(5x + y) - (5x + y)$

22. $4w(y - 8x) + 5z(y - 8x) + (y - 8x)$

23. $2a(a - 6b) - 3b(a - 6b) - 2(a - 6b)$

24. $3a(a + 4b) - 5b(a + 4b) - 9(a + 4b)$

Factor.

25. $x^3 + 5x^2 + 3x + 15$

26. $x^3 + 8x^2 + 2x + 16$

27. $2x + 6 - 3ax - 9a$

28. $2bc + 4b - 5c - 10$

29. $ab - 4a + 12 - 3b$

30. $2m^2 - 8mn - 5m + 20n$

31. $5x - 20 + 3xy - 12y$

32. $3x - 21 + 4xy - 28y$

33. $9y + 2x - 6 - 3xy$

34. $10y + 3x - 6 - 5xy$

35. $ab^3 + c + b^2 + abc$

36. $25x^2z - 14y - 10yz + 35x^2$

Applications

37. Stacked Oranges The total number of oranges stacked in a pile of x rows is given by the polynomial $\frac{1}{3}x^3 + \frac{1}{2}x^2 + \frac{1}{6}x$. Write this polynomial in factored form.

▲ **38. Geometry** The volume of the box pictured below is given by the polynomial $4x^3 + 2x^2 - 6x$. Write this polynomial in factored form.

$x - 1$

$2x$

$2x + 3$

Cumulative Review *Graph the equations in 39 and 40.*

39. [3.1.1] $6x - 2y = -12$

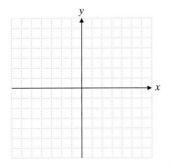

40. [3.3.1] $y = \frac{2}{3}x - 2$

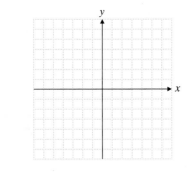

41. [3.2.1] Find the slope of the line that passes through $(5, -3)$ and $(2, 6)$.

42. [3.3.1] Find the slope and y-intercept of $2y + 6x = -3$.

Quick Quiz 5.4 Factor completely.

1. $15a^2b^3 - 10a^3b^3 - 25a^2b^4$

2. $2x(5x - 3y) - 5(5x - 3y)$

3. $3ac + 20b^2 - 4ab - 15bc$

4. Concept Check Explain how you would rearrange the order of $8xy - 15 - 10y + 12x$ in order to factor the polynomial.

5.5 FACTORING TRINOMIALS

Student Learning Objectives

After studying this section, you will be able to:

1 Factor trinomials of the form $x^2 + bx + c$.

2 Factor trinomials of the form $ax^2 + bx + c$.

1 Factoring Trinomials of the Form $x^2 + bx + c$

If we multiply $(x + 4)(x + 5)$, we obtain $x^2 + 9x + 20$. But suppose that we already have the polynomial $x^2 + 9x + 20$ and need to factor it. In other words, suppose we need to find the expressions that, when multiplied, give us the polynomial. Let's use this example to find a general procedure.

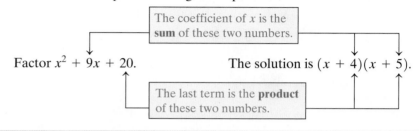

The coefficient of x is the **sum** of these two numbers.

Factor $x^2 + 9x + 20$.   The solution is $(x + 4)(x + 5)$.

The last term is the **product** of these two numbers.

> **FACTORING TRINOMIALS OF THE FORM $x^2 + bx + c$**
>
> **1.** The answer has the form $(x + m)(x + n)$, when m and n are real numbers.
>
> **2.** The numbers m and n are chosen so that
>   **(a)** $m \cdot n = c$ and   **(b)** $m + n = b$.

If the last term of the trinomial is positive and the middle term is negative, the two numbers m and n will be negative numbers.

EXAMPLE 1   Factor $x^2 - 14x + 24$.

Solution   We want to find two numbers whose product is 24 and whose sum is -14. They will both be negative numbers.

Factor Pairs of 24	Sum of the Factors
$(-24)(-1)$	$-24 - 1 = -25$
$(-12)(-2)$	$-12 - 2 = -14$ ✓
$(-6)(-4)$	$-6 - 4 = -10$
$(-8)(-3)$	$-8 - 3 = -11$

The numbers whose product is 24 and whose sum is -14 are -12 and -2. Thus,

$$x^2 - 14x + 24 = (x - 12)(x - 2).$$

Practice Problem 1   Factor $x^2 - 10x + 21$.

If the last term of the trinomial is negative, the two numbers m and n will be opposite in sign.

EXAMPLE 2   Factor $x^2 + 11x - 26$.

Solution   We want to find two numbers whose product is -26 and whose sum is 11. One number will be positive and the other negative.

Factor Pairs of -26	Sum of the Factors
$(-26)(1)$	$-26 + 1 = -25$
$(26)(-1)$	$26 - 1 = 25$
$(-13)(2)$	$-13 + 2 = -11$
$(13)(-2)$	$13 - 2 = 11$ ✓

The numbers whose product is -26 and whose sum is 11 are 13 and -2. Thus,

$$x^2 + 11x - 26 = (x + 13)(x - 2).$$

Practice Problem 2 Factor $x^2 - 13x - 48$.

Sometimes we can make a substitution that makes a polynomial easier to factor, as shown in the following example.

EXAMPLE 3 Factor $x^4 - 2x^2 - 24$.

Solution We need to recognize that we can write this as $(x^2)^2 - 2(x^2) - 24$. We can make this polynomial easier to factor if we substitute y for x^2. Then we have $y^2 + (-2)y + (-24)$. So the factors will be $(y + m)(y + n)$. The two numbers whose product is -24 and whose sum is -2 are -6 and 4. Therefore, we have $(y - 6)(y + 4)$. But $y = x^2$, so our answer is

$$x^4 - 2x^2 - 24 = (x^2 - 6)(x^2 + 4).$$

Practice Problem 3 Factor $x^4 + 9x^2 + 8$.

FACTS ABOUT SIGNS

Suppose $x^2 + bx + c = (x + m)(x + n)$. We know certain facts about m and n.

1. m and n have the same sign if c is positive. (*Note:* We did *not* say that they will have the same sign as c.)
   (a) They are positive if b is positive.
   (b) They are negative if b is negative.

2. m and n have opposite signs if c is negative. The larger number is positive if b is positive and negative if b is negative.

If you understand these sign facts, continue on to Example 4. If not, review Examples 1 through 3.

EXAMPLE 4 Factor.
(a) $y^2 + 5y - 36$ (b) $x^4 - 4x^2 - 12$

Solution
(a) $y^2 + 5y - 36 = (y + 9)(y - 4)$ The larger number (9) is positive because $b = 5$ is positive.

(b) $x^4 - 4x^2 - 12 = (x^2 - 6)(x^2 + 2)$ The larger number (6) is negative because $b = -4$ is negative.

Practice Problem 4 Factor.
(a) $a^2 - 2a - 48$ (b) $x^4 + 2x^2 - 15$

Does the order in which we write the factors make any difference? In other words, is it true that $x^2 + bx + c = (x + n)(x + m)$? Since multiplication is commutative,

$$x^2 + bx + c = (x + n)(x + m) = (x + m)(x + n).$$

The order of the factors is not important.

We can also factor certain kinds of trinomials that have more than one variable.

EXAMPLE 5 Factor. (a) $x^2 - 21xy + 20y^2$ (b) $x^2 + 4xy - 21y^2$

Solution (a) $x^2 - 21xy + 20y^2 = (x - 20y)(x - y)$

The last terms in each factor contain the variable y.

(b) $x^2 + 4xy - 21y^2 = (x + 7y)(x - 3y)$

Practice Problem 5 Factor.

(a) $x^2 - 16xy + 15y^2$ (b) $x^2 + xy - 42y^2$

NOTE TO STUDENT: Fully worked-out solutions to all of the Practice Problems can be found at the back of the text starting at page SP-1

If the terms of a trinomial have a common factor, you should remove the greatest common factor from the terms first. Then you will be able to follow the factoring procedure we have used in the previous examples.

EXAMPLE 6 Factor $3x^2 - 30x + 48$.

Solution The factor 3 is common to all three terms of the polynomial. Factoring out the 3 gives us the following:

$$3x^2 - 30x + 48 = 3(x^2 - 10x + 16)$$

Now we continue to factor the trinomial in the usual fashion.

$$3(x^2 - 10x + 16) = 3(x - 8)(x - 2)$$

Practice Problem 6 Factor $4x^2 - 44x + 72$.

② Factoring Trinomials of the Form $ax^2 + bx + c$

Using the Grouping Number Method. One way to factor a trinomial $ax^2 + bx + c$ is to write it as four terms and factor it by grouping as we did in Section 5.4. For example, the trinomial $2x^2 + 11x + 12$ can be written as $2x^2 + 3x + 8x + 12$.

$$2x^2 + 3x + 8x + 12 = x(2x + 3) + 4(2x + 3)$$
$$= (2x + 3)(x + 4)$$

We can factor all factorable trinomials of the form $ax^2 + bx + c$ in this way. Use the following procedure.

> **GROUPING NUMBER METHOD FOR FACTORING TRINOMIALS OF THE FORM $ax^2 + bx + c$**
>
> 1. Obtain the grouping number ac.
> 2. Find the factor pair of the grouping number whose sum is b.
> 3. Use those two factors to write bx as the sum of two terms.
> 4. Factor by grouping.

EXAMPLE 7 Factor $2x^2 + 19x + 24$.

Solution

1. The grouping number is $(a)(c) = (2)(24) = 48$.
2. The factor pairs of 48 are as follows:

$$48 \cdot 1 \qquad 24 \cdot 2 \qquad 16 \cdot 3 \qquad 12 \cdot 4 \qquad 8 \cdot 6$$

Now $b = 19$ so we want the factor pair of 48 whose sum is 19. Therefore, we select the factors 16 and 3.

3. We use the numbers 16 and 3 to write $19x$ as the sum of $16x$ and $3x$.

$$2x^2 + 19x + 24 = 2x^2 + 16x + 3x + 24$$

4. Factor by grouping.

$$2x^2 + 16x + 3x + 24 = 2x(x + 8) + 3(x + 8)$$
$$= (x + 8)(2x + 3)$$

Practice Problem 7 Factor $10x^2 - 9x + 2$.

EXAMPLE 8 Factor $6x^2 + 7x - 5$.

Solution

1. The grouping number is -30.
2. We want the factor pair of -30 whose sum is 7.

$$
\begin{aligned}
-30 &= (-30)(1) & -30 &= (5)(-6)\\
&= (30)(-1) & &= (-5)(6)\\
&= (15)(-2) & &= (3)(-10)\\
&= (-15)(2) & &= (-3)(10)
\end{aligned}
$$

3. Since $-3 + 10 = 7$, use -3 and 10 to write $6x^2 + 7x - 5$ with four terms.

$$6x^2 + 7x - 5 = 6x^2 - 3x + 10x - 5$$

4. Factor by grouping.

$$6x^2 - 3x + 10x - 5 = 3x(2x - 1) + 5(2x - 1)$$
$$= (2x - 1)(3x + 5)$$

Practice Problem 8 Factor $3x^2 + 2x - 8$.

If the three terms have a common factor, then prior to using the four-step grouping number procedure, we first factor out the greatest common factor from the terms of the trinomial.

EXAMPLE 9 Factor $6x^3 - 26x^2 + 24x$.

Solution First we factor out the greatest common factor $2x$ from each term.

$$6x^3 - 26x^2 + 24x = 2x(3x^2 - 13x + 12)$$

Next we follow the four steps to factor $3x^2 - 13x + 12$.

1. The grouping number is 36.
2. We want the factor pair of 36 whose sum is -13. The two factors are -4 and -9.
3. We use -4 and -9 to write $3x^2 - 13x + 12$ with four terms.

$$3x^2 - 13x + 12 = 3x^2 - 4x - 9x + 12$$

4. Factor by grouping. Remember that we first factored out the factor $2x$. This factor must be part of the answer.

$$2x(3x^2 - 4x - 9x + 12) = 2x[x(3x - 4) - 3(3x - 4)]$$
$$= 2x(3x - 4)(x - 3)$$

Practice Problem 9 Factor $9x^3 - 15x^2 - 6x$.

Using the Trial-and-Error Method. Another way to factor trinomials of the form $ax^2 + bx + c$ is by trial and error. This method has an advantage if the grouping number is large and we would have to list many factors. In the trial-and-error method, we try different values and see which ones can be multiplied out to obtain the original expression.

If the last term is negative, there are many more sign possibilities.

EXAMPLE 10 Factor by trial and error $10x^2 - 49x - 5$.

Solution The first terms in the factors could be $10x$ and x or $5x$ and $2x$. The second terms could be $+1$ and -5 or -1 and $+5$. We list all the possibilities and look for one that will yield a middle term of $-49x$.

Possible Factors	Middle Term of Product
$(2x - 1)(5x + 5)$	$+5x$
$(2x + 1)(5x - 5)$	$-5x$
$(2x + 5)(5x - 1)$	$+23x$
$(2x - 5)(5x + 1)$	$-23x$
$(10x - 5)(x + 1)$	$+5x$
$(10x + 5)(x - 1)$	$-5x$
$(10x - 1)(x + 5)$	$+49x$
$(10x + 1)(x - 5)$	$-49x$

Thus,

$$10x^2 - 49x - 5 = (10x + 1)(x - 5)$$

As a check, it is always a good idea to multiply the two binomials to see whether you obtain the original expression.

$$(10x + 1)(x - 5) = 10x^2 - 50x + 1x - 5$$
$$= 10x^2 - 49x - 5$$

NOTE TO STUDENT: Fully worked-out solutions to all of the Practice Problems can be found at the back of the text starting at page SP-1

Practice Problem 10 Factor by trial and error.

$$8x^2 - 6x - 5$$

EXAMPLE 11 Factor by trial and error $6x^4 + x^2 - 12$.

Solution The first term of each factor must contain x^2. Suppose that we try the following:

Possible Factors	Middle Term of Product
$(2x^2 - 3)(3x^2 + 4)$	$-x^2$

The middle term we get is $-x^2$, but we need its opposite, x^2. In this case, we just need to reverse the signs of -3 and 4. Do you see why? Therefore,

$$6x^4 + x^2 - 12 = (2x^2 + 3)(3x^2 - 4).$$

Practice Problem 11 Factor by trial and error.

$$6x^4 + 13x^2 - 5$$

Factor each polynomial.

1. $x^2 + 8x + 7$

2. $x^2 + 12x + 11$

3. $x^2 - 8x + 15$

4. $x^2 - 10x + 16$

5. $x^2 - 10x + 24$

6. $x^2 - 9x + 18$

7. $a^2 + 4a - 45$

8. $a^2 + 2a - 35$

9. $x^2 - xy - 42y^2$

10. $x^2 - xy - 56y^2$

11. $x^2 - 15xy + 14y^2$

12. $x^2 + 10xy + 9y^2$

13. $x^4 - 3x^2 - 40$

14. $x^4 + 6x^2 + 5$

15. $x^4 + 16x^2y^2 + 63y^4$

16. $x^4 - 6x^2 - 55$

Factor out the greatest common factor from the terms of the trinomial. Then factor the remaining trinomial.

17. $2x^2 + 26x + 44$

18. $2x^2 + 30x + 52$

19. $x^3 + x^2 - 20x$

20. $x^3 - 4x^2 - 45x$

Factor each polynomial. You may use the grouping number method or the trial-and-error method.

21. $2x^2 - x - 1$

22. $3x^2 + x - 2$

23. $6x^2 - 7x - 5$

24. $5x^2 - 13x - 28$

25. $3a^2 - 8a + 5$

26. $6a^2 + 11a + 3$

27. $4a^2 + a - 14$

28. $3a^2 - 20a + 12$

29. $2x^2 + 13x + 15$

30. $5x^2 - 8x - 4$

31. $3x^4 - 8x^2 - 3$

32. $6x^4 - 7x^2 - 5$

33. $6x^2 + 35xy + 11y^2$

34. $5x^2 + 12xy + 7y^2$

35. $7x^2 + 11xy - 6y^2$

36. $4x^2 - 13xy + 3y^2$

Factor out the greatest common factor from the terms of the trinomial. Then factor the remaining trinomial.

37. $8x^3 - 2x^2 - x$

38. $9x^3 - 9x^2 - 10x$

39. $10x^4 + 15x^3 + 5x^2$

40. $16x^4 + 48x^3 + 20x^2$

Mixed Practice *Factor each polynomial.*

41. $x^2 - 2x - 63$

42. $x^2 + 6x - 40$

43. $6x^2 + x - 2$

44. $5x^2 + 17x + 6$

45. $x^2 - 20x + 51$

46. $x^2 - 20x + 99$

47. $15x^2 + x - 2$

48. $12x^2 - 5x - 3$

49. $2x^2 + 4x - 96$

50. $3x^2 + 9x - 84$

51. $18x^2 + 21x + 6$

52. $24x^2 + 26x + 6$

53. $40ax^2 + 72ax - 16a$

54. $60bx^2 - 84bx - 72b$

55. $6x^3 + 26x^2 - 20x$

56. $12x^3 - 14x^2 + 4x$

57. $7x^4 + 13x^2 - 2$

58. $8x^4 + 2x^2 - 3$

59. $9a^2 - 18ab - 7b^2$

60. $13a^2 - 8ab - 5b^2$

61. $x^6 - 10x^3 - 39$

62. $x^6 - 3x^3 - 70$

63. $10x^3y - 15x^2y - 10xy$

64. $8x^3y + 20x^2y - 12xy$

Cumulative Review

▲ **65.** **[1.6.2]** Find the area of a circle of radius 3 inches. (Use $\pi \approx 3.14$.)

66. **[2.2.1]** Solve for b. $A = \dfrac{1}{3}(3b + 4a)$

67. **[4.3.1]** *Airplane Seating* A large commercial jetliner flies from Atlanta to San Francisco. The jetliner can carry a total of 184 passengers. There are two types of seats on the aircraft: first class and coach. The number of coach seats is sixteen more than six times the number of first-class seats. How many of each type of seat are there on the airplane?

68. **[3.3.1]** Graph $6x + 4y = -12$.

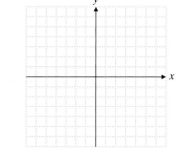

Quick Quiz 5.5 Factor each polynomial.

1. $x^2 + 5x - 84$

2. $6x^2 - 17x + 12$

3. $4x^3 + 6x^2 - 40x$

4. **Concept Check** Explain what the first step would be in factoring the following polynomial.
$3x^2y^2 + 6xy^2 - 72y^2$

How are you doing with your homework assignments on Sections 5.1 to 5.5? Do you feel you have mastered the material so far? Do you understand the concepts you have covered? Before you go further in the textbook, take some time to do each of the following problems.

5.1

Simplify.

1. $(5x^2 - 3x + 2) + (-3x^2 - 5x - 8) - (x^2 + 3x - 10)$

2. $(x^2 - 3x - 4)(2x - 3)$ **3.** $(5a - 8)(a - 7)$

4. $(3y - 5)(3y + 5)$ **5.** $(3x^2 + 4)^2$

6. Evaluate the polynomial function $p(x) = 2x^3 - 5x^2 - 6x + 1$ for $p(-3)$.

5.2 and 5.3

Divide.

7. $(25x^3y^2 - 30x^2y^3 - 50x^2y^2) \div 5x^2y^2$

8. $(3y^3 - 5y^2 + 2y - 1) \div (y - 2)$

9. $(6x^4 - x^3 + 11x^2 - 2x - 2) \div (3x + 1)$

10. Use synthetic division to do the following:
$(2x^4 + 10x^3 + 11x^2 - 6x - 9) \div (x + 3)$

5.4

Factor completely.

11. $24a^3b^2 + 36a^4b^2 - 60a^3b^3$ **12.** $3x(4x - 3y) - 2(4x - 3y)$

13. $10wx + 6xz - 15yz - 25wy$ **14.** $18a^2 + 6ab - 15a - 5b$

5.5

Factor completely.

15. $x^2 - 7x + 10$ **16.** $4y^2 - 4y - 15$

17. $28x^2 - 19xy + 3y^2$ **18.** $2x^2 + 21x + 40$

19. $3x^2 - 6x - 72$ **20.** $8x^2 - 18x + 9$

Now turn to page SA-14 for the answers to each of these problems. Each answer also includes a reference to the objective in which the problem is first taught. If you missed any of these problems, you should stop and review the Examples and Practice Problems in the referenced objective. A little review now will help you master the material in the upcoming sections of the text.

1. _____
2. _____
3. _____
4. _____
5. _____
6. _____
7. _____
8. _____
9. _____
10. _____
11. _____
12. _____
13. _____
14. _____
15. _____
16. _____
17. _____
18. _____
19. _____
20. _____

1 Factoring the Difference of Two Squares

Recall the special product formula: $(a + b)(a - b) = a^2 - b^2$. We can use it now as a factoring formula.

> **FACTORING THE DIFFERENCE OF TWO SQUARES**
> $$a^2 - b^2 = (a + b)(a - b)$$

EXAMPLE 1 Factor $x^2 - 16$.

Solution In this case $a = x$ and $b = 4$ in the formula.

$$
\begin{array}{cccccc}
a^2 & - & b^2 & = & (a + b) & (a - b) \\
\downarrow & & \downarrow & & \downarrow \quad \downarrow & \downarrow \quad \downarrow \\
(x)^2 & - & (4)^2 & = & (x + 4) & (x - 4)
\end{array}
$$

Practice Problem 1 Factor $x^2 - 9$.

EXAMPLE 2 Factor $25x^2 - 36$.

Solution Here we will use the formula $a^2 - b^2 = (a + b)(a - b)$.

$$25x^2 - 36 = (5x)^2 - (6)^2 = (5x + 6)(5x - 6)$$

Practice Problem 2 Factor $64x^2 - 121$.

EXAMPLE 3 Factor $100w^4 - 9z^4$.

Solution $100w^4 - 9z^4 = (10w^2)^2 - (3z^2)^2 = (10w^2 + 3z^2)(10w^2 - 3z^2)$

Practice Problem 3 Factor $49x^2 - 25y^4$.

Whenever possible, a common factor should be factored out in the first step. Then the formula can be applied.

EXAMPLE 4 Factor $75x^2 - 3$.

Solution We factor out a common factor of 3 from each term.

$$
\begin{aligned}
75x^2 - 3 &= 3(25x^2 - 1) \\
&= 3(5x + 1)(5x - 1)
\end{aligned}
$$

Practice Problem 4 Factor $7x^2 - 28$.

2 Factoring Perfect Square Trinomials

Recall the formulas for squaring a binomial.

$$(a - b)^2 = a^2 - 2ab + b^2$$
$$(a + b)^2 = a^2 + 2ab + b^2$$

We can use these formulas to factor perfect square trinomials.

PERFECT SQUARE FACTORING FORMULAS
$$a^2 - 2ab + b^2 = (a - b)^2$$
$$a^2 + 2ab + b^2 = (a + b)^2$$

Recognizing these special cases will save you a lot of time when factoring. How can we recognize a perfect square trinomial?

1. The first and last terms are perfect squares. (The numerical values are 1, 4, 9, 16, 25, 36, ..., and the variables have an exponent that is an even whole number.)
2. The middle term is twice the product of the values that, when squared, give the first and last terms.

EXAMPLE 5   Factor $25x^2 - 20x + 4$.

Solution   Is this trinomial a perfect square? Yes.

1. The first and last terms are perfect squares.
$$25x^2 - 20x + 4 = (5x)^2 - 20x + (2)^2$$

2. The middle term is twice the product of the value $5x$ and the value 2. In other words, $2(5x)(2) = 20x$.
$$(5x)^2 - 2(5x)(2) + (2)^2 = (5x - 2)^2$$
Therefore, we can use the formula $a^2 - 2ab + b^2 = (a - b)^2$. Thus,
$$25x^2 - 20x + 4 = (5x - 2)^2.$$

Practice Problem 5   Factor $9x^2 - 30x + 25$.

EXAMPLE 6   Factor $16x^2 - 24x + 9$.

Solution

1. The first and last terms are perfect squares: $16x^2 = (4x)^2$ and $9 = (3)^2$.
2. The middle term is twice the product $(4x)(3)$. Therefore, we have the following:
$$a^2 - 2ab + b^2 = (a - b)^2$$
$$16x^2 - 24x + 9 = (4x)^2 - 2(4x)(3) + (3)^2$$
$$= (4x - 3)^2.$$

Practice Problem 6   Factor $25x^2 - 70x + 49$.

EXAMPLE 7   Factor $200x^2 + 360x + 162$.

Solution   First we factor out the common factor of 2.
$$200x^2 + 360x + 162 = 2(100x^2 + 180x + 81)$$
$$a^2 + 2ab + b^2 = (a + b)^2$$
$$2(100x^2 + 180x + 81) = 2[(10x)^2 + (2)(10x)(9) + (9)^2]$$
$$= 2(10x + 9)^2$$

Practice Problem 7   Factor $242x^2 + 88x + 8$.

EXAMPLE 8 Factor.

(a) $x^4 + 14x^2 + 49$ (b) $9x^4 + 30x^2y^2 + 25y^4$

Solution

(a) $x^4 + 14x^2 + 49 = (x^2)^2 + 2(x^2)(7) + (7)^2$
$$= (x^2 + 7)^2$$

(b) $9x^4 + 30x^2y^2 + 25y^4 = (3x^2)^2 + 2(3x^2)(5y^2) + (5y^2)^2$
$$= (3x^2 + 5y^2)^2$$

Practice Problem 8 Factor.

(a) $49x^4 + 28x^2 + 4$ (b) $36x^4 + 84x^2y^2 + 49y^4$

③ Factoring the Sum or Difference of Two Cubes

There are also special formulas for factoring cubic binomials. We see that the factors of $x^3 + 27$ are $(x + 3)(x^2 - 3x + 9)$, and that the factors of $x^3 - 64$ are $(x - 4)(x^2 + 4x + 16)$. We can generalize these patterns and derive the following factoring formulas.

> **SUM AND DIFFERENCE OF CUBES FACTORING FORMULAS**
> $$a^3 + b^3 = (a + b)(a^2 - ab + b^2)$$
> $$a^3 - b^3 = (a - b)(a^2 + ab + b^2)$$

EXAMPLE 9 Factor $125x^3 + y^3$.

Solution Here $a = 5x$ and $b = y$.

$$a^3 + b^3 = (a + b)(a^2 - ab + b^2)$$
$$125x^3 + y^3 = (5x)^3 + (y)^3 = (5x + y)(25x^2 - 5xy + y^2)$$

Practice Problem 9 Factor $8x^3 + 125y^3$.

EXAMPLE 10 Factor $64x^3 + 27$.

Solution Here $a = 4x$ and $b = 3$.

$$a^3 + b^3 = (a + b)(a^2 - ab + b^2)$$
$$64x^3 + 27 = (4x)^3 + (3)^3 = (4x + 3)(16x^2 - 12x + 9)$$

Practice Problem 10 Factor $64x^3 + 125y^3$.

EXAMPLE 11 Factor $125w^3 - 8z^6$.

Solution Here $a = 5w$ and $b = 2z^2$.

$$a^3 - b^3 = (a - b)(a^2 + ab + b^2)$$
$$125w^3 - 8z^6 = (5w)^3 - (2z^2)^3 = (5w - 2z^2)(25w^2 + 10wz^2 + 4z^4)$$

Practice Problem 11 Factor $27w^3 - 125z^6$.

EXAMPLE 12 Factor $250x^3 - 2$.

Solution First we factor out the common factor of 2.

$$250x^3 - 2 = 2(125x^3 - 1)$$
$$= 2(5x - 1)\underbrace{(25x^2 + 5x + 1)}$$
$$\uparrow$$

Note that this trinomial cannot be factored.

Practice Problem 12 Factor $54x^3 - 16$.

What should you do if a polynomial is the difference of two cubes *and* the difference of two squares? Usually, it's easier to use the difference of two squares formula first. Then apply the difference of two cubes formula.

EXAMPLE 13 Factor $x^6 - y^6$.

Solution We can write this binomial as $(x^2)^3 - (y^2)^3$ or as $(x^3)^2 - (y^3)^2$. Therefore, we can use either the difference of two cubes formula or the difference of two squares formula. It's usually better to use the difference of two squares formula first, so we'll do that.

$$x^6 - y^6 = (x^3)^2 - (y^3)^2$$

Here $a = x^3$ and $b = y^3$. Therefore,

$$(x^3)^2 - (y^3)^2 = (x^3 + y^3)(x^3 - y^3).$$

Now we use the sum of two cubes formula for the first factor and the difference of two cubes formula for the second factor.

$$x^3 + y^3 = (x + y)(x^2 - xy + y^2)$$
$$x^3 - y^3 = (x - y)(x^2 + xy + y^2)$$

Hence,

$$x^6 - y^6 = (x + y)(x^2 - xy + y^2)(x - y)(x^2 + xy + y^2).$$

Practice Problem 13 Factor $64a^6 - 1$.

You'll see these special cases of factoring often. You should be very familiar with the following formulas. Be sure you understand how to use each one of them.

SPECIAL CASES OF FACTORING

Difference of Two Squares

$$a^2 - b^2 = (a + b)(a - b)$$

Perfect Square Trinomial

$$a^2 - 2ab + b^2 = (a - b)^2$$
$$a^2 + 2ab + b^2 = (a + b)^2$$

Sum and Difference of Cubes

$$a^3 + b^3 = (a + b)(a^2 - ab + b^2)$$
$$a^3 - b^3 = (a - b)(a^2 + ab + b^2)$$

Verbal and Writing Skills

1. How do you determine if a factoring problem will use the difference of two squares?

2. How do you determine if a factoring problem will use the perfect square trinomial formula?

3. How do you determine if a factoring problem will use the sum of two cubes formula?

4. How do you determine if a factoring problem will use the difference of two cubes formula?

Use the difference of two squares formula to factor. Be sure to factor out any common factors.

5. $a^2 - 64$

6. $y^2 - 49$

7. $16x^2 - 81$

8. $4x^2 - 25$

9. $64x^2 - 1$

10. $81x^2 - 1$

11. $49m^2 - 9n^2$

12. $36x^2 - 25y^2$

13. $100y^2 - 81$

14. $49y^2 - 144$

15. $1 - 36x^2y^2$

16. $1 - 64x^2y^2$

17. $32x^2 - 18$

18. $50x^2 - 8$

19. $5x - 20x^3$

20. $49x^3 - 36x$

Use the perfect square trinomial formulas to factor. Be sure to factor out any common factors.

21. $9x^2 - 6x + 1$

22. $16y^2 - 8y + 1$

23. $49x^2 - 14x + 1$

24. $100y^2 - 20y + 1$

25. $81w^2 + 36wt + 4t^2$

26. $25w^2 + 20wt + 4t^2$

27. $36x^2 + 60xy + 25y^2$

28. $64x^2 + 48xy + 9y^2$

29. $8x^2 + 40x + 50$

30. $108x^2 + 36x + 3$

31. $3x^3 - 24x^2 + 48x$

32. $50x^3 - 20x^2 + 2x$

Use the sum and difference of cubes formulas to factor. Be sure to factor out any common factors.

33. $x^3 - 27$

34. $x^3 - 8$

35. $x^3 + 125$

36. $x^3 + 64$

37. $64x^3 - 1$

38. $125x^3 - 1$

39. $8x^3 - 125$

40. $27x^3 - 64$

41. $1 - 27x^3$

42. $1 - 8x^3$

43. $64x^3 + 125$

44. $27x^3 + 125$

45. $64s^6 + t^6$

46. $125s^6 + t^6$

47. $5y^3 - 40$

48. $54y^3 - 2$

49. $250x^3 + 2$ **50.** $128y^3 + 2$ **51.** $x^5 - 8x^2y^3$ **52.** $x^5 - 27x^2y^3$

Mixed Practice *Factor by the methods of this section.*

53. $25w^4 - 1$ **54.** $16m^4 - 25$ **55.** $b^4 + 6b^2 + 9$ **56.** $a^4 - 10a^2 + 25$

57. $9m^6 - 64$ **58.** $144 - m^6$ **59.** $36y^6 - 60y^3 + 25$ **60.** $100n^6 - 140n^3 + 49$

61. $45z^8 - 5$ **62.** $2a^8 - 98$ **63.** $125m^3 + 8n^3$

64. $64z^3 - 27w^3$ **65.** $24a^3 - 3b^3$ **66.** $54w^3 + 250$

67. $4w^2 - 20wz + 25z^2$ **68.** $81x^4 - 36x^2 + 4$ **69.** $36a^2 - 81b^2$

70. $400x^4 - 36y^2$ **71.** $16x^4 - 81y^4$ **72.** $256x^4 - 1$

73. $125m^6 + 8$ **74.** $27n^6 + 125$

Try to factor the following four exercises by using the formulas for the perfect square trinomial. Why can't the formulas be used? Then factor each exercise correctly using an appropriate method.

75. $25x^2 + 25x + 4$ **76.** $16x^2 + 40x + 9$

77. $49x^2 - 35x + 4$ **78.** $4x^2 - 25x + 36$

Applications

▲ **79.** *Carpentry* Find the area of a maple cabinet surface that is constructed by a carpenter as a large square with sides of $4x$ feet and has a square cut-out region whose sides are y feet. Factor the expression.

▲ **80.** *Base of a Lamp* A copper base for a lamp consists of a large circle of radius $2y$ inches with a cut-out area in the center of radius x inches. Write an expression for the area of the top of this copper base. Write your answer in factored form.

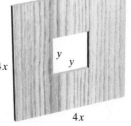

Cumulative Review

81. **[2.1.1]** *Public School Salaries* The average annual salary y, in thousands of dollars, paid to public school principals in the United States can be approximated by the equation $y = 1.57x + 84$, where x is the number of years since 2002. The average annual salary y, in thousands of dollars, paid to public high school classroom teachers can be approximated by the equation $y = 1.04x + 44$, where x is the number of years since 2002. If this trend continues, in what year will a public school principal be paid $50,000 more than a public school classroom teacher? (*Source:* www.census.gov)

82. **[2.4.1]** *MP3 Player Prices* Three friends each bought a portable MP3 player. The total for the three purchases was $387. George paid $34 more than Marcel. Rebecca paid $19 less than Marcel. How much did each person pay?

Quick Quiz 5.6 Factor completely.

1. $25x^2 - 64y^2$

2. $49x^2 - 56xy + 16y^2$

3. $64x^3 - 27y^3$

4. **Concept Check** Explain why the formula $(a + b)^2 = a^2 + 2ab + b^2$ does not work when factoring the following expression.

$$36x^2 + 66xy + 121y^2$$

Developing Your Study Skills

Keep Trying

You may be one of those students who has had much difficulty with mathematics in the past and who is sure that you cannot do well in this course. Perhaps you are thinking, "I have never been any good at mathematics" or "I have always hated mathematics." You may even have picked up on the label "math anxiety" and attached it to yourself.

It is time for you to reprogram your thinking. Replace those negative thoughts with more positive ones. You need to say things like, "I will give this math class my best shot" or "I can learn mathematics if I work at it." You will be pleasantly surprised at the difference a positive attitude makes!

We live in a highly technical world, and you cannot afford to give up on the study of mathematics. Dropping mathematics may prevent you from entering certain career fields that you find interesting. Such courses as finite math, college algebra, and statistics may be necessary. Learning mathematics can open new doors for you.

1 Factoring Factorable Polynomials

Not all polynomials have the convenient form of one of the special formulas. Most do not. The following procedure will help you handle these common cases. You must practice this procedure until you can *recognize the various forms* and *determine which factoring method to use*.

Student Learning Objectives

After studying this section, you will be able to:

1. Factor any factorable polynomial.

2. Recognize polynomials that are prime.

COMPLETELY FACTORING A POLYNOMIAL

1. Check for a common factor. Factor out the greatest common factor (if there is one) before doing anything else.

2. **(a)** If the remaining polynomial has two terms, try to factor it as one of the following.
   (1) The difference of two squares: $a^2 - b^2 = (a + b)(a - b)$
   (2) The difference of two cubes: $a^3 - b^3 = (a - b)(a^2 + ab + b^2)$
   (3) The sum of two cubes: $a^3 + b^3 = (a + b)(a^2 - ab + b^2)$

   (b) If the remaining polynomial has three terms, try to factor it as one of the following.
   (1) A perfect square trinomial: $a^2 + 2ab + b^2 = (a + b)^2$ or $a^2 - 2ab + b^2 = (a - b)^2$
   (2) A general trinomial of the form $x^2 + bx + c$ or the form $ax^2 + bx + c$

   (c) If the remaining polynomial has four terms, try to factor by grouping.

3. Check to see whether the factors can be factored further.

EXAMPLE 1 Factor completely.

(a) $2x^2 - 18$

(b) $27x^4 - 8x$

(c) $27x^2 + 36xy + 12y^2$

(d) $2x^2 - 100x + 98$

(e) $6x^3 + 11x^2 - 10x$

(f) $5ax + 5ay - 20x - 20y$

Solution

(a) $2x^2 - 18 = 2(x^2 - 9)$ Factor out the common factor.
$= 2(x + 3)(x - 3)$ Use $a^2 - b^2 = (a + b)(a - b)$.

(b) $27x^4 - 8x = x(27x^3 - 8)$ Factor out the common factor.
$= x(3x - 2)(9x^2 + 6x + 4)$ Use $a^3 - b^3 = (a - b)(a^2 + ab + b^2)$.

(c) $27x^2 + 36xy + 12y^2 = 3(9x^2 + 12xy + 4y^2)$ Factor out the common factor.
$= 3(3x + 2y)^2$ Use $(a + b)^2 = a^2 + 2ab + b^2$.

(d) $2x^2 - 100x + 98 = 2(x^2 - 50x + 49)$ Factor out the common factor.
$= 2(x - 49)(x - 1)$ The trinomial has the form $x^2 + bx + c$.

(e) $6x^3 + 11x^2 - 10x = x(6x^2 + 11x - 10)$ Factor out the common factor.
$= x(3x - 2)(2x + 5)$ The trinomial has the form $ax^2 + bx + c$.

(f) $5ax + 5ay - 20x - 20y = 5(ax + ay - 4x - 4y)$ Factor out the common factor.

$= 5[a(x + y) - 4(x + y)]$ Factor by grouping.

$= 5(x + y)(a - 4)$

NOTE TO STUDENT: *Fully worked-out solutions to all of the Practice Problems can be found at the back of the text starting at page SP-1*

Practice Problem 1 Factor completely.

(a) $7x^5 + 56x^2$

(b) $125x^2 + 50xy + 5y^2$

(c) $12x^2 - 75$

(d) $3x^2 - 39x + 126$

(e) $6ax + 6ay + 18bx + 18by$

(f) $6x^3 - x^2 - 12x$

Recognizing Polynomials That Are Prime

Can all polynomials be factored? No. Many polynomials cannot be factored. If a polynomial cannot be factored using rational numbers, it is said to be **prime.**

EXAMPLE 2 If possible, factor $6x^2 + 10x + 3$.

Solution The trinomial has the form $ax^2 + bx + c$. The grouping number is 18. If the trinomial can be factored, we must find two numbers whose product is 18 and whose sum is 10.

Factor Pairs of 18	Sum of the Factors
(18)(1)	19
(9)(2)	11
(6)(3)	9

There are no numbers meeting the necessary conditions. Thus, the polynomial is prime. (If you use the trial-and-error method, try all the possible factors and show that none of them has a product with a middle term of $10x$.)

Practice Problem 2 If possible, factor $3x^2 - 10x + 4$.

EXAMPLE 3 If possible, factor $25x^2 + 49$.

Solution Unless there is a common factor that can be factored out, binomials of the form $a^2 + b^2$ cannot be factored. Therefore, $25x^2 + 49$ is prime.

Practice Problem 3 If possible, factor $16x^2 + 81$.

Verbal and Writing Skills

1. In any factoring problem the first step is

_____ .

2. If $x^2 + bx + c = (x + e)(x + f)$ and c is positive and b is negative, what can you say about the signs of e and of f?

3. If you were asked to factor a problem of the form $49x^2 + 9y^2$, how would you know immediately that this polynomial is prime?

4. If you were asked to factor a problem of the form $x^2 + 9x + 12$, how would you know very quickly that this polynomial is prime?

Factor, if possible. These problems will require only one step.

5. $3xy - 6yz$

6. $33a - 44a^2b$

7. $y^2 + 7y - 18$

8. $b^2 - 7b + 12$

9. $3x^2 - 8x + 5$

10. $3x^2 + 4x - 7$

11. $4x^2 + 20x + 25$

12. $16x^2 - 24x + 9$

13. $8x^3 - 125y^3$

14. $27x^3 + 64y^3$

15. $a^3 - 3ab - ac$

16. $3a^3 + a^2b - 2a^2c$

17. $x^2 + 16$

18. $4x^4 + 25$

19. $64y^2 - 25z^2$

20. $81m^2 - 49n^2$

Mixed Practice   *Factor if possible. Be sure to factor completely.*

21. $6x^2 - 23x - 4$

22. $5x^2 + x - 4$

23. $3x^2 - x - 1$

24. $5x^2 - x - 2$

25. $x^3 + 9x^2 + 14x$

26. $x^3 + x^2 - 20x$

27. $25x^2 - 40x + 16$

28. $9r^2 + 48r + 64$

29. $6a^2 - 6a - 36$

30. $9a^2 + 18a - 72$

31. $3x^2 - 3x - xy + y$

32. $xb - x - yb + y$

33. $81a^4 - 1$

34. $1 - 16x^4$

35. $2x^5 - 16x^3 - 18x$

36. $2x^4 - 2x^2 - 24$ **37.** $8a^3b - 50ab^3$ **38.** $50x^2y^2 - 32y^2$

39. $2ax - 8xy + aw - 4wy$ **40.** $3ax - xy + 6aw - 2wy$

Applications

Cattle Farming *A cattle pen is constructed with solid wood walls. The pen is divided into four rectangular compartments. Each compartment is x feet long and y feet wide. The walls are x − 10 feet high.*

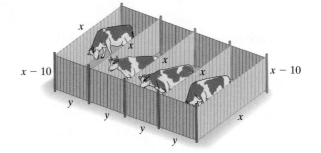

▲ **41.** Find the total surface area of the walls used in the cattle pen. Express the answer in factored form and in the form with the factors multiplied.

▲ **42.** The rancher who owns the cattle pen wants to increase the length x by 3 feet. Find the new total surface area of the walls that would be used in this enlarged cattle pen. Express the answer in factored form and in the form with the factors multiplied.

Cumulative Review *Solve the following inequalities.*

43. **[2.6.3]** $3x - 2 \le -5 + 2(x - 3)$ **44.** **[2.8.1]** $|2 + 5x - 3| < 2$

45. **[2.8.2]** $\left|\dfrac{1}{3}(5 - 4x)\right| > 4$ **46.** **[2.7.2]** $x - 4 \ge 7$ or $4x + 1 \le 17$

Quick Quiz 5.7 Factor if possible.

1. $49x^3 + 84x^2y + 36xy^2$ **2.** $6x^4 + x^2 - 12$

3. $9x^2 + 12x - 12$ **4.** **Concept Check** Explain why $4x^2 + 3x + 1$ is prime.

1 Factoring to Find the Roots of a Quadratic Equation

Up until now, we have solved only first-degree equations. In this section we will solve quadratic, or second-degree, equations.

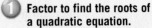
> **DEFINITION OF QUADRATIC EQUATION**
>
> A second-degree equation of the form $ax^2 + bx + c = 0$, where a, b, c are real numbers and $a \neq 0$, is a **quadratic equation.** $ax^2 + bx + c = 0$ is the **standard form** of a quadratic equation.

Before solving a quadratic equation, we will first write the equation in standard form. Although it is not necessary that a, b, and c be integers, the equation is usually written this way.

The key to solving quadratic equations by factoring is called the **zero factor property.** When we multiply two real numbers, the resulting product will be zero if one or both of the factors is zero. Thus, if the product of two real numbers is zero, at least one of the factors must be zero. We state this property of real numbers formally.

> **ZERO FACTOR PROPERTY**
>
> For all real numbers a and b,
>
> if $a \cdot b = 0$, then $a = 0$, $b = 0$, or both $= 0$.

EXAMPLE 1 Solve the equation $x^2 + 15x = 100$.

Solution When we say "solve the equation" or "find the roots," we mean "find the values of x that satisfy the equation."

$x^2 + 15x - 100 = 0$	Subtract 100 from both sides so that one side is 0.
$(x + 20)(x - 5) = 0$	Factor the trinomial.
$x + 20 = 0$ or $x - 5 = 0$	Set each factor equal to 0.
$x = -20$ $x = 5$	Solve each equation.

Check: Use the *original* equation $x^2 + 15x = 100$.

$$x = -20: \quad (-20)^2 + 15(-20) \stackrel{?}{=} 100$$
$$400 - 300 \stackrel{?}{=} 100$$
$$100 = 100 \quad \checkmark$$
$$x = 5: \quad (5)^2 + 15(5) \stackrel{?}{=} 100$$
$$25 + 75 \stackrel{?}{=} 100$$
$$100 = 100 \quad \checkmark$$

Thus, 5 and -20 are both roots of the quadratic equation $x^2 + 15x = 100$.

Practice Problem 1 Find the roots of $x^2 + x = 56$.

For convenience, on the next page we list the steps we have employed to solve the quadratic equation.

Graphing Calculator

Finding Roots

Find the roots of

$$10x(x + 1) = 83x + 12{,}012.$$

This can be written in the form

$$10x^2 - 73x - 12{,}012 = 0$$

and factored to obtain the following:

$$(5x + 156)(2x - 77) = 0$$

$$x = -\frac{156}{5} \text{ and } x = \frac{77}{2}.$$

In decimal form the solutions are −31.2 and 38.5.

However, you can use the graphing calculator to graph

$$y = 10x^2 - 73x - 12{,}012.$$

By setting an appropriate viewing window and then using the Zoom and Trace features of your calculator, you can find the two places where $y = 0$.

Some calculators have a command that will find the zeros (roots) of a graph.

In a similar fashion graph each equation using the form $y = ax^2 + bx + c$ and use the graph to find the roots.

(a) $10x^2 - 189x - 12{,}834 = 0$

(b) $10x(x + 2) = 11{,}011 - 193x$

SOLVING A QUADRATIC EQUATION BY FACTORING

1. Rewrite the quadratic equation in standard form (so that one side of the equation is 0) and, if possible, *factor* the quadratic expression.
2. Set each factor equal to zero.
3. Solve the resulting equations to find both roots. (A quadratic equation has two roots.)
4. Check your solutions.

It is extremely important to remember that when you are placing the quadratic equation in standard form, one side of the equation must be zero. Several algebraic operations may be necessary to obtain that desired result before you can factor the polynomial.

EXAMPLE 2   Find the roots of $6x^2 + 4 = 7(x + 1)$.

Solution

$6x^2 + 4 = 7x + 7$	Apply the distributive property.
$6x^2 - 7x - 3 = 0$	Rewrite the equation in standard form.
$(2x - 3)(3x + 1) = 0$	Factor the trinomial.
$2x - 3 = 0 \quad \text{or} \quad 3x + 1 = 0$	Set each factor equal to 0.
$2x = 3 \qquad\qquad 3x = -1$	Solve the equations.
$x = \dfrac{3}{2} \qquad\qquad x = -\dfrac{1}{3}$	

Check: Use the *original* equation $6x^2 + 4 = 7(x + 1)$.

$$x = \frac{3}{2}: \quad 6\left(\frac{3}{2}\right)^2 + 4 \stackrel{?}{=} 7\left(\frac{3}{2} + 1\right)$$

$$6\left(\frac{9}{4}\right) + 4 \stackrel{?}{=} 7\left(\frac{5}{2}\right)$$

$$\frac{27}{2} + 4 \stackrel{?}{=} \frac{35}{2}$$

$$\frac{27}{2} + \frac{8}{2} \stackrel{?}{=} \frac{35}{2}$$

$$\frac{35}{2} = \frac{35}{2} \quad \checkmark$$

It checks, so $\frac{3}{2}$ is a root. Verify that $-\frac{1}{3}$ is also a root.

If you are using a calculator to check your roots, you can complete the check more rapidly using the decimal values 1.5 and −0.33333333. The latter value is approximate, so some rounding error is expected.

Practice Problem 2   Find the roots of $12x^2 - 9x + 2 = 2x$.

EXAMPLE 3   Find the roots of $3x^2 - 5x = 0$.

Solution

$3x^2 - 5x = 0$	The equation is already in standard form.
$x(3x - 5) = 0$	Factor.
$x = 0 \quad \text{or} \quad 3x - 5 = 0$	Set each factor equal to 0.
$\qquad\qquad 3x = 5$	
$\qquad\qquad x = \dfrac{5}{3}$	

Check: Verify that 0 and $\frac{5}{3}$ are roots of $3x^2 - 5x = 0$.

Practice Problem 3   Find the roots of $7x^2 - 14x = 0$.

NOTE TO STUDENT: Fully worked-out solutions to all of the Practice Problems can be found at the back of the text starting at page SP-1

EXAMPLE 4 Solve $9x(x - 1) = 3x - 4$.

Solution

$$9x^2 - 9x = 3x - 4 \quad \text{Remove parentheses.}$$
$$9x^2 - 9x - 3x + 4 = 0 \quad \text{Get 0 on one side.}$$
$$9x^2 - 12x + 4 = 0 \quad \text{Combine like terms.}$$
$$(3x - 2)^2 = 0 \quad \text{Factor.}$$
$$3x - 2 = 0 \quad \text{or} \quad 3x - 2 = 0$$
$$3x = 2 \qquad\qquad 3x = 2$$
$$x = \frac{2}{3} \qquad\qquad x = \frac{2}{3}$$

We obtain one solution twice. This value is called a **double root.**

Practice Problem 4 Solve $16x(x - 2) = 8x - 25$.

The zero factor property can be extended to polynomial equations of degree greater than 2. In the following example, we will find the three roots of a third-degree polynomial equation.

EXAMPLE 5 Solve $2x^3 = 24x - 8x^2$.

Solution

$$2x^3 + 8x^2 - 24x = 0 \qquad\qquad \text{Get 0 on one side of the equation.}$$
$$2x(x^2 + 4x - 12) = 0 \qquad\qquad \text{Factor out the common factor } 2x.$$
$$2x(x + 6)(x - 2) = 0 \qquad\qquad \text{Factor the trinomial.}$$
$$2x = 0 \quad \text{or} \quad x + 6 = 0 \quad \text{or} \quad x - 2 = 0 \quad \text{Zero factor property.}$$
$$x = 0 \qquad\qquad x = -6 \qquad\qquad x = 2 \quad \text{Solve for } x.$$

The solutions are 0, -6, and 2.

Practice Problem 5 Solve $3x^3 + 6x^2 = 45x$.

 Solving Applications That Involve Factorable Quadratic Equations

Some applied exercises lead to factorable quadratic equations. Using the methods developed in this section, we can solve these types of exercises.

▲ **EXAMPLE 6** A racing sailboat has a triangular sail. Find the base and altitude of the triangular sail that has an area of 35 square meters and a base that is 3 meters shorter than the altitude.

Solution

1. **Understand the problem.** We draw a sketch and recall the formula for the area of a triangle.

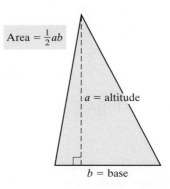

Area $= \frac{1}{2}ab$

$a =$ altitude

$b =$ base

Let $x =$ the length of the altitude in meters.

Then $x - 3 =$ the length of the base in meters.

2. **Write an equation.**

$$A = \frac{1}{2}ab$$

$$35 = \frac{1}{2}x(x - 3) \quad \text{Replace } A \text{ (area) by 35, } a \text{ (altitude) by } x,\text{ and } b \text{ (base) by } x - 3.$$

3. **Solve the equation and state the answer.**

$$70 = x(x - 3) \qquad \text{Multiply each side by 2.}$$
$$70 = x^2 - 3x \qquad \text{Remove parentheses.}$$
$$0 = x^2 - 3x - 70 \qquad \text{Subtract 70 from each side.}$$
$$0 = (x - 10)(x + 7)$$
$$x = 10 \quad \text{or} \quad x = -7$$

The altitude of a triangle must be a positive number, so we disregard -7. Thus,

$$\text{altitude} = x = 10 \text{ meters and}$$
$$\text{base} = x - 3 = 7 \text{ meters.}$$

The altitude of the triangular sail measures 10 meters, and the base of the sail measures 7 meters.

4. **Check.** Is the base 3 meters shorter than the altitude?

$$10 - 3 = 7 \quad \checkmark$$

Is the area of the triangle 35 square meters?

$$A = \frac{1}{2}ab$$

$$A = \frac{1}{2}(10)(7) = 5(7) = 35 \quad \checkmark$$

▲ **Practice Problem 6** A racing sailboat has a triangular sail. Find the base and the altitude of the triangular sail if the area is 52 square feet and the altitude is 5 feet longer than the base.

NOTE TO STUDENT: Fully worked-out solutions to all of the Practice Problems can be found at the back of the text starting at page SP-1

▲ **EXAMPLE 7**   A car manufacturer uses a square panel that holds fuses. The square panel that is used this year has an area that is 72 square centimeters greater than the area of the square panel used last year. The length of each side of the new panel is 3 centimeters more than double the length of last year's panel. Find the dimensions of each panel.

Solution

1. **Understand the problem.** We draw a sketch of each square panel and recall that the area of each panel is obtained by squaring its side.

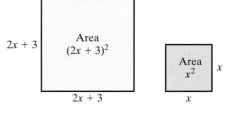

Let x = the length in centimeters of last year's panel.

Then $2x + 3$ = the length in centimeters of this year's panel.

The area of this year's panel is 72 square centimeters greater than the area of last year's panel.

2. **Write an equation.**

Area of the larger square is 72 square centimeters greater than the area of the smaller square.

$$(2x + 3)^2 \qquad = 72 \qquad\qquad + \qquad x^2$$

3. **Solve the equation and state the answer.**

$$4x^2 + 12x + 9 = 72 + x^2 \qquad \text{Remove parentheses.}$$
$$4x^2 + 12x + 9 - x^2 - 72 = 0 \qquad \text{Get 0 on one side of the equation.}$$
$$3x^2 + 12x - 63 = 0 \qquad \text{Simplify.}$$
$$3(x^2 + 4x - 21) = 0 \qquad \text{Factor out the common factor of 3.}$$
$$3(x + 7)(x - 3) = 0 \qquad \text{Factor the trinomial.}$$
$$x + 7 = 0 \quad \text{or} \quad x - 3 = 0 \qquad \text{Use the zero factor property.}$$
$$x = -7 \qquad\quad x = 3 \qquad\qquad \text{Solve for } x.$$

A fuse box cannot measure -7 centimeters, so we reject the negative answer. We use $x = 3$; then $2x + 3 = 2(3) + 3 = 6 + 3 = 9$.

Thus, the old fuse panel measured 3 centimeters on a side. The new fuse panel measures 9 centimeters on a side.

4. **Check.** Verify that each answer is correct.

▲ **Practice Problem 7**   Last year Grandpa Jones had a small square garden. This year he has a square garden that is 112 square feet larger in area. Each side of the new garden is 2 feet longer than triple each side of the old garden. Find the dimensions of each garden.

Developing Your Study Skills

Getting Help

Getting the right kind of help at the right time can be a key ingredient in being successful in mathematics. Even if you have gone to class on a regular basis, taken careful notes, methodically read your textbook, and diligently done your homework—all of which means making every effort possible to learn the mathematics—you may find that you are still having difficulty. If this is the case, then you need to seek help. Make an appointment with your instructor to find out what help is available to you. The instructor, tutoring services, a mathematics lab, videotapes, and computer software may be among the resources you can draw on.

Once you discover the resources available in your school, you need to take advantage of them. Do not put it off, or you will find yourself getting behind. You cannot afford that. When studying mathematics, you must keep up with your work.

Find all the roots and check your answers.

1. $x^2 - x - 6 = 0$

2. $x^2 + 5x - 14 = 0$

3. $5x^2 - 6x = 0$

4. $3x^2 + 5x = 0$

5. $9x^2 - 16 = 0$

6. $49x^2 - 25 = 0$

7. $3x^2 - 2x - 8 = 0$

8. $4x^2 - 13x + 3 = 0$

9. $8x^2 - 3 = 2x$

10. $4x^2 - 15 = 4x$

11. $8x^2 = 11x - 3$

12. $5x^2 = 11x - 2$

13. $x^2 + \dfrac{5}{3}x = \dfrac{2}{3}x$

14. $x^2 - \dfrac{5}{2}x = \dfrac{x}{2}$

15. $25x^2 + 10x + 1 = 0$

16. $36x^2 - 12x + 1 = 0$

Find all the roots and check your answers.

17. $x^3 + 7x^2 + 12x = 0$

18. $x^3 + 7x^2 + 10x = 0$

19. $\dfrac{x^3}{6} - 8x = \dfrac{x^2}{3}$

20. $\dfrac{x^3}{24} - \dfrac{x^2}{12} = x$

21. $3x^3 - 10x = 17x$

22. $5x^3 - 8x = 12x$

23. $2x^3 + 4x^2 = 30x$

24. $3x^3 + 12x^2 = 36x$

25. $\dfrac{7x^2 - 3}{2} = 2x$

26. $\dfrac{3x^2 + 3x}{2} = \dfrac{2}{3}$

27. $2(x + 3) = -3x + 2(x^2 - 3)$

28. $2(x^2 - 4) - 3x = 4x - 11$

29. $7x^2 + 6 = 2x^2 + 2(4x + 3)$

30. $11x^2 - 3x + 1 = 2(x^2 - 5x) + 1$

To Think About

31. The equation $2x^2 - 3x + c = 0$ has a solution of $-\frac{1}{2}$. What is the value of c? What is the other solution to the equation?

32. The equation $x^2 + bx - 12 = 0$ has a solution of -4. What is the value of b? What is the other solution to the equation?

Solve the following applied exercises.

▲ **33.** *Warning Road Sign* An orange triangular warning sign by the side of the road has an area of 180 square inches. The base of the sign is 2 inches longer than the altitude. Find the measurements of the base and altitude.

▲ **34.** *Baseball Banner* A triangular Boston Red Sox banner has an area of 150 square inches. The altitude of the triangle is 3 times the base. Find the measurements of the base and altitude.

▲ **35.** *Neon Billboard* The area of a triangular neon billboard advertising the local mall is 104 square feet. The base of the triangle is 2 feet longer than triple the length of the altitude.

   (a) What are the dimensions of the triangular billboard in feet?

   (b) What are the dimensions of the triangular billboard in yards?

▲ **36.** *Entertainment Platform* During halftime at the Super Bowl, one of the performers will sing on a triangular platform that measures 119 square yards. The base of the triangular stage is 6 yards longer than four times the length of the altitude. What are the dimensions of the triangular stage?

▲ **37.** *Desk Telephone* The area of the base of a rectangular desk telephone is 896 square centimeters. The length of the rectangular telephone is 4 centimeters longer than its width.

   (a) What are the length and width, in centimeters, of the desk telephone?

   (b) What are the length and width, in millimeters, of the desk telephone?

▲ **38.** *Mouse Pad* The area of a rectangular mouse pad is 480 square centimeters. Its length is 16 centimeters shorter than twice its width.

   (a) What are the length and width of the mouse pad in centimeters?

   (b) What are the length and width of the mouse pad in millimeters?

▲ **39.** *Turkish Rug* A rare square Turkish rug, belonging to the family of President John F. Kennedy, was auctioned off. The area of the Turkish rug in square feet is 96 more than its perimeter in feet. Find the length in feet of the side.

▲ **40.** *Circus* The backstage dressing room of the most famous circus in the world is in the shape of a square. The area of the square dressing room in square feet is 165 more than its perimeter in feet. Find the length of the side.

▲ **41.** *Cereal Box* The volume of a rectangular solid can be written as $V = LWH$, where L is the length of the solid, W is the width, and H is the height. A box of cereal has a width of 2 inches. Its height is 2 inches longer than its length. If the volume of the box is 198 cubic inches, what are the length and height of the box?

▲ **42.** *College Catalog* The volume of a rectangular solid can be written as $V = LWH$, where L is the length of the solid, W is the width, and H is the height. The North Shore Community College catalog is $\frac{1}{2}$ inch wide. Its height is 3 inches longer than its length. The volume of the catalog is 27 cubic inches. What are the height and length of the catalog?

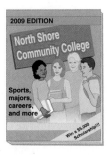

▲ **43.** *Moon Exploration* In planning for the first trip to the moon, NASA surveyed a rectangular area of 54 square miles. The length of the rectangle was 3 miles less than double the width. What were the dimensions of this potential landing area?

▲ **44.** *Vegetable Garden* The Patel family has a square vegetable garden. They increased the size of the garden so that the new, larger square garden is 144 square feet larger than the old garden. The new garden has a side that is 3 feet longer than double the length of the side of the old garden. What are the dimensions of the old and new gardens?

Debit Card Transactions The number N (in billions) of debit card transactions in the United States can be approximated by the equation $N = 3.2x + 9.8$, where x is the number of years since 2000. Use this equation to solve exercises 45–48. (Source: Statistical Abstract of the United States: 2007)

45. Approximately how many debit card transactions were there in the year 2000?

46. Approximately how many debit card transactions were there in the year 2007?

47. In what year were there 29,000,000,000 debit card transactions?

48. In what year will there be 57.8 billion debit card transactions?

Cumulative Review *Simplify. Do not leave negative exponents in your answers.*

49. **[1.4.2]** $(2x^3y^2)^3(5xy^2)^2$

50. **[1.4.3]** $\dfrac{(2a^3b^2)^3}{16a^5b^8}$

51. **[4.1.3]** Solve the system.
$x - 2y = 8$
$x + y = -1$

52. **[3.3.2]** Find an equation of the line passing through $(5, 2)$ and $(6, 1)$.

Quick Quiz 5.8 Solve the following equations.

1. $x^2 = -4x + 32$

2. $15x^2 - 11x + 2 = 0$

3. The area of a triangular field is 160 square yards. The altitude of the triangle is 4 yards shorter than the base of the triangle. Find the altitude and the base of the triangle.

4. **Concept Check** Explain how you would solve the following equation.
$$\frac{3}{8}x^2 + \frac{1}{4}x = 1$$

Putting Your Skills to Work: Use Math to Save Money

EXPENSES THAT COST MORE THAN YOU THINK

Do you find yourself running out of money each month? Do you wish you could find a little extra cash for yourself? Is there some habit that costs money that you might able give up? How about that cup of special coffee in the morning or in the afternoon? Now consider the story of Tracy and Max.

Tracy and Max liked their special blended coffee; however, they really wanted to save for an exciting vacation this year on their third wedding anniversary. Tracy and Max decided they would try making their own coffee. It was pretty tedious, and at first they just focused on the additional time and effort it took. But then they noticed they were having more money left over at the end of the month. They decided to estimate how much they were saving each month. Tracy had

coffee once a day, and the special blend she bought at Donna's Coffee Stop cost about $3.00 a cup. Max bought coffee at Donna's Coffee Stop twice a day. He bought several different types of special blends that cost $4.00 a cup on average.

1. If Tracy and Max estimated their coffee purchases using a 30-day month, find the total amount they both spend on coffee in one month at Donna's Coffee Stop.

2. If Tracy and Max make their own specialty blend of coffee, they estimate that it costs about $1.00 for every 2 cups of coffee. Determine how much they are saving by making their own coffee.

3. Tracy and Max determined that the vacation they would like to take for their anniversary would cost $2250. If they put the money saved by making their own coffee into a vacation account, would there be enough money in eight months to take their vacation?

4. Can you think of one extra expense you could eliminate so you could save money for a vacation? Calculate the savings.

Topic	Procedure	Examples
Adding and subtracting polynomials, p. 249.	Combine like terms to add and subtract polynomials.	$(5x^2 - 6x - 8) + (-2x^2 - 5x + 3) = 3x^2 - 11x - 5$ $(3a^2 - 2ab - 5b^2) - (-7a^2 + 6ab - b^2)$ $= (3a^2 - 2ab - 5b^2) + (7a^2 - 6ab + b^2)$ $= 10a^2 - 8ab - 4b^2$
Multiplying polynomials, p. 250.	**1.** Multiply each term of the first polynomial by each term of the second polynomial. If two binomials are being multiplied, use the FOIL (First, Outer, Inner, Last) method. **2.** Combine like terms.	$2x^2(3x^2 - 5x - 6) = 6x^4 - 10x^3 - 12x^2$ $(3x + 4)(2x - 7) = 6x^2 - 21x + 8x - 28$ $= 6x^2 - 13x - 28$ $(x - 3)(x^2 + 5x + 8)$ $= x^3 + 5x^2 + 8x - 3x^2 - 15x - 24$ $= x^3 + 2x^2 - 7x - 24$
Division of a polynomial by a monomial, p. 255.	**1.** Write the division as the sum of separate fractions. **2.** If possible, reduce the separate fractions.	$(16x^3 - 24x^2 + 56x) \div (-8x)$ $= \dfrac{16x^3}{-8x} + \dfrac{-24x^2}{-8x} + \dfrac{56x}{-8x}$ $= -2x^2 + 3x - 7$
Dividing a polynomial by a binomial or a trinomial, p. 256.	**1.** Write the division exercise in long division form. Write both polynomials in descending order; write any missing terms with a coefficient of zero. **2.** Divide the *first* term of the divisor into the first term of the dividend. The result is the first term of the quotient. **3.** Multiply the first term of the quotient by *every* term in the divisor. **4.** Write this product under the dividend (align like terms) and subtract. **5.** Treat this difference as a new dividend. Repeat steps 2 to 4. Continue until the remainder is zero or is a polynomial of lower degree than the divisor. **6.** If there is a remainder, write it as the numerator of a fraction with the divisor as the denominator. Add this fraction to the quotient.	Divide $(6x^3 + 5x^2 - 2x + 1) \div (3x + 1)$. $\begin{array}{r} 2x^2 + x - 1 \\ 3x+1\overline{)6x^3 + 5x^2 - 2x + 1} \\ \underline{6x^3 + 2x^2} \\ 3x^2 - 2x \\ \underline{3x^2 + x} \\ -3x + 1 \\ \underline{-3x - 1} \\ 2 \end{array}$ The quotient is $2x^2 + x - 1 + \dfrac{2}{3x + 1}$.
Synthetic division, p. 260.	Synthetic division can be used if the divisor is in the form $(x + b)$. **1.** Write the coefficients of the terms in descending order of the dividend. Write any missing terms with a coefficient of zero. **2.** The divisor must be of the form $x + b$. Write down the opposite of b to the left. **3.** Bring down the first coefficient to the bottom row. **4.** Multiply the coefficient in the bottom row by the opposite of b. Write the result in the second row under the next coefficient in the top row. **5.** Add the values in the top and second rows and place the result in the bottom row. **6.** Repeat steps 4 and 5 until the bottom row is filled.	Divide $(3x^5 - 2x^3 + x^2 - x + 7) \div (x + 2)$. $\begin{array}{r|rrrrrr} -2 & 3 & 0 & -2 & 1 & -1 & 7 \\ & & -6 & 12 & -20 & 38 & -74 \\ \hline & 3 & -6 & 10 & -19 & 37 & \underline{-67} \end{array}$ The quotient is $3x^4 - 6x^3 + 10x^2 - 19x + 37 + \dfrac{-67}{x + 2}$.
Factoring out a common factor, p. 264.	Remove the greatest common factor from each term. Many factoring problems are two steps, of which this is the first.	$5x^3 - 25x^2 - 10x = 5x(x^2 - 5x - 2)$ $20a^3b^2 - 40a^4b^3 + 30a^3b^3 = 10a^3b^2(2 - 4ab + 3b)$
Factoring the difference of two squares, p. 278.	$a^2 - b^2 = (a + b)(a - b)$	$9x^2 - 1 = (3x + 1)(3x - 1)$ $8x^2 - 50 = 2(4x^2 - 25) = 2(2x + 5)(2x - 5)$

Topic	Procedure	Examples
Factoring a perfect square trinomial, p. 279.	$a^2 + 2ab + b^2 = (a + b)^2$ $a^2 - 2ab + b^2 = (a - b)^2$	$16x^2 + 40x + 25 = (4x + 5)^2$ $18x^2 + 120xy + 200y^2 = 2(9x^2 + 60xy + 100y^2)$ $\qquad = 2(3x + 10y)^2$ $4x^2 - 36x + 81 = (2x - 9)^2$ $25a^3 - 10a^2b + ab^2 = a(25a^2 - 10ab + b^2)$ $\qquad = a(5a - b)^2$
Factoring the sum and difference of two cubes, p. 280.	$a^3 + b^3 = (a + b)(a^2 - ab + b^2)$ $a^3 - b^3 = (a - b)(a^2 + ab + b^2)$	$8x^3 + 27 = (2x + 3)(4x^2 - 6x + 9)$ $250x^3 + 2y^3 = 2(125x^3 + y^3)$ $\qquad = 2(5x + y)(25x^2 - 5xy + y^2)$ $27x^3 - 64 = (3x - 4)(9x^2 + 12x + 16)$ $125y^4 - 8y = y(125y^3 - 8)$ $\qquad = y(5y - 2)(25y^2 + 10y + 4)$
Factoring trinomials of the form $x^2 + bx + c$, p. 270.	The factors will be of the form $(x + m)(x + n)$, where $m \cdot n = c$ and $m + n = b$.	$x^2 - 7x + 12 = (x - 4)(x - 3)$ $3x^2 - 36x + 60 = 3(x^2 - 12x + 20)$ $\qquad = 3(x - 2)(x - 10)$ $x^2 + 2x - 15 = (x + 5)(x - 3)$ $2x^2 - 44x - 96 = 2(x^2 - 22x - 48)$ $\qquad = 2(x - 24)(x + 2)$
Factoring trinomials of the form $ax^2 + bx + c$, p. 272.	Use the trial-and-error method or the grouping number method.	$2x^2 + 7x + 3 = (2x + 1)(x + 3)$ $8x^2 - 26x + 6 = 2(4x^2 - 13x + 3)$ $\qquad = 2(4x - 1)(x - 3)$ $7x^2 + 20x - 3 = (7x - 1)(x + 3)$ $5x^3 - 18x^2 - 8x = x(5x^2 - 18x - 8)$ $\qquad = x(5x + 2)(x - 4)$
Factoring by grouping, p. 266.	1. Make sure that the first two terms have a common factor and the last two terms have a common factor; otherwise, rearrange the terms. 2. Factor out the common factor in each group. 3. Factor out the common binomial factor.	$6xy - 8y + 3xw - 4w$ $= 2y(3x - 4) + w(3x - 4)$ $= (3x - 4)(2y + w)$
Solving a quadratic equation by factoring, p. 290.	1. Write the equation in standard form. 2. Factor, if possible. 3. Set each factor equal to 0. 4. Solve each of the resulting equations.	Solve $(x + 3)(x - 2) = 5(x + 3)$. $x^2 - 2x + 3x - 6 = 5x + 15$ $x^2 + x - 6 = 5x + 15$ $x^2 + x - 5x - 6 - 15 = 0$ $x^2 - 4x - 21 = 0$ $(x - 7)(x + 3) = 0$ $x - 7 = 0 \quad \text{or} \quad x + 3 = 0$ $x = 7 \qquad\qquad x = -3$

Chapter 5 Review Problems

Perform the indicated operations.

1. $(x^2 - 3x + 5) + (-2x^2 - 7x + 8)$

2. $(-4x^2y - 7xy + y) + (5x^2y + 2xy - 9y)$

3. $(-6x^2 + 7xy - 3y^2) - (5x^2 - 3xy - 9y^2)$

4. $(-13x^2 + 9x - 14) - (-2x^2 - 6x + 1)$

5. $(5x + 2) - (6 - x) + (2x + 3)$

6. $(4x - 5) - (3x^2 + x) + (x^2 - 2)$

For the polynomial function $p(x) = 3x^3 - 2x^2 - 6x + 1$ find the following.

7. $p(-4)$　　　　　**8.** $p(-1)$　　　　　**9.** $p(3)$

For the polynomial function $g(x) = -x^4 + 2x^3 - x + 5$ find the following.

10. $g(-1)$　　　　　**11.** $g(3)$　　　　　**12.** $g(0)$

For the polynomial function $h(x) = -x^3 - 6x^2 + 12x - 4$ find the following.

13. $h(3)$　　　　　**14.** $h(-2)$　　　　　**15.** $h(0)$

Multiply.

16. $3xy(x^2 - xy + y^2)$　　　　　**17.** $(3x^2 + 1)(2x - 1)$　　　　　**18.** $(5x^2 + 3)^2$

19. $(x - 3)(2x - 5)(x + 2)$　　　　　**20.** $(x^2 - 3x + 1)(-2x^2 + x - 2)$　　　　　**21.** $(3x - 5)(3x^2 + 2x - 4)$

22. $(5ab - 2)(5ab + 2)$　　　　　**23.** $(3a - b^2)(6a - 5b^2)$

Divide.

24. $(25x^3y - 15x^2y - 100xy) \div (-5xy)$　　　　　**25.** $(12x^2 - 5x - 2) \div (3x - 2)$

26. $(2x^3 + x^2 - x + 1) \div (2x + 3)$　　　　　**27.** $(3y^3 - 2y + 5) \div (y - 3)$

28. $(8a^4 + 2a^3 - 10a^2 - a + 3) \div (2a^2 - 1)$　　　　　**29.** $(2x^4 - x^2 + 6x + 3) \div (x - 1)$

30. $(2x^4 - 13x^3 + 16x^2 - 9x + 20) \div (x - 5)$　　　　　**31.** $(3x^4 + 5x^3 - x^2 + x - 2) \div (x + 2)$

Remove the greatest common factor.

32. $15a^2b + 5ab^2 - 10ab$

33. $x^5 - 3x^4 + 2x^2$

34. $12mn - 8m$

Factor using the grouping method.

35. $xy - 6y + 3x - 18$

36. $8x^2y + x^2b + 8y + b$

37. $3ab - 15a - 2b + 10$

Factor the trinomials.

38. $x^2 - 9x - 22$

39. $4x^2 - 5x - 6$

40. $6x^2 + 5x - 21$

Factor using one of the formulas from Section 5.6.

41. $100x^2 - 49$

42. $4x^2 - 28x + 49$

43. $8a^3 - 27$

Mixed Practice *Factor, if possible. Be sure to factor completely.*

44. $9x^2 - 121$

45. $5x^2 - 11x + 2$

46. $x^3 + 8x^2 + 12x$

47. $x^2 - 8wy + 4wx - 2xy$

48. $36x^2 + 25$

49. $2x^2 - 7x - 3$

50. $x^2 + 6xy - 27y^2$

51. $27x^4 - x$

52. $21a^2 + 20ab + 4b^2$

53. $-3a^3b^3 + 2a^2b^4 - a^2b^3$

54. $a^4b^4 + a^3b^4 - 6a^2b^4$

55. $3x^4 - 5x^2 - 2$

56. $9a^2b + 15ab - 14b$

57. $2x^2 + 7x - 6$

58. $3x^2 + 5x + 4$

59. $4y^4 - 13y^3 + 9y^2$

60. $y^4 + 2y^3 - 35y^2$

61. $10x^2y^2 - 20x^2y + 5x^2$

62. $4x^4 + 4x^2 - 15$

63. $a^2 + 5ab^3 + 4b^6$

64. $3x^2 - 12 - 8x + 2x^3$

65. $2x^4 - 12x^2 - 54$

66. $8a + 8b - 4bx - 4ax$

67. $8x^4 + 34x^2y^2 + 21y^4$

68. $4x^3 + 10x^2 - 6x$ **69.** $2a^2x - 15ax + 7x$ **70.** $9x^4y^2 - 30x^2y + 25$

71. $27x^3y - 3xy$ **72.** $5bx - 28y + 4by - 35x$ **73.** $27abc^2 - 12ab$

Solve the following equations.

74. $5x^2 - 9x - 2 = 0$ **75.** $2x^2 - 11x + 12 = 0$

76. $(6x + 5)(x + 2) = -2$ **77.** $6x^2 = 24x$

78. $3x^2 + 14x + 3 = -1 + 4(x + 1)$ **79.** $x^3 + 7x^2 = -12x$

Use a quadratic equation to solve each of the following exercises.

▲ **80.** *Geometry* The area of a triangle is 75 square meters. The altitude of the triangle is 5 meters longer than the base of the triangle. Find the base and the altitude of the triangle.

▲ **81.** *Ping-Pong Table* A regulation ping-pong table has an area of 45 square feet. The length of the table is one foot less than double the width. Find the dimensions of a regulation ping-pong table.

82. *Manufacturing Profit* The hourly profit in dollars made by a scientific calculator manufacturing plant is given by the equation $P = 3x^2 - 7x - 10$, where x is the number of calculators assembled in 1 hour. Find the number of calculators that should be made in 1 hour if the hourly profit is to be $30.

▲ **83.** *Sound Insulators* A square sound insulator is constructed for a restaurant. It does not provide enough insulation, so a larger square is constructed. The larger square has 24 square yards more insulation. The side of the larger square is 3 yards longer than double the side of the smaller square. Find the dimensions of each square.

Remember to use your Chapter Test Prep Video CD to see the worked-out solutions to the test problems you want to review.

Combine.

1. $(3x^2y - 2xy^2 - 6) + (5 + 2xy^2 - 7x^2y)$

2. $(5a^2 - 3) - (2 + 5a) - (4a - 3)$

Multiply.

3. $-2x(x + 3y - 4)$

4. $(2x - 3y^2)^2$

5. $(x - 2)(2x^2 + x - 1)$

Divide.

6. $(-15x^3 - 12x^2 + 21x) \div (-3x)$

7. $(2x^4 - 7x^3 + 7x^2 - 9x + 10) \div (2x - 5)$

8. $(x^3 - x^2 - 5x + 2) \div (x + 2)$

Use synthetic division to perform this division.

9. $(x^4 + x^3 - x - 3) \div (x + 1)$

Factor, if possible.

10. $121x^2 - 25y^2$

11. $9x^2 + 30xy + 25y^2$

12. $x^3 - 26x^2 + 48x$

13. $4x^3y + 8x^2y^2 + 4x^2y$

14. $x^2 - 6wy + 3xy - 2wx$

15. $2x^2 - 3x + 2$

1. _____

2. _____

3. _____

4. _____

5. _____

6. _____

7. _____

8. _____

9. _____

10. _____

11. _____

12. _____

13. _____

14. _____

15. _____

16. _____

16. $18x^2 + 3x - 15$

17. _____

17. $54a^4 - 16a$

18. _____

18. $9x^5 - 6x^3y + xy^2$

19. _____

19. $3x^4 + 17x^2 + 10$

20. _____

20. $3x - 10ay + 6y - 5ax$

Find the following if $p(x) = -2x^3 - x^2 + 6x - 10.$
21. $p(2)$

21. _____

22. $p(-3)$

22. _____

Solve the following equations.
23. $x^2 = 5x + 14$

23. _____

24. $3x^2 - 11x - 4 = 0$

24. _____

25. $7x^2 + 6x = 8x$

25. _____

Use a quadratic equation to solve the following exercise.

▲ **26.** The area of a triangular road sign is 70 square inches. The altitude of the triangle is 4 inches less than the base of the triangle. Find the altitude and the base of the triangle.

26. _____

Cumulative Test for Chapters 1–5

Approximately one-half of this test covers the content of Chapters 1–4. The remainder covers the content of Chapter 5.

1. What property is illustrated by the equation below?

$$3(5 \cdot 2) = (3 \cdot 5)2$$

2. Evaluate. $\dfrac{2 + 6(-2)}{(2 - 4)^3 + 3}$

3. Simplify. $7x - 3[1 + 2(x - y)]$

4. Solve for x. $5x + 7y = 2$

5. Solve for x. $\dfrac{1}{3}(x - 1) + \dfrac{x}{4} = x - 2$

6. Find the slope of the line passing through $(-2, -3)$ and $(1, 5)$.

7. Graph $y = -\dfrac{2}{3}x + 4$.

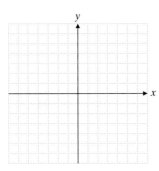

8. Graph the region $3x - 4y \geq -12$.

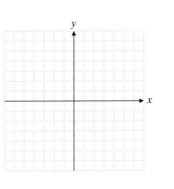

9. Solve the inequality. $-3(x + 2) < 5x - 2(4 + x)$

▲ 10. What are the dimensions of a rectangle with a perimeter of 46 meters if the length is 5 meters longer than twice the width?

11. Evaluate. $x^3 - 2x^2 + 5xy - 6$ for $x = -1, y = 0$

12. Simplify. $a[ab - 2b(a + 4)]$

13. Multiply and simplify your answer.

$(3x + 1)(x^2 - x + 4)$

1. _____

2. _____

3. _____

4. _____

5. _____

6. _____

7. _____

8. _____

9. _____

10. _____

11. _____

12. _____

13. _____

14. _____

15. _____

16. _____

17. _____

18. _____

19. _____

20. _____

21. _____

22. _____

23. _____

24. _____

25. _____

26. _____

27. _____

28. _____

Divide.

14. $(-21x^3 + 14x^2 - 28x) \div (7x)$

15. $(2x^3 - 3x^2 + 3x - 4) \div (x - 2)$

Factor, if possible.

16. $2x^3 - 10x^2$

17. $64x^2 - 49$

18. $3x^2 - 2x - 8$

19. $49x^2 + 42x + 9$

20. $3x^2 - 9x - 54$

21. $2x^2 + 24x + 40$

22. $16x^2 + 9$

23. $4x^3 + 8x^2 - 5x$

24. $27x^4 + 64x$

25. $2x - 6 - 5xy + 15y$

Solve for x.

26. $3x^2 - 4x - 4 = 0$

27. $x^2 - 8x = 33$

▲ **28.** A hospital has paved a triangular parking lot for emergency helicopter landings. The area of the triangle is 68 square meters. The altitude of the triangle is 1 meter longer than double the base of the triangle. Find the altitude and the base of this triangular region.

Watching the snow removal team go to work in an airport in Boston, Chicago, or Denver during a snowstorm is an amazing thing to see. Huge trucks, plows, and snow blowers are all working together in an all-out effort to keep the runways clear. But how do you assign large snow plows and small snow plows to work together most effectively? How long will it take to clear the snow if they all work together? The mathematics you will learn in this chapter allows airport managers to deploy the right equipment and to correctly predict the amount of time needed for snow removal.

Rational Expressions and Equations

6.1 RATIONAL EXPRESSIONS AND FUNCTIONS: SIMPLIFYING, MULTIPLYING, AND DIVIDING   308

6.2 ADDING AND SUBTRACTING RATIONAL EXPRESSIONS   316

6.3 COMPLEX RATIONAL EXPRESSIONS   323

HOW AM I DOING? SECTION 6.1–6.3   329

6.4 RATIONAL EQUATIONS   330

6.5 APPLICATIONS: FORMULAS AND ADVANCED RATIO EXERCISES   335

CHAPTER 6 ORGANIZER   344

CHAPTER 6 REVIEW PROBLEMS   346

HOW AM I DOING? CHAPTER 6 TEST   350

CUMULATIVE TEST FOR CHAPTERS 1–6   351

Student Learning Objectives

After studying this section, you will be able to:

1 Simplify a rational expression or function.

2 Simplify the product of two or more rational expressions.

3 Simplify the quotient of two or more rational expressions.

Graphing Calculator

Finding Domains

Find the domain in Example 1.

Since the domain will be values except those that make the denominator equal to zero, we can graph $y = x^2 + 8x - 20$ in order to identify the values that are not in the domain.

Display:

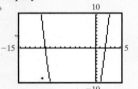

Use the Zoom and Trace features or the zero command to find where $y = 0$.

Try to find the domain of the following function. Round any values where the function is not defined to the nearest tenth.

$$y = \frac{x + 9}{2.1x^2 + 5.2x - 3.1}$$

1 Simplifying a Rational Expression or Function

You may recall from Chapter 1 that a *rational number* is an exact quotient $\frac{a}{b}$ of two integers a and b with $b \neq 0$. A **rational expression** is an expression of the form $\frac{P}{Q}$, where P and Q are polynomials and Q is not zero. For example, $\frac{7}{x + 2}$ and $\frac{x + 5}{x - 3}$ are rational expressions. Since the denominator cannot equal zero, the first expression is undefined if $x + 2 = 0$. This occurs when $x = -2$. We say that x can be any real number except -2. Look at the second expression. When is this expression undefined? Why? For what values of x is the expression defined? Note that the numerator can be zero because any fraction $\frac{0}{a}(a \neq 0)$ is just 0.

A function defined by a rational expression is a **rational function.** The domain of a rational function is the set of values that can be used to replace the variable. Thus, the domain of $f(x) = \frac{7}{x + 2}$ is all real numbers except -2. The domain of $g(x) = \frac{x + 5}{x - 3}$ is all real numbers except 3.

EXAMPLE 1 Find the domain of the function $f(x) = \frac{x - 7}{x^2 + 8x - 20}$.

Solution The domain will be all real numbers except those that make the denominator equal to zero. What value(s) will make the denominator equal to zero? We determine this by solving the equation $x^2 + 8x - 20 = 0$.

$$(x + 10)(x - 2) = 0 \quad \text{Factor.}$$
$$x + 10 = 0 \quad \text{or} \quad x - 2 = 0 \quad \text{Use the zero factor property.}$$
$$x = -10 \qquad x = 2 \quad \text{Solve for } x.$$

The domain of $y = f(x)$ is all real numbers except -10 and 2.

Practice Problem 1 Find the domain of $f(x) = \frac{4x + 3}{x^2 - 9x - 22}$.

We have learned that we can simplify fractions (or reduce them to lowest terms) by factoring the numerator and denominator into prime factors and dividing by the common factors. For example,

$$\frac{15}{25} = \frac{3 \cdot \cancel{5}}{5 \cdot \cancel{5}} = \frac{3}{5}.$$

We can do this by using the **basic rule of fractions.**

What we are doing is factoring out a common factor of 1 ($\frac{c}{c} = 1$). We have

$$\frac{ac}{bc} = \frac{a}{b} \cdot \frac{c}{c} = \frac{a}{b} \cdot 1 = \frac{a}{b}.$$

Note that c must be a factor of the numerator *and* the denominator. Thus, the basic rule of fractions simply says that we may divide out a *common* factor from the numerator and the denominator. This nonzero factor can be a number or an algebraic expression.

EXAMPLE 2 Simplify. $\dfrac{2a^2 - ab - b^2}{a^2 - b^2}$

Solution

$$\frac{(2a + b)(a - b)}{(a + b)(a - b)} = \frac{2a + b}{a + b} \cdot 1 = \frac{2a + b}{a + b}$$

As you become more familiar with this basic rule, you won't have to write out every step. We did so here to show the application of the rule. We cannot simplify this fraction any further.

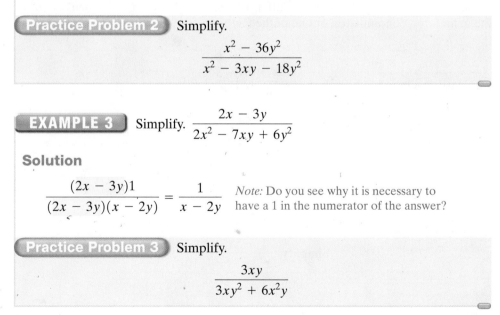

Practice Problem 2 Simplify.

$$\frac{x^2 - 36y^2}{x^2 - 3xy - 18y^2}$$

EXAMPLE 3 Simplify. $\dfrac{2x - 3y}{2x^2 - 7xy + 6y^2}$

Solution

$$\frac{(2x - 3y)1}{(2x - 3y)(x - 2y)} = \frac{1}{x - 2y} \qquad \text{\textit{Note:} Do you see why it is necessary to}$$
$$\text{have a 1 in the numerator of the answer?}$$

Practice Problem 3 Simplify.

$$\frac{3xy}{3xy^2 + 6x^2y}$$

EXAMPLE 4 Simplify. $\dfrac{2x^2 + 2x - 12}{x^3 + 7x^2 + 12x}$

Solution

$$\frac{2x^2 + 2x - 12}{x^3 + 7x^2 + 12x} = \frac{2(x^2 + x - 6)}{x(x^2 + 7x + 12)} = \frac{2(x + 3)(x - 2)}{x(x + 3)(x + 4)} = \frac{2(x - 2)}{x(x + 4)}$$

We usually leave the answer in factored form.

Practice Problem 4 Simplify. $\dfrac{x^3 - 4x^2 - 5x}{3x^2 - 30x + 75}$

Be alert for situations in which one factor in the numerator is the opposite of another factor in the denominator. In such cases you should factor -1 from one of the factors.

EXAMPLE 5 Simplify. $\dfrac{25y^2 - 16x^2}{8x^2 - 14xy + 5y^2}$

Solution

$$\frac{(5y + 4x)(5y - 4x)}{(4x - 5y)(2x - y)} = \frac{(5y + 4x)(5y - 4x)}{-1(-4x + 5y)(2x - y)} = \frac{5y + 4x}{-1(2x - y)} = -\frac{5y + 4x}{2x - y}$$

Observe that $4x - 5y = -1(-4x + 5y)$.

Practice Problem 5 Simplify. $\dfrac{7a^2 - 23ab + 6b^2}{4b^2 - 49a^2}$

Sometimes you will need to factor a different negative number (other than -1) from the numerator, denominator, or one of the factors.

EXAMPLE 6 Simplify. $\dfrac{-2x + 14y}{x^2 - 5xy - 14y^2}$

Solution

$$\frac{-2x + 14y}{x^2 - 5xy - 14y^2} = \frac{-2(x - 7y)}{(x + 2y)(x - 7y)} \qquad \text{Factor } -2 \text{ from each term of the numerator and factor the denominator.}$$

$$= \frac{-2}{x + 2y} \qquad \text{Use the basic rule of fractions.}$$

Practice Problem 6 Simplify. $\dfrac{-3x + 6y}{x^2 - 7xy + 10y^2}$

② Simplifying the Product of Two or More Rational Expressions

Multiplication of rational expressions follows the same rule as multiplication of integer fractions. However, it is particularly helpful to use the basic rule of fractions to simplify whenever possible.

MULTIPLYING RATIONAL EXPRESSIONS

For any polynomials a, b, c, and d,

$$\frac{a}{b} \cdot \frac{c}{d} = \frac{ac}{bd}, \qquad \text{where } b \text{ and } d \neq 0.$$

EXAMPLE 7   Multiply. $\dfrac{2x^2 - 4x}{x^2 - 5x + 6} \cdot \dfrac{x^2 - 9}{2x^4 + 14x^3 + 24x^2}$

Solution   We first use the basic rule of fractions; that is, we factor (if possible) the numerator and denominator and divide out common factors.

$$\frac{2x(x-2)}{(x-2)(x-3)} \cdot \frac{(x+3)(x-3)}{2x^2(x^2 + 7x + 12)} = \frac{2x(x-2)(x+3)(x-3)}{(2x)x(x-2)(x-3)(x+3)(x+4)}$$

$$= \frac{2x}{2x} \cdot \frac{1}{x} \cdot \frac{x-2}{x-2} \cdot \frac{x+3}{x+3} \cdot \frac{x-3}{x-3} \cdot \frac{1}{x+4}$$

$$= 1 \cdot \frac{1}{x} \cdot 1 \cdot 1 \cdot 1 \cdot \frac{1}{x+4}$$

$$= \frac{1}{x(x+4)} \quad \text{or} \quad \frac{1}{x^2 + 4x}$$

Although either form of the answer is correct, we usually use the factored form.

Practice Problem 7   Multiply. $\dfrac{2x^2 + 5x + 2}{4x^2 - 1} \cdot \dfrac{10x^2 + 5x - 5}{x^2 + x - 2}$

EXAMPLE 8   Multiply. $\dfrac{7x + 7y}{4ax + 4ay} \cdot \dfrac{8a^2x^2 - 8b^2x^2}{35ax^3 - 35bx^3}$

Solution $\dfrac{7(x+y)}{4a(x+y)} \cdot \dfrac{8x^2(a^2 - b^2)}{35x^3(a-b)} = \dfrac{7(x+y)}{4a(x+y)} \cdot \dfrac{8x^2(a+b)(a-b)}{35x^3(a-b)}$

$$= \frac{7(x+y)}{4a(x+y)} \cdot \frac{\overset{2}{8}x^2(a+b)(a-b)}{\underset{5}{35}\,\underset{x}{x^3}(a-b)}$$

$$= \frac{2(a+b)}{5ax} \quad \text{or} \quad \frac{2a + 2b}{5ax}$$

Note that we shortened our steps by not writing out every factor of 1 as we did in Example 7. Either way is correct.

Practice Problem 8   Multiply. $\dfrac{9x + 9y}{5ax + 5ay} \cdot \dfrac{10ax^2 - 40x^2b^2}{27ax^2 - 54bx^2}$

③ Simplifying the Quotient of Two or More Rational Expressions

When we divide fractions, we take the **reciprocal** of the second fraction and then multiply the fractions. (Remember that the reciprocal of a fraction $\dfrac{m}{n}$ is $\dfrac{n}{m}$. Thus, the reciprocal of $\dfrac{2}{3}$ is $\dfrac{3}{2}$, and the reciprocal of $\dfrac{3x}{11y^2}$ is $\dfrac{11y^2}{3x}$.) We divide rational expressions in the same way.

DIVIDING RATIONAL EXPRESSIONS

For any polynomials a, b, c, and d,

$$\frac{a}{b} \div \frac{c}{d} = \frac{a}{b} \cdot \frac{d}{c}, \qquad \text{where } b, c, \text{ and } d \neq 0.$$

EXAMPLE 9 Divide. $\dfrac{4x^2 - y^2}{x^2 + 4xy + 4y^2} \div \dfrac{4x - 2y}{3x + 6y}$

Solution We take the reciprocal of the second fraction and multiply the fractions.

$$\frac{4x^2 - y^2}{x^2 + 4xy + 4y^2} \cdot \frac{3x + 6y}{4x - 2y} = \frac{(2x + y)\cancel{(2x - y)}}{\cancel{(x + 2y)}(x + 2y)} \cdot \frac{3\cancel{(x + 2y)}}{2\cancel{(2x - y)}}$$

$$= \frac{3(2x + y)}{2(x + 2y)} \quad \text{or} \quad \frac{6x + 3y}{2x + 4y}$$

Practice Problem 9 Divide.

$$\frac{8x^3 + 27y^3}{64x^3 - y^3} \div \frac{4x^2 - 9y^2}{16x^2 + 4xy + y^2}$$

EXAMPLE 10 Divide. $\dfrac{24 + 10x - 4x^2}{2x^2 + 13x + 15} \div (2x - 8)$

Solution We take the reciprocal of the second fraction and multiply the fractions. The reciprocal of $(2x - 8)$ is $\dfrac{1}{2x - 8}$.

$$\frac{-4x^2 + 10x + 24}{2x^2 + 13x + 15} \cdot \frac{1}{2x - 8} = \frac{-2(2x^2 - 5x - 12)}{(2x + 3)(x + 5)} \cdot \frac{1}{2(x - 4)}$$

$$= \frac{\overset{-1}{\cancel{-2}}\cancel{(x - 4)}\cancel{(2x + 3)}}{\cancel{(2x + 3)}(x + 5)} \cdot \frac{1}{\underset{1}{\cancel{2}}\cancel{(x - 4)}}$$

$$= \frac{-1}{x + 5} \quad \text{or} \quad -\frac{1}{x + 5}$$

NOTE TO STUDENT: Fully worked-out solutions to all of the Practice Problems can be found at the back of the text starting at page SP-1

Practice Problem 10 Divide.

$$\frac{4x^2 - 9}{2x^2 + 11x + 12} \div (-6x + 9)$$

Find the domain of each of the following rational functions.

1. $f(x) = \dfrac{5x + 6}{2x - 6}$

2. $f(x) = \dfrac{3x - 8}{4x + 20}$

3. $g(x) = \dfrac{-3x + 5}{x^2 + 5x - 24}$

4. $g(x) = \dfrac{-2x + 7}{x^2 - x - 20}$

Simplify completely.

5. $\dfrac{-18x^4 y}{12x^2 y^6}$

6. $\dfrac{-7xy^2}{28x^5 y}$

7. $\dfrac{3x^3 - 24x^2}{6x - 48}$

8. $\dfrac{10x^2 + 15x}{35x^2 - 5x}$

9. $\dfrac{9x^2}{12x^2 - 15x}$

10. $\dfrac{20x^2}{28x^2 - 12x}$

11. $\dfrac{5x^2 y^2 - 15xy^2}{10x^3 y - 20x^3 y^2}$

12. $\dfrac{6m^2 n^4 + 2mn^3}{4mn^2 - 8n^2}$

13. $\dfrac{2x + 10}{2x^2 - 50}$

14. $\dfrac{6x^2 - 15x}{3x}$

15. $\dfrac{3y^2 - 27}{3y + 9}$

16. $\dfrac{x - 2}{8x^2 - 32}$

17. $\dfrac{2x^2 - x^3 - x^4}{x^4 - x^3}$

18. $\dfrac{30x - x^2 - x^3}{x^3 - x^2 - 20x}$

$\dfrac{2y^2 + y - 10}{4 - y^2}$

20. $\dfrac{36 - b^2}{3b^2 - 16b - 12}$

$\dfrac{n^5}{3} \cdot \dfrac{9m^3 n^3}{6mn}$

22. $\dfrac{35x^2 y^6}{10x^7 y^2} \cdot \dfrac{-15y^3}{21xy}$

$\dfrac{a^2}{a + 4} \cdot \dfrac{a^2 - 4}{3a}$

24. $\dfrac{8x^2}{x^2 - 9} \cdot \dfrac{x^2 + 6x + 9}{16x^3}$

$\dfrac{+ 7}{+ 6} \cdot \dfrac{3x - 6}{x^2 + 5x + 7}$

26. $\dfrac{x - 5}{10x - 2} \cdot \dfrac{25x^2 - 1}{x^2 - 10x + 25}$

$\dfrac{- 10y^2}{y} \cdot \dfrac{x^2 + 7xy + 6y^2}{x + 2y}$

28. $\dfrac{x - 4y}{x^2 + 3xy - 28y^2} \cdot \dfrac{x^2 + 10xy + 21y^2}{x + 3y}$

29. $\dfrac{y^2 - y - 12}{2y^2 + y - 1} \cdot \dfrac{2y^2 + 7y - 4}{2y^2 - 32}$

30. $\dfrac{6a^2 - a - 1}{a^2 - 3a - 4} \cdot \dfrac{5a^2 - 5}{3a^2 - 2a - 1}$

31. $\dfrac{x^3 - 125}{x^5 y} \cdot \dfrac{x^3 y^2}{x^2 + 5x + 25}$

32. $\dfrac{3a^3 b^2}{8a^3 - b^3} \cdot \dfrac{4a^2 + 2ab + b^2}{12ab^4}$

Divide.

33. $\dfrac{2mn - m}{15m^3} \div \dfrac{2n - 1}{3m^2}$

34. $\dfrac{3y + 12}{8y^3} \div \dfrac{9y + 36}{16y^3}$

35. $\dfrac{b^2 - 6b + 9}{5b^2 - 16b + 3} \div \dfrac{6b - 3}{15b - 3}$

36. $\dfrac{a^2 - 3a}{a^2 - 9} \div \dfrac{a^3 - a^2}{4a^2 - a - 3}$

37. $\dfrac{x^2 - xy - 6y^2}{x^2 + 2} \div (x^2 + 2xy)$

38. $\dfrac{x^2 - 5x + 4}{2x - 8} \div (3x^2 - 3x)$

Mixed Practice *Perform the indicated operation. When no operation is indicated, simplify the rational expression completely.*

39. $\dfrac{7x}{y^2} \div 21x^3$

40. $\dfrac{10m^4}{9a^2} \cdot 3a^2 b$

41. $\dfrac{5x^2 - 2x}{15x - 6}$

42. $\dfrac{3x^2 - 5x}{18x - 30}$

43. $\dfrac{x^2 y - 49y}{x^2 y^3} \cdot \dfrac{3x^2 y - 21xy}{x^2 - 14x + 49}$

44. $\dfrac{x^2 + 6x + 9}{2x^2 y - 18y} \div \dfrac{6xy + 18y}{3x^2 y - 27y}$

45. $\dfrac{x^2 - 9x + 14}{x^3 y^4} \div \dfrac{x^2 - 6x - 7}{x^2 y^2}$

46. $\dfrac{x^2 - 10x + 16}{x^5 y} \div \dfrac{x^2 - 6x - 16}{x^2 y^5}$

47. $\dfrac{a^2 - a - 12}{2a^2 + 5a - 12}$

48. $\dfrac{3y^2 - 3y - 36}{2y^2 - y - 3}$

Optional Graphing Calculator Problems *Find the domain of the following functions. Round any values that you exclude to the nearest tenth.*

49. $f(x) = \dfrac{2x + 5}{3.6x^2 + 1.8x - 4.3}$

50. $f(x) = \dfrac{5x - 4}{1.6x^2 - 1.3x - 5.9}$

Applications

Tropical Fish *The total number of fish will depend on the size of a pond or aquarium. At the Reading Mandarin Restaurant a total of 90 tropical fish were placed in an aquarium. For the next 12 months the total number of fish could be predicted by the equation*

$$P(x) = \frac{90(1 + 1.5x)}{1 + 0.5x},$$

where P is the number of fish and x is the number of months since the fish were placed in the aquarium.

51. One month after the fish are first placed in the aquarium, what is the total number of fish?

52. Three months after the fish are first placed in the aquarium, what is the total number of fish?

53. Six months after the fish are first placed in the aquarium, what is the total number of fish?

54. One year after the fish are first placed in the aquarium, what is the total number of fish? Round your answer to the nearest whole number.

55. Using your answers for exercises 51 to 54, graph the function $P(x)$.

56. Based on your graph, what would you estimate to be the total number of fish eleven months after the fish are first placed in the aquarium?

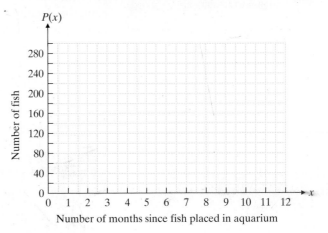

Cumulative Review

57. **[4.1.3]** Solve the system. $2x - y = 8$
$$x + 3y = 4$$

58. **[2.6.3]** Solve. $3x - (x + 5) < 7x + 10$

59. **[1.2.2]** ***Motorcyclist Deaths*** Between 1996 and 2006, the number of annual deaths from motorcycle accidents went from 2160 to 4810. (This may have been the result of many states loosening laws requiring riders to wear helmets.) What was the percent increase in the number of deaths between 1996 and 2006? In 2006, 42% of riders killed were not wearing helmets. Of the number of people killed in 2006 in motorcycle accidents, how many were not wearing helmets? (*Source:* www.dot.gov)

Quick Quiz 6.1 Simplify.

1. $\dfrac{x^3 + 11x^2 + 30x}{3x^3 + 17x^2 - 6x}$

2. $\dfrac{2x^2 - 128}{x^2 + 16x + 64} \cdot \dfrac{3x^2 + 30x + 48}{x^2 - 6x - 16}$

3. $\dfrac{x^2 + x - 6}{x - 5} \div \dfrac{x^2 + 8x + 15}{15 - 3x}$

4. **Concept Check** Explain how you would simplify the following.

$$\frac{9 - x^2}{x^2 - 7x + 12}$$

1 Finding the LCD of Two or More Rational Expressions

Recall that if we wish to add or subtract fractions, the denominators must be the same. If the denominators are not the same, we use the basic rule of fractions to rewrite one or both fractions so that the denominators are the same. For example,

$$\frac{3}{7} + \frac{2}{7} = \frac{5}{7} \quad \text{and} \quad \frac{2}{3} + \frac{3}{4} = \frac{2}{3} \cdot \frac{4}{4} + \frac{3}{4} \cdot \frac{3}{3} = \frac{8}{12} + \frac{9}{12} = \frac{17}{12}.$$

How did we know that 12 was the least common denominator (LCD)? The least common denominator of two or more fractions is the product of the different prime factors in each denominator. If a factor is repeated, we use the highest power that appears on that factor in the denominators.

$$3 = 3$$
$$4 = 2 \cdot 2$$
$$LCD = 3 \cdot 2 \cdot 2 = 12$$

This same technique is used to add or subtract rational expressions.

HOW TO FIND THE LCD

1. Factor each denominator completely into prime factors.

2. List all the different prime factors.

3. The LCD is the product of these factors, each of which is raised to the highest power that appears on that factor in the denominators.

Now we'll do some sample problems.

EXAMPLE 1 Find the LCD of the rational expressions $\dfrac{7}{x^2 - 4}$ and $\dfrac{2}{x - 2}$.

Solution

Step 1 We factor each denominator completely (into prime factors).

$$x^2 - 4 = (x + 2)(x - 2)$$
$$x - 2 \text{ cannot be factored.}$$

Step 2 We list all the *different* prime factors. The different factors are $x + 2$ and $x - 2$.

Step 3 Since no factor occurs more than once in the denominators, the LCD is the product $(x + 2)(x - 2)$.

Practice Problem 1 Find the LCD of the rational expressions

$$\frac{8}{x^2 - x - 12} \quad \text{and} \quad \frac{3}{x - 4}.$$

EXAMPLE 2 Find the LCD of the rational expressions $\dfrac{7}{12xy^2}$ and $\dfrac{4}{15x^3y}$.

Solution

Step 1 We factor each denominator.

$$12xy^2 = 2 \cdot 2 \cdot 3 \cdot x \cdot y \cdot y$$
$$15x^3y = 3 \cdot 5 \cdot x \cdot x \cdot x \cdot y$$

Step 2 Our LCD will require each of the different factors that appear in either denominator. They are $2, 3, 5, x$, and y.

Step 3 The factor 2 and the factor y each occur twice in $12xy^2$.
The factor x occurs three times in $15x^3y$.
Thus, the LCD $= 2 \cdot 2 \cdot 3 \cdot 5 \cdot x \cdot x \cdot x \cdot y \cdot y$
$= 60x^3y^2$

Practice Problem 2 Find the LCD of $\dfrac{2}{15x^3y^2}$ and $\dfrac{13}{25xy^3}$.

NOTE TO STUDENT: Fully worked-out solutions to all of the Practice Problems can be found at the back of the text starting at page SP-1

② Adding or Subtracting Two or More Rational Expressions

We can add and subtract rational expressions with the same denominator just as we do in arithmetic: We simply add or subtract the numerators.

ADDITION AND SUBTRACTION OF RATIONAL EXPRESSIONS

For any polynomials a, b, and c we have the following:

$$\frac{a}{b} + \frac{c}{b} = \frac{a + c}{b}, \qquad b \neq 0$$

$$\frac{a}{b} - \frac{c}{b} = \frac{a - c}{b}, \qquad b \neq 0.$$

EXAMPLE 3 Subtract. $\dfrac{5x + 2}{(x + 3)(x - 4)} - \dfrac{6x}{(x + 3)(x - 4)}$

Solution $\dfrac{5x + 2}{(x + 3)(x - 4)} - \dfrac{6x}{(x + 3)(x - 4)} = \dfrac{-x + 2}{(x + 3)(x - 4)}$

Practice Problem 3 Subtract.
$$\frac{4x}{(x + 6)(2x - 1)} - \frac{3x + 1}{(x + 6)(2x - 1)}$$

If the two rational expressions have different denominators, we first need to find the LCD. Then we must rewrite each fraction as an equivalent fraction that has the LCD as the denominator.

EXAMPLE 4 Add the rational expressions. $\dfrac{7}{(x + 2)(x - 2)} + \dfrac{2}{x - 2}$

Solution The LCD $= (x + 2)(x - 2)$.
Before we can add the fractions, we must rewrite our fractions as fractions with the LCD. The first fraction needs no change, but the second fraction does.

$$\frac{7}{(x + 2)(x - 2)} + \frac{2}{x - 2} \cdot \frac{x + 2}{x + 2}$$

Since $\dfrac{x + 2}{x + 2} = 1$, we have not changed the *value* of the fraction. We are simply writing it in an equivalent form. Thus, we now have the following:

$$\frac{7}{(x + 2)(x - 2)} + \frac{2(x + 2)}{(x + 2)(x - 2)} = \frac{7 + 2(x + 2)}{(x + 2)(x - 2)}$$

$$= \frac{7 + 2x + 4}{(x + 2)(x - 2)} = \frac{2x + 11}{(x + 2)(x - 2)}$$

Practice Problem 4 Add the rational expressions.
$$\frac{8}{(x - 4)(x + 3)} + \frac{3}{x - 4}$$

EXAMPLE 5 Add. $\dfrac{4}{x+6} + \dfrac{5}{6x}$

Solution The LCD $= 6x(x+6)$. We need to multiply the fractions by 1 to obtain two equivalent fractions that have the LCD for the denominator. We multiply the first fraction by $1 = \dfrac{6x}{6x}$. We multiply the second fraction by $1 = \dfrac{x+6}{x+6}$.

$$\dfrac{4(6x)}{(x+6)(6x)} + \dfrac{5(x+6)}{6x(x+6)}$$

$$= \dfrac{24x}{6x(x+6)} + \dfrac{5x+30}{6x(x+6)} = \dfrac{29x+30}{6x(x+6)}$$

Practice Problem 5 Add. $\dfrac{5}{x+4} + \dfrac{3}{4x}$

EXAMPLE 6 Add. $\dfrac{7}{2x^2y} + \dfrac{3}{xy^2}$

Solution You should be able to see that the LCD of these fractions is $2x^2y^2$.

$$\dfrac{7}{2x^2y} \cdot \dfrac{y}{y} + \dfrac{3}{xy^2} \cdot \dfrac{2x}{2x} = \dfrac{7y}{2x^2y^2} + \dfrac{6x}{2x^2y^2} = \dfrac{6x+7y}{2x^2y^2}$$

Practice Problem 6 Add. $\dfrac{7}{4ab^3} + \dfrac{1}{3a^3b^2}$

When two rational expressions are subtracted, we must be very careful with the signs in the numerator of the second fraction.

EXAMPLE 7 Subtract. $\dfrac{2}{x^2+3x+2} - \dfrac{4}{x^2+4x+3}$

Solution $x^2 + 3x + 2 = (x+1)(x+2)$

$x^2 + 4x + 3 = (x+1)(x+3)$

Therefore, the LCD is $(x+1)(x+2)(x+3)$. We now have the following:

$$\dfrac{2}{(x+1)(x+2)} \cdot \dfrac{x+3}{x+3} - \dfrac{4}{(x+1)(x+3)} \cdot \dfrac{x+2}{x+2}$$

$$= \dfrac{2x+6}{(x+1)(x+2)(x+3)} - \dfrac{4x+8}{(x+1)(x+2)(x+3)} = \dfrac{2x+6-4x-8}{(x+1)(x+2)(x+3)}$$

$$= \dfrac{-2x-2}{(x+1)(x+2)(x+3)} = \dfrac{-2\cancel{(x+1)}}{\cancel{(x+1)}(x+2)(x+3)}$$

$$= \dfrac{-2}{(x+2)(x+3)}$$

Study this problem carefully. Be sure you understand the reason for each step. You'll see this type of problem often.

Practice Problem 7 Subtract.

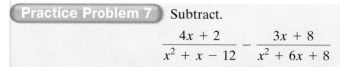

$$\dfrac{4x+2}{x^2+x-12} - \dfrac{3x+8}{x^2+6x+8}$$

The following example involves repeated factors in the denominator. Read it through carefully to be sure you understand each step.

EXAMPLE 8 Subtract. $\dfrac{2x + 1}{25x^2 + 10x + 1} - \dfrac{6x}{25x + 5}$

Solution

Step 1 Factor each denominator into prime factors.

$$25x^2 + 10x + 1 = (5x + 1)^2$$
$$25x + 5 = 5(5x + 1)$$

Step 2 The different factors are 5 and $5x + 1$. However, $5x + 1$ appears to the first power *and* to the second power. So we need step 3.

Step 3 We must use the *highest* power of each factor. In this example the highest power of the factor 5 is 1. The highest power of the factor $5x + 1$ is 2, so we use $(5x + 1)^2$. Thus, the LCD is $5(5x + 1)^2$.

So first we write our problem in factored form.

$$\frac{2x + 1}{(5x + 1)^2} - \frac{6x}{5(5x + 1)}$$

Next, we must multiply our fractions by the appropriate factor to change them to equivalent fractions with the LCD.

$$\frac{2x + 1}{(5x + 1)^2} \cdot \frac{5}{5} - \frac{6x}{5(5x + 1)} \cdot \frac{5x + 1}{5x + 1} = \frac{5(2x + 1) - 6x(5x + 1)}{5(5x + 1)^2}$$

$$= \frac{10x + 5 - 30x^2 - 6x}{5(5x + 1)^2}$$

$$= \frac{-30x^2 + 4x + 5}{5(5x + 1)^2}$$

Practice Problem 8 Subtract.

$$\frac{7x - 3}{4x^2 + 20x + 25} - \frac{3x}{4x + 10}$$

CAUTION: Adding and subtracting rational expressions is somewhat difficult. You should take great care in finding the LCD. Students sometimes make careless errors when finding the LCD.

Likewise, great care should be taken to copy correctly all + and − signs. It is very easy to make a sign error when combining the equivalent fractions. Try to work very neatly and very carefully. A little extra diligence will result in greater accuracy.

Verbal and Writing Skills

1. Explain how to find the LCD of the fractions $\dfrac{3}{5xy}$ and $\dfrac{11}{y^3}$.

2. Explain how to find the LCD of the fractions $\dfrac{8}{7xy}$ and $\dfrac{3}{x^2}$.

Find the LCD.

3. $\dfrac{3}{x-1}, \dfrac{4}{x^2-2x+1}$

4. $\dfrac{5}{x-6}, \dfrac{7}{x^2-12x+36}$

5. $\dfrac{7}{2m^3n}, \dfrac{3}{2mn^2}$

6. $\dfrac{5}{7ab^5}, \dfrac{2}{7a^3b^3}$

7. $\dfrac{3x}{(2x+5)^3}, \dfrac{x-2}{(x+1)(2x+5)^2}$

8. $\dfrac{10x}{(x+5)(x-3)^2}, \dfrac{12xy}{(x-3)^3}$

9. $\dfrac{15xy}{3x^2+2x}, \dfrac{17y}{18x^2+9x-2}$

10. $\dfrac{8x}{3x^2-4x}, \dfrac{10xy}{3x^2+5x-12}$

Add or subtract and simplify your answers.

11. $\dfrac{3}{x+4} + \dfrac{2}{x^2-16}$

12. $\dfrac{7}{x^2-x-6} + \dfrac{5}{x-3}$

13. $\dfrac{9}{4xy} + \dfrac{3}{4y^2}$

14. $\dfrac{5}{6a^2} + \dfrac{1}{6ab}$

15. $\dfrac{3}{x^2-7x+12} + \dfrac{5}{x^2-4x}$

16. $\dfrac{7}{x^2-1} + \dfrac{5}{3x^2+3x}$

17. $\dfrac{6x}{2x-5} + 4$

18. $\dfrac{15}{7a+3} + 5$

19. $\dfrac{-5y}{y^2-1} + \dfrac{6}{y^2-2y+1}$

20. $\dfrac{7y}{y^2+6y+9} + \dfrac{5}{y^2-9}$

Mixed Practice *Add or subtract and simplify your answers.*

21. $\dfrac{a+4}{3a-6} + \dfrac{a-1}{a^2-4}$

22. $\dfrac{4b}{b^2-b-6} + \dfrac{b-1}{3b-9}$

23. $\dfrac{5}{x-4} - \dfrac{3}{x+1}$

24. $\dfrac{7}{x+2} - \dfrac{4}{2x-3}$

25. $\dfrac{1}{x^2-x-2} - \dfrac{3}{x^2+2x+1}$

26. $\dfrac{3x}{x^2+3x-10} - \dfrac{2x}{x^2+x-6}$

27. $\dfrac{4y}{y^2+3y+2} - \dfrac{y-3}{y+2}$

28. $\dfrac{3y^2}{y^2-1} - \dfrac{y+2}{y+1}$

29. $a+2+\dfrac{3}{2a-5}$

30. $a-4+\dfrac{5}{3a+1}$

Applications

31. **Artificial Lung** If an artificial lung company manufactures more than five machines per day, the revenue function in thousands of dollars to manufacture and sell x machines is given by

$$R(x) = \frac{80-24x}{2-x}.$$

The cost function in thousands of dollars to manufacture x machines is given by

$$C(x) = \frac{60-12x}{3-x}.$$

Determine the profit function in thousands of dollars for this company when more than five machines per day are manufactured by obtaining $P(x) = R(x) - C(x)$.

32. **Artificial Heart** If an artificial heart company manufactures more than five machines per day, the revenue function in thousands of dollars to manufacture and sell x machines is given by

$$R(x) = \frac{150-38x}{3-x}.$$

The cost function in thousands of dollars to manufacture x machines is given by

$$C(x) = \frac{120-25x}{2-x}.$$

Determine the profit function in thousands of dollars for this company when more than five machines per day are manufactured by obtaining $P(x) = R(x) - C(x)$.

33. *Artificial Lung* Determine the daily profit of the artificial lung company in exercise 31 if ten machines per day are manufactured. Round your answer to the nearest dollar.

34. *Artificial Heart* Determine the daily profit of the artificial heart company in exercise 32 if twenty machines per day are manufactured. Round your answer to the nearest dollar.

Cumulative Review

35. **[1.3.3]** Evaluate. $\dfrac{-6 \cdot 9 \div |1 - 10|}{4^2 - 5 \cdot 2 + 3\sqrt{25}}$

36. **[1.4.5]** Write using scientific notation. 0.000351

37. **[4.3.2]** *Purchase Price of Automobiles* Jamie, Denise, and Amanda each purchased a car. The total cost of the cars was $24,000. Jamie purchased a car that cost twice as much as the one Amanda purchased. Denise purchased a car that cost $2000 more than the cost of Amanda's car. How much did each car cost?

38. **[2.5.1]** *Chemical Mixtures* A chemist at Argonne Laboratories must combine a mixture that is 15% acid with a mixture that is 30% acid to obtain 60 liters of a mixture that is 20% acid. How much of each kind should he use?

Quick Quiz 6.2 Add or subtract and simplify your answers.

1. $\dfrac{5}{x} - \dfrac{4}{x + 3}$

2. $\dfrac{4}{x^2 + 3x + 2} + \dfrac{3}{x^2 + 6x + 8}$

3. $\dfrac{6x}{3x - 7} + 5$

4. **Concept Check** Explain how you would find the LCD of $\dfrac{2}{x^2 - 16}$ and $\dfrac{3}{x^2 - 6x + 8}$.

1 Simplifying Complex Rational Expressions

A **complex rational expression** is a large fraction that has at least one rational expression in the numerator, in the denominator, or in both the numerator and the denominator. The following are three examples of complex rational expressions.

$$\frac{7 + \dfrac{1}{x}}{x + 2}, \qquad \frac{2}{\dfrac{x}{y} + 3}, \qquad \frac{\dfrac{a + b}{7}}{\dfrac{1}{x} + \dfrac{1}{x + a}}$$

There are two ways to simplify complex rational expressions. You can use whichever method you like.

Student Learning Objective

After studying this section, you will be able to:

1 Simplify complex rational expressions.

EXAMPLE 1 Simplify.
$$\frac{x + \dfrac{1}{x}}{\dfrac{1}{x} + \dfrac{3}{x^2}}$$

Solution

Method 1

1. Simplify numerator and denominator.
$$x + \frac{1}{x} = \frac{x^2 + 1}{x}$$
$$\frac{1}{x} + \frac{3}{x^2} = \frac{x + 3}{x^2}$$

2. Divide the numerator by the denominator.
$$\frac{\dfrac{x^2 + 1}{x}}{\dfrac{x + 3}{x^2}} = \frac{x^2 + 1}{x} \div \frac{x + 3}{x^2}$$
$$= \frac{x^2 + 1}{x} \cdot \frac{x^2}{x + 3}$$
$$= \frac{x^2 + 1}{\cancel{x}} \cdot \frac{\cancel{x^2}^{\,x}}{x + 3}$$
$$= \frac{x(x^2 + 1)}{x + 3}$$

3. The result is already simplified.

Method 2

1. Find the LCD of all the fractions in the numerator and denominator. The LCD is x^2.

2. Multiply the numerator and denominator by the LCD. Use the distributive property.
$$\frac{x + \dfrac{1}{x}}{\dfrac{1}{x} + \dfrac{3}{x^2}} \cdot \frac{x^2}{x^2} = \frac{x^3 + x}{x + 3}$$

3. The result is already simplified, but we will write it in factored form.
$$\frac{x^3 + x}{x + 3} = \frac{x(x^2 + 1)}{x + 3}$$

Practice Problem 1 Simplify.
$$\frac{y + \dfrac{3}{y}}{\dfrac{2}{y^2} + \dfrac{5}{y}}$$

Graphing Calculator

 Exploration

You can verify the answer for Example 1 on a graphing calculator. Graph y_1 and y_2 on the same set of axes.

$$y_1 = \frac{x + \dfrac{1}{x}}{\dfrac{1}{x} + \dfrac{3}{x^2}}$$

$$y_2 = \frac{x(x^2 + 1)}{x + 3}$$

The domain of y_1 is more restrictive than that of y_2. If we use the domain of y_1, then the graphs of y_1 and y_2 should be identical.

Show on your graphing calculator whether y_1 is or is not equivalent to y_2 in the following:

$$y_1 = \frac{1 + \dfrac{3}{x + 2}}{1 + \dfrac{6}{x - 1}}$$

$$y_2 = \frac{x - 1}{x + 2}$$

METHOD 1: COMBINING FRACTIONS IN BOTH NUMERATOR AND DENOMINATOR

1. Simplify the numerator and denominator, if possible, by combining quantities to obtain one fraction in the numerator and one fraction in the denominator.

2. Divide the numerator by the denominator (that is, multiply the numerator by the reciprocal of the denominator).

3. Simplify the expression.

EXAMPLE 2   Simplify. $\dfrac{\dfrac{1}{2x+6}+\dfrac{3}{2}}{\dfrac{3}{x^2-9}+\dfrac{x}{x-3}}$

Solution

Method 1

1. Simplify the numerator.

$$\frac{1}{2x+6}+\frac{3}{2}=\frac{1}{2(x+3)}+\frac{3}{2}$$

$$=\frac{1}{2(x+3)}+\frac{3(x+3)}{2(x+3)}$$

$$=\frac{1+3x+9}{2(x+3)}$$

$$=\frac{3x+10}{2(x+3)}$$

Simplify the denominator.

$$\frac{3}{x^2-9}+\frac{x}{x-3}=\frac{3}{(x+3)(x-3)}+\frac{x}{x-3}$$

$$=\frac{3}{(x+3)(x-3)}+\frac{x(x+3)}{(x+3)(x-3)}$$

$$=\frac{x^2+3x+3}{(x+3)(x-3)}$$

2. Divide the numerator by the denominator.

$$\frac{3x+10}{2(x+3)}\div\frac{x^2+3x+3}{(x+3)(x-3)}=\frac{3x+10}{2\cancel{(x+3)}}\cdot\frac{\cancel{(x+3)}(x-3)}{x^2+3x+3}$$

$$=\frac{(3x+10)(x-3)}{2(x^2+3x+3)}$$

3. Simplify. The answer is already simplified.

Before we continue the example, we state Method 2 in the following box.

> **METHOD 2: MULTIPLYING EACH TERM OF THE NUMERATOR AND DENOMINATOR BY THE LCD OF ALL INDIVIDUAL FRACTIONS**
>
> 1. Find the LCD of all the rational expressions in the numerator and denominator.
> 2. Multiply the numerator and denominator of the complex fraction by the LCD.
> 3. Simplify the result.

We will now proceed to do Example 2 by Method 2.

Method 2

1. To find the LCD, we factor.

$$\frac{\dfrac{1}{2x+6}+\dfrac{3}{2}}{\dfrac{3}{x^2-9}+\dfrac{x}{x-3}} = \frac{\dfrac{1}{2(x+3)}+\dfrac{3}{2}}{\dfrac{3}{(x+3)(x-3)}+\dfrac{x}{x-3}}$$

The LCD of the two fractions in the numerator and the two fractions in the denominator is

$$2(x+3)(x-3).$$

2. Multiply the numerator and the denominator by the LCD.

$$\frac{\dfrac{1}{2(x+3)}+\dfrac{3}{2}}{\dfrac{3}{(x+3)(x-3)}+\dfrac{x}{x-3}} \cdot \frac{2(x+3)(x-3)}{2(x+3)(x-3)}$$

$$= \frac{\dfrac{1}{2\cancel{(x+3)}}\cdot 2\cancel{(x+3)}(x-3) + \dfrac{3}{\cancel{2}}\cdot \cancel{2}(x+3)(x-3)}{\dfrac{3}{\cancel{(x+3)}\cancel{(x-3)}}\cdot 2\cancel{(x+3)}\cancel{(x-3)} + \dfrac{x}{\cancel{x-3}}\cdot 2(x+3)\cancel{(x-3)}}$$

$$= \frac{x-3+3(x+3)(x-3)}{6+2x(x+3)}$$

$$= \frac{3x^2+x-30}{2x^2+6x+6} = \frac{(3x+10)(x-3)}{2(x^2+3x+3)}$$

3. Simplify. The answer is already simplified.

Whether we use Method 1 or Method 2, we can leave the answer in factored form, or we can multiply it out to obtain

$$\frac{3x^2+x-30}{2x^2+6x+6}.$$

Practice Problem 2   Simplify.

$$\frac{\dfrac{4}{16x^2-1}+\dfrac{3}{4x+1}}{\dfrac{x}{4x-1}+\dfrac{5}{4x+1}}$$

EXAMPLE 3 Simplify by Method 1. $\dfrac{x+3}{\dfrac{9}{x}-x}$

Solution

$$\frac{x+3}{\dfrac{9}{x}-\dfrac{x}{1}\cdot\dfrac{x}{x}}=\frac{x+3}{\dfrac{9}{x}-\dfrac{x^2}{x}}=\frac{\dfrac{x+3}{1}}{\dfrac{9-x^2}{x}}=\frac{x+3}{1}\div\frac{9-x^2}{x}=\frac{x+3}{1}\cdot\frac{x}{9-x^2}$$

$$=\frac{\cancel{(x+3)}}{1}\cdot\frac{x}{\cancel{(3+x)}(3-x)}=\frac{x}{3-x}$$

Practice Problem 3 Simplify by Method 1.

$$\frac{4+x}{x-\dfrac{16}{x}}$$

EXAMPLE 4 Simplify by Method 2. $\dfrac{\dfrac{3}{x+2}+\dfrac{1}{x}}{\dfrac{3}{y}-\dfrac{2}{x}}$

Solution The LCD of the numerator is $x(x+2)$. The LCD of the denominator is xy. Thus, the LCD of the complex fraction is $xy(x+2)$.

$$\frac{\dfrac{3}{x+2}+\dfrac{1}{x}}{\dfrac{3}{y}-\dfrac{2}{x}}\cdot\frac{xy(x+2)}{xy(x+2)}=\frac{3xy+xy+2y}{3x(x+2)-2y(x+2)}$$

$$=\frac{4xy+2y}{(x+2)(3x-2y)}$$

$$=\frac{2y(2x+1)}{(x+2)(3x-2y)}$$

Practice Problem 4 Simplify by Method 2.

$$\frac{\dfrac{7}{y+3}-\dfrac{3}{y}}{\dfrac{2}{y}+\dfrac{5}{y+3}}$$

Simplify the complex fractions by any method.

1. $\dfrac{\dfrac{7}{x}}{\dfrac{3}{xy}}$

✓ **2.** $\dfrac{-\dfrac{2}{ab^2}}{\dfrac{5}{a}}$

3. $\dfrac{\dfrac{2x}{x+5}}{\dfrac{x^2}{x-1}}$

✓ **4.** $\dfrac{\dfrac{3y+2}{7y^3}}{\dfrac{y-1}{y^2}}$

5. $\dfrac{1-\dfrac{4}{3y}}{\dfrac{2}{y}+1}$

6. $\dfrac{3-\dfrac{1}{2y}}{\dfrac{1}{4y}+1}$

7. $\dfrac{\dfrac{y}{6}-\dfrac{1}{2y}}{\dfrac{3}{2y}-\dfrac{1}{y}}$

✓ **8.** $\dfrac{\dfrac{1}{3y}+\dfrac{1}{6y}}{\dfrac{1}{2y}+\dfrac{3}{4y}}$

9. $\dfrac{\dfrac{2}{y^2-9}}{\dfrac{3}{y+3}+1}$

✓ **10.** $\dfrac{\dfrac{2}{y+4}}{\dfrac{3}{y-4}-\dfrac{1}{y^2-16}}$

11. $\dfrac{\dfrac{3}{2x+4}+\dfrac{1}{2}}{\dfrac{2}{x^2-4}+\dfrac{x}{x+2}}$

12. $\dfrac{\dfrac{5}{2x+8}+\dfrac{3}{2}}{\dfrac{3}{x^2-16}+\dfrac{1}{x+4}}$

13. $\dfrac{-8}{\dfrac{6x}{x-1}-4}$

14. $\dfrac{6}{2x-\dfrac{10}{x-4}}$

15. $\dfrac{\dfrac{1}{2x+1}+\dfrac{4}{4x^2+4x+1}}{\dfrac{6}{2x^2+x}}$

16. $\dfrac{\dfrac{4}{3x-2}-\dfrac{1}{9x^2-4}}{\dfrac{5x}{3x^2-2x}}$

17. $\dfrac{\dfrac{4}{x+y}+\dfrac{1}{3}}{\dfrac{x}{x+y}-1}$

18. $\dfrac{\dfrac{2}{y+3}+1}{\dfrac{1}{3}-\dfrac{2}{y+3}}$

19. $\dfrac{\dfrac{1}{x-a}-\dfrac{1}{x}}{a}$

20. $\dfrac{\dfrac{1}{x+a}-\dfrac{1}{x}}{a}$

Cumulative Review *Solve for x.*

21. [2.3.1] $|2-3x|=4$

22. [2.3.1] $\left|\dfrac{1}{2}(5-x)\right|=5$

23. [2.8.1] $|7x-3-2x|<6$

24. [2.8.2] $|0.6x+0.3|\geq 1.2$

Quick Quiz 6.3 Simplify.

1. $\dfrac{\dfrac{1}{4x}+\dfrac{1}{2x}}{\dfrac{1}{3y}+\dfrac{5}{6y}}$

2. $\dfrac{\dfrac{3}{x+4}-2}{4-\dfrac{3}{x+4}}$

3. $\dfrac{\dfrac{3}{x+3}-\dfrac{1}{x}}{\dfrac{5}{x^2+3x}}$

4. Concept Check Explain how you would simplify the following.

$$\dfrac{\dfrac{1}{x}+\dfrac{2}{y}}{\dfrac{3}{x}-\dfrac{4}{y}}$$

How are you doing with your homework assignments in Sections 6.1 to 6.3? Do you feel you have mastered the material so far? Do you understand the concepts you have covered? Before you go further in the textbook, take some time to do each of the following problems.

6.1

Simplify.

1. $\dfrac{49x^2 - 9}{7x^2 + 4x - 3}$

2. $\dfrac{x^2 - 4x - 21}{x^2 + x - 56}$

3. $\dfrac{2x^3 - 5x^2 - 3x}{x^3 - 8x^2 + 15x}$

4. $\dfrac{6a - 30}{3a + 3} \cdot \dfrac{9a^2 + a - 8}{2a^2 - 15a + 25}$

5. $\dfrac{5x^3y^2}{x^2y + 10xy^2 + 25y^3} \div \dfrac{2x^4y^5}{3x^3 - 75xy^2}$

6. $\dfrac{8x^3 + 1}{4x^2 + 4x + 1} \cdot \dfrac{6x + 3}{4x^2 - 2x + 1}$

6.2

Add or subtract. Simplify your answers.

7. $\dfrac{x}{2x - 4} - \dfrac{5}{2x}$

8. $\dfrac{2}{x + 5} + \dfrac{3}{x - 5} + \dfrac{7x}{x^2 - 25}$

9. $\dfrac{y - 1}{y^2 - 2y - 8} - \dfrac{y + 2}{y^2 + 6y + 8}$

10. $\dfrac{x + 1}{x + 4} + \dfrac{4 - x^2}{x^2 - 16}$

6.3

Simplify.

11. $\dfrac{\dfrac{1}{12x} + \dfrac{5}{3x}}{\dfrac{2}{3x^2}}$

12. $\dfrac{\dfrac{x}{4x^2 - 1}}{3 - \dfrac{2}{2x + 1}}$

13. $\dfrac{\dfrac{5}{x} + 3}{\dfrac{6}{x} - 2}$

14. $\dfrac{\dfrac{1}{x} + \dfrac{x}{x + 3}}{\dfrac{x + 1}{x + 3} + \dfrac{4}{x}}$

Now turn to page SA-16 for the answers to each of these problems. Each answer also includes a reference to the objective in which the problem is first taught. If you missed any of these problems, you should stop and review the Examples and Practice Problems in the referenced objective. A little review now will help you master the material in the upcoming sections of the text.

1. _____

2. _____

3. _____

4. _____

5. _____

6. _____

7. _____

8. _____

9. _____

10. _____

11. _____

12. _____

13. _____

14. _____

6.4 RATIONAL EQUATIONS

Student Learning Objectives

After studying this section, you will be able to:

1. Solve a rational equation that has a solution and be able to check the solution.

2. Identify those rational equations that have no solution.

1 Solving a Rational Equation

A **rational equation** is an equation that has one or more rational expressions as terms. To solve a rational equation, we find the LCD of all fractions in the equation and multiply each side of the equation by the LCD. We then solve the resulting equation.

EXAMPLE 1 Solve and check your solution. $\dfrac{9}{4} - \dfrac{1}{2x} = \dfrac{4}{x}$

Solution First we multiply each side of the equation by the LCD, which is $4x$.

$$4x\left(\frac{9}{4} - \frac{1}{2x}\right) = 4x\left(\frac{4}{x}\right)$$

$$\cancel{4}x\left(\frac{9}{\cancel{4}}\right) - \overset{2}{\cancel{4x}}\left(\frac{1}{\cancel{2x}}\right) = 4\cancel{x}\left(\frac{4}{\cancel{x}}\right) \quad \text{Use the distributive property.}$$

$$9x - 2 = 16 \quad \text{Simplify.}$$

$$9x = 18 \quad \text{Combine like terms.}$$

$$x = 2 \quad \text{Divide each side by the coefficient of } x.$$

Check:

$$\frac{9}{4} - \frac{1}{2(2)} \overset{?}{=} \frac{4}{2}$$

$$\frac{9}{4} - \frac{1}{4} \overset{?}{=} 2$$

$$\frac{8}{4} \overset{?}{=} 2$$

$$2 = 2 \checkmark$$

Practice Problem 1 Solve and check. $\dfrac{4}{3x} + \dfrac{x+1}{x} = \dfrac{1}{2}$

Usually, we combine the first two steps of the example and show only the step of multiplying each term of the equation by the LCD. We will follow this approach in the remaining examples in this section.

This is another illustration of the need to understand a mathematical principle rather than merely copying down a step without understanding it. Because we understand the distributive property, we can move directly to simplifying a rational equation by multiplying each term of the equation by the LCD.

EXAMPLE 2 Solve and check. $\dfrac{2}{3x+6} = \dfrac{1}{6} - \dfrac{1}{2x+4}$

Solution

$$\frac{2}{3(x+2)} = \frac{1}{6} - \frac{1}{2(x+2)} \quad \text{Factor each denominator.}$$

$$\overset{2}{\cancel{6}}\,\cancel{(x+2)}\left[\frac{2}{\cancel{3(x+2)}}\right] = \cancel{6}(x+2)\left[\frac{1}{\cancel{6}}\right] - \overset{3}{\cancel{6}}\,\cancel{(x+2)}\left[\frac{1}{\cancel{2(x+2)}}\right] \quad \begin{array}{l}\text{Multiply}\\\text{each term}\\\text{by the LCD}\\6(x+2).\end{array}$$

$$4 = x + 2 - 3 \quad \text{Simplify.}$$

$$4 = x - 1 \quad \text{Combine like terms.}$$

$$5 = x \quad \text{Solve for } x.$$

Check: Verify that 5 is the solution.

Practice Problem 2 Solve and check. $\dfrac{1}{3x-9} = \dfrac{1}{2x-6} - \dfrac{5}{6}$

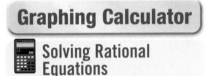

Graphing Calculator

Solving Rational Equations

To solve Example 2 on your graphing calculator, find the point of intersection of

$$y_1 = \frac{2}{3x+6}$$

and $y_2 = \dfrac{1}{6} - \dfrac{1}{2x+4}$.

Use the Zoom and Trace features or the intersection command to find that the solution is 5.00 (to the nearest hundredth). What two difficulties do you observe in the graph that make this exploration more challenging? How can these be overcome?

EXAMPLE 3 Solve. $\dfrac{y^2 - 10}{y^2 - y - 20} = 1 + \dfrac{7}{y - 5}$

Solution

$\dfrac{y^2 - 10}{(y - 5)(y + 4)} = 1 + \dfrac{7}{y - 5}$

Factor each denominator. Multiply each term by the LCD $(y - 5)(y + 4)$.

$\cancel{(y - 5)}\cancel{(y + 4)}\left[\dfrac{y^2 - 10}{\cancel{(y - 5)}\cancel{(y + 4)}}\right] = (y - 5)(y + 4)(1) + \cancel{(y - 5)}(y + 4)\left[\dfrac{7}{\cancel{(y - 5)}}\right]$

$y^2 - 10 = (y - 5)(y + 4)(1) + 7(y + 4)$ Divide out common factors.

$y^2 - 10 = y^2 - y - 20 + 7y + 28$ Simplify.

$y^2 - 10 = y^2 + 6y + 8$ Combine like terms.

$-10 = 6y + 8$ Subtract y^2 from each side.

$-18 = 6y$ Add -8 to each side.

$-3 = y$ Divide each side by the coefficient of y.

Check:

$$\dfrac{(-3)^2 - 10}{(-3)^2 - (-3) - 20} \overset{?}{=} 1 + \dfrac{7}{-3 - 5}$$

$$\dfrac{9 - 10}{9 + 3 - 20} \overset{?}{=} 1 + \dfrac{7}{-8}$$

$$\dfrac{-1}{-8} \overset{?}{=} 1 - \dfrac{7}{8}$$

$$\dfrac{1}{8} = \dfrac{1}{8} \quad \checkmark$$

Practice Problem 3 Solve. $\dfrac{y^2 + 4y - 2}{y^2 - 2y - 8} = 1 + \dfrac{4}{y - 4}$

② Identifying Equations with No Solution

Some rational equations have no solution. This can happen in two distinct ways. In the first case, when you attempt to solve the equation, you obtain a contradiction, such as $0 = 1$. This occurs because the variable "drops out" of the equation. No solution can be obtained. In the second case, we may solve an equation to get an *apparent* solution, but it may not satisfy the original equation. We call the apparent solution an **extraneous solution.** An equation that yields only an extraneous solution has no solution.

Case 1: The Variable Drops Out. In this case, when you attempt to solve the equation, the coefficient of the variable term becomes zero. Thus, you are left with a statement such as $0 = 1$, which is false. In such a case you know that there is no value for the variable that could make $0 = 1$; hence, there cannot be any solution.

EXAMPLE 4 Solve. $\dfrac{z + 1}{z^2 - 3z + 2} + \dfrac{3}{z - 1} = \dfrac{4}{z - 2}$

Solution

$\dfrac{z + 1}{(z - 2)(z - 1)} + \dfrac{3}{z - 1} = \dfrac{4}{z - 2}$ Factor to find the LCD $(z - 2)(z - 1)$. Then multiply each term by the LCD.

$\cancel{(z - 2)}\cancel{(z - 1)}\left[\dfrac{z + 1}{\cancel{(z - 2)}\cancel{(z - 1)}}\right] + (z - 2)\cancel{(z - 1)}\left[\dfrac{3}{\cancel{z - 1}}\right] = \cancel{(z - 2)}(z - 1)\left[\dfrac{4}{\cancel{(z - 2)}}\right]$

$$z + 1 + 3(z - 2) = 4(z - 1) \quad \text{Divide out common factors.}$$

$$z + 1 + 3z - 6 = 4z - 4 \quad \text{Simplify.}$$

$$4z - 5 = 4z - 4 \quad \text{Combine like terms.}$$

$$4z - 4z = -4 + 5 \quad \text{Move variable terms to one side and constant values to the other.}$$

$$0 = 1$$

Of course, $0 \neq 1$. Therefore, no value of z makes the original equation true. Hence, the equation has **no solution.**

Practice Problem 4   Solve. $\dfrac{2x - 1}{x^2 - 7x + 10} + \dfrac{3}{x - 5} = \dfrac{5}{x - 2}$

Case 2: The Obtained Value of the Variable Leads to a Denominator of Zero

EXAMPLE 5   Solve. $\dfrac{4y}{y + 3} - \dfrac{12}{y - 3} = \dfrac{4y^2 + 36}{y^2 - 9}$

Solution

$$\frac{4y}{y + 3} - \frac{12}{y - 3} = \frac{4y^2 + 36}{(y + 3)(y - 3)} \quad \begin{array}{l}\text{Factor each denominator to find the LCD } (y + 3)(y - 3).\\ \text{Multiply each term by the LCD.}\end{array}$$

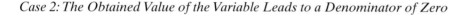

$$4y(y - 3) - 12(y + 3) = 4y^2 + 36 \qquad \text{Divide out common factors.}$$

$$4y^2 - 12y - 12y - 36 = 4y^2 + 36 \qquad \text{Remove parentheses.}$$

$$4y^2 - 24y - 36 = 4y^2 + 36 \qquad \text{Combine like terms.}$$

$$-24y - 36 = 36 \qquad \text{Subtract } 4y^2 \text{ from each side.}$$

$$-24y = 72 \qquad \text{Add 36 to each side.}$$

$$y = \frac{72}{-24} \qquad \text{Divide each side by } -24.$$

$$y = -3$$

Check:
$$\frac{4(-3)}{-3 + 3} - \frac{12}{-3 - 3} \overset{?}{=} \frac{4(-3)^2 + 36}{(-3)^2 - 9}$$

$$\frac{-12}{0} - \frac{12}{-6} \overset{?}{=} \frac{36 + 36}{0}$$

You cannot divide by zero. Division by zero is not defined. A value of a variable that makes a denominator in the original equation zero is not a solution to the equation. Thus, this equation has **no solution.**

TO THINK ABOUT: Quick Solution Checks   Sometimes you may find that you do not have sufficient time for a complete check, but you still wish to make sure that you do not have a "no solution" situation. In those instances you can do a quick analysis to be sure that your obtained value for the variable does not make a denominator zero. If you were solving the equation $\dfrac{4x - 1}{x^2 + 5x - 14} = \dfrac{1}{x - 2} - \dfrac{2}{x + 7}$, you would know immediately that you could not have 2 or -7 as a solution. Do you see why?

Practice Problem 5   Solve and check. $\dfrac{y}{y - 2} - 3 = 1 + \dfrac{2}{y - 2}$

Solve the equations and check your solutions. If there is no solution, say so.

1. $\dfrac{2}{x} + \dfrac{3}{2x} = \dfrac{7}{6}$

2. $\dfrac{1}{x} + \dfrac{2}{3x} = \dfrac{1}{3}$

3. $2 - \dfrac{1}{x} = \dfrac{1}{5x}$

4. $\dfrac{5}{4x} + 2 = \dfrac{1}{x}$

5. $\dfrac{5}{2x + 3} + \dfrac{1}{x} = \dfrac{3}{x}$

6. $\dfrac{4}{3x - 1} + \dfrac{1}{2x} = \dfrac{3}{2x}$

7. $\dfrac{2}{y} = \dfrac{5}{y - 3}$

8. $\dfrac{10}{3x - 8} = \dfrac{2}{x}$

9. $\dfrac{y + 6}{y + 3} - 2 = \dfrac{3}{y + 3}$

10. $4 - \dfrac{8x}{x + 1} = \dfrac{8}{x + 1}$

11. $\dfrac{1}{3x} - \dfrac{2}{x} = \dfrac{-5}{x + 4}$

12. $\dfrac{2}{x} - \dfrac{1}{5x} = \dfrac{3}{2x - 3}$

13. $\dfrac{2x + 3}{x + 3} = \dfrac{2x}{x + 1}$

14. $\dfrac{3x}{x - 2} = \dfrac{3x + 5}{x - 1}$

15. $\dfrac{3}{x^2 - 4} = \dfrac{2}{2x^2 + 4x}$

16. $\dfrac{4}{y^2 + y} = \dfrac{2}{y^2 - 1}$

Mixed Practice *Solve the equations and check your solutions. If there is no solution, say so.*

17. $\dfrac{1}{x^2 + x} + \dfrac{5}{x} = \dfrac{3}{x + 1}$

18. $\dfrac{2}{3x - 1} + \dfrac{2}{3x + 1} = \dfrac{4x}{9x^2 - 1}$

19. $\dfrac{5}{y - 3} + 2 = \dfrac{3}{3y - 9}$

20. $\dfrac{2}{3} + \dfrac{5}{y - 4} = \dfrac{y + 6}{3y - 12}$

21. $1 - \dfrac{10}{z - 3} = \dfrac{-5}{3z - 9}$

22. $\dfrac{3}{2} + \dfrac{2}{2z - 8} = \dfrac{1}{z - 4}$

23. $\dfrac{8}{3x + 2} - \dfrac{7x + 4}{3x^2 + 5x + 2} = \dfrac{2}{x + 1}$

24. $\dfrac{1}{x - 1} - \dfrac{2x + 1}{2x^2 + 5x - 7} = \dfrac{6}{2x + 7}$

25. $\dfrac{4}{z^2 - 9} = \dfrac{2}{z^2 - 3z}$

26. $\dfrac{z^2 + 16}{z^2 - 16} = \dfrac{z}{z + 4} - \dfrac{4}{z - 4}$

27. $\dfrac{3x + 1}{3} + \dfrac{1}{x + 2} = x$

28. $\dfrac{4x - 3}{4} - x = -\dfrac{3}{x + 4}$

Verbal and Writing Skills

29. In what situations will a rational equation have no solution?

30. What does "extraneous solution" mean? What must we do to determine whether a solution is an extraneous solution?

Cumulative Review *Factor completely.*

31. [5.7.1] $7x^2 - 63$

32. [5.7.1] $2x^2 + 20x + 50$

33. [5.7.1] $64x^3 - 27y^3$

34. [5.7.1] $3x^2 - 13x + 14$

Quick Quiz 6.4 Solve.

1. $2 + \dfrac{x}{x + 3} = \dfrac{3x}{x - 3}$

2. $\dfrac{1}{x + 2} - \dfrac{1}{3} = \dfrac{-2}{3x + 6}$

3. $\dfrac{1}{3x - 2} + \dfrac{2x}{x + 1} = 2$

4. **Concept Check** Why does the equation $\dfrac{1}{x - 4} - \dfrac{2}{2x - 8} = \dfrac{3}{2}$ have no solution? Explain how this can be determined.

6.5 APPLICATIONS: FORMULAS AND ADVANCED RATIO EXERCISES

 Solving a Formula for a Particular Variable

In science, economics, business, and mathematics, we use formulas that contain rational expressions. We often have to solve these formulas for a specific variable in terms of the other variables.

Student Learning Objectives

After studying this section, you will be able to:

1 Solve a formula for a particular variable.

2 Solve advanced exercises involving ratio and rate.

EXAMPLE 1 Solve for a. $\dfrac{1}{f} = \dfrac{1}{a} + \dfrac{1}{b}$

Solution This formula is used in optics in the study of light passing through a lens. It relates the focal length f of the lens to the distance a of an object from the lens and the distance b of the image from the lens.

$ab\!\!\!/\!f\left[\dfrac{1}{\not f}\right] = \not abf\left[\dfrac{1}{\not a}\right] + ab\!\!\!/\!f\left[\dfrac{1}{\not b}\right]$ Multiply each term by the LCD abf.

$ab = bf + af$ Simplify.

$ab - af = bf$ Obtain all the terms containing the variable a on one side of the equation.

$a(b - f) = bf$ Factor.

$a = \dfrac{bf}{b - f}$ Divide each side by $b - f$.

Practice Problem 1 Solve for t. $\dfrac{1}{t} = \dfrac{1}{c} + \dfrac{1}{d}$

This formula relates the total amount of time t in hours that is required for two workers to complete a job working together if one worker can complete it alone in c hours and the other worker in d hours.

EXAMPLE 2 The gravitational force F between two masses m_1 and m_2 a distance d apart is represented by the following formula. Solve for m_2.

$$F = \dfrac{Gm_1m_2}{d^2}$$

Solution The subscripts on the variable m mean that m_1 and m_2 are *different*. (The m stands for "mass.")

$F = \dfrac{Gm_1m_2}{d^2}$

$d^2[F] = \not d^{\,2}\left[\dfrac{Gm_1m_2}{\not d^{\,2}}\right]$ Multiply each side by the LCD d^2.

$d^2F = Gm_1m_2$ Simplify.

$\dfrac{d^2F}{Gm_1} = \dfrac{Gm_1m_2}{Gm_1}$ Divide each side by the coefficient of m_2, which is Gm_1.

$\dfrac{d^2F}{Gm_1} = m_2$

Practice Problem 2 The number of telephone calls C between two cities of populations p_1 and p_2 that are a distance d apart may be represented by the formula

$$C = \dfrac{Bp_1p_2}{d^2}.$$

Solve this equation for p_1.

② Solving Advanced Exercises Involving Ratio and Rate

You have already encountered the idea of proportions in a previous course. In that course you learned that a **proportion** is an equation that says that two ratios are equal.

For example, if Wendy's car traveled 180 miles on 7 gallons of gas, how many miles can the car travel on 11 gallons of gas? You probably remember that you can solve this type of exercise quickly if you let x be the number of miles the car can travel. Then to find x you solve the proportion $\dfrac{7}{180} = \dfrac{11}{x}$. Rounded to the nearest mile, the answer is 283 miles. Can you obtain that answer? So far in Chapters 1–5 in this book in the Cumulative Review Exercises, there have been several ratio and proportion exercises for you to solve at this elementary level.

Now we proceed with more advanced exercises in which the ratio of two quantities is more difficult to establish. Study the next two examples carefully. See whether you can follow the reasoning in each case.

EXAMPLE 3 A company plans to employ 910 people with a ratio of two managers for every eleven workers. How many managers should be hired? How many workers?

Solution If we let $x =$ the number of managers, then $910 - x =$ the number of workers. We are given the ratio of managers to workers, so let's set up our proportion in that way.

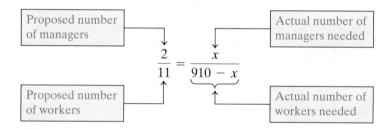

The LCD is $11(910 - x)$. Multiplying by the LCD, we get the following:

$$\cancel{11}(910 - x)\left[\frac{2}{\cancel{11}}\right] = 11\cancel{(910 - x)}\left[\frac{x}{\cancel{910 - x}}\right]$$

$$2(910 - x) = 11x$$
$$1820 - 2x = 11x$$
$$1820 = 13x$$
$$140 = x$$
$$910 - x = 910 - 140 = 770$$

The number of managers needed is 140. The number of workers needed is 770.

NOTE TO STUDENT: Fully worked-out solutions to all of the Practice Problems can be found at the back of the text starting at page SP-1

Practice Problem 3 Western University has 1932 faculty and students. The university always maintains a student-to-faculty ratio of 21:2. How many students are enrolled? How many faculty members are there?

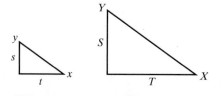

The next example concerns **similar triangles.** These triangles have corresponding angles that are equal and corresponding sides that are *proportional* (not equal). Similar triangles are frequently used to determine distances that cannot be conveniently measured. For example, in the following sketch, x and X are corresponding angles, y and Y are corresponding angles, and s and S are corresponding sides.

Hence, angle x = angle X, angle y = angle Y, and side s is proportional to side S. (Again, note that we did not say that side s is equal to side S.) Also, side t is proportional to side T. So, really, one triangle is just a magnification of the other triangle. If the sides are proportional, they are in the same ratio.

▲ **EXAMPLE 4**   A helicopter is hovering an unknown distance above an 850-foot building. A man watching the helicopter is 500 feet from the base of the building and 11 feet from a flagpole that is 29 feet tall. The man's line of sight to the helicopter is directly above the flagpole, as you can see in the sketch. How far above the building is the helicopter? Round your answer to the nearest foot.

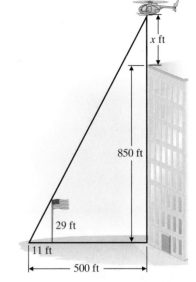

Solution

1. Understand the problem.

Can you see the two triangles in the diagram in the margin? For convenience, we separate them out in the sketch on the right. We want to find the distance x. Are the triangles similar? The angles at the bases of the triangles are equal. (Why?) It follows, then, that the top angles must also be equal. (Remember that the angles of any triangle add up to 180°.) Since the angles are equal, the triangles are similar and the sides are proportional.

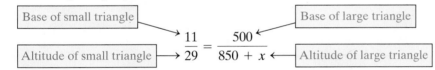

2. Write an equation.

We can set up our proportion like this:

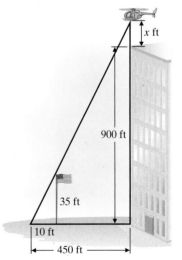

$$\frac{11}{29} = \frac{500}{850 + x}$$

Base of small triangle → 11
Altitude of small triangle → 29

Base of large triangle → 500
Altitude of large triangle → 850 + x

3. Solve the equation and state the answer.

The LCD is $29(850 + x)$.

$$\cancel{29}(850 + x)\left[\frac{11}{\cancel{29}}\right] = 29\cancel{(850 + x)}\left[\frac{500}{\cancel{850 + x}}\right]$$

$$11(850 + x) = 29(500)$$

$$9350 + 11x = 14{,}500$$

$$11x = 5150$$

$$x = \frac{5150}{11} = 468.\overline{18}$$

So the helicopter is about 468 feet above the building.

▲ **Practice Problem 4**   Solve the exercise in Example 4 for a man watching 450 feet from the base of a 900-foot building as shown in the figure in the margin. The flagpole is 35 feet tall, and the man is 10 feet from the flagpole.

We will see some challenging exercises involving similar triangles in exercises 51 and 52.

We will sometimes encounter exercises in which two or more people or machines are working together to complete a certain task. These types of exercises are sometimes called *work problems*. In general, these types of exercises can be analyzed by using the following concept.

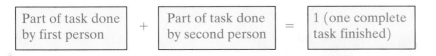

Part of task done by first person $+$ Part of task done by second person $=$ 1 (one complete task finished)

We will also use a general idea about the rate at which something is done. If Robert can do a task in 3 hours, then he can do $\frac{1}{3}$ of the task in 1 hour. If Susan can do the same task in 2 hours, then she can do $\frac{1}{2}$ of the task in 1 hour. In general, if a person can do a task in t hours, then that person can do $\frac{1}{t}$ of the task in 1 hour.

EXAMPLE 5 Robert can paint the kitchen in 3 hours. Susan can paint the kitchen in 2 hours. How long will it take Robert and Susan to paint the kitchen if they work together?

Solution

1. Understand the problem.

Robert can paint $\frac{1}{3}$ of the kitchen in 1 hour.

Susan can paint $\frac{1}{2}$ of the kitchen in 1 hour.

We do not know how long it will take them working together, so we let $x =$ the number of hours it takes them to paint the kitchen working together. To assist us, we will construct a table that relates the data. We will use the concept that (rate)(time) = fraction of task done.

	Rate of Work per Hour	Time Worked in Hours	Fraction of Task Done
Robert	$\frac{1}{3}$	x	$\frac{x}{3}$
Susan	$\frac{1}{2}$	x	$\frac{x}{2}$

2. Write an equation.

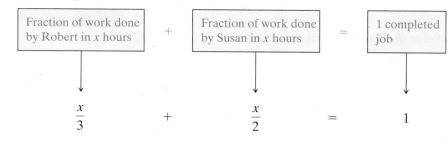

Fraction of work done by Robert in x hours	+	Fraction of work done by Susan in x hours	=	1 completed job
$\frac{x}{3}$	+	$\frac{x}{2}$	=	1

3. Solve the equation and state the answer.

Multiply each side of the equation by the LCD and use the distributive property.

$$6\left(\frac{x}{3}\right) + 6\left(\frac{x}{2}\right) = 6(1)$$

$$2x + 3x = 6$$

$$5x = 6$$

$$x = \frac{6}{5} = 1\frac{1}{5} = 1.2 \text{ hours}$$

Working together, Robert and Susan can paint the kitchen in 1.2 hours.

NOTE TO STUDENT: *Fully worked-out solutions to all of the Practice Problems can be found at the back of the text starting at page SP-1*

Practice Problem 5 Alfred can mow the huge lawn at his Vermont farm with his new lawn mower in 4 hours. His young son can mow the lawn with the old lawn mower in 5 hours. If they work together, how long will it take them to mow the lawn?

Solve for the indicated variable.

1. $x = \dfrac{y - b}{m}$; for m

2. $m = \dfrac{y - b}{x}$; for x

3. $\dfrac{1}{f} = \dfrac{1}{a} + \dfrac{1}{b}$; for b

4. $\dfrac{1}{f} = \dfrac{1}{a} + \dfrac{1}{b}$; for f

5. $\dfrac{V}{lh} = w$; for h

6. $r = \dfrac{I}{pt}$; for p

7. $G = \dfrac{ab + ac}{3}$; for a

8. $H = \dfrac{xy + xz}{4}$; for x

9. $\dfrac{r^3}{V} = \dfrac{3}{4\pi}$; for V

10. $\dfrac{V}{\pi r^2} = h$; for r^2

11. $\dfrac{E}{e} = \dfrac{R + r}{r}$; for e

12. $\dfrac{E}{e} = \dfrac{R + r}{r}$; for r

13. $\dfrac{QR_1}{S_1} = \dfrac{QR_2}{S_2}$; for S_1

14. $\dfrac{QR_1}{S_1} = \dfrac{QR_2}{S_2}$; for S_2

15. $\dfrac{S - 2lw}{2w + 2l} = h$; for w

16. $\dfrac{S - 2lw}{2w + 2l} = h$; for l

Mixed Practice *Solve for the indicated variable.*

17. $E = T_1 - \dfrac{T_1}{T_2}$; for T_1

18. $E = T_1 - \dfrac{T_1}{T_2}$; for T_2

19. $m = \dfrac{y_2 - y_1}{x_2 - x_1}$; for x_1

20. $m = \dfrac{y_2 - y_1}{x_2 - x_1}$; for x_2

21. $\dfrac{2D - at^2}{2t} = V$; for D

22. $\dfrac{2D - at^2}{2t} = V$; for a

23. $Q = \dfrac{kA(t_1 - t_2)}{L}$; for t_2

24. $Q = \dfrac{kA(t_1 - t_2)}{L}$; for A

25. $\dfrac{T_2 W}{T_2 - T_1} = q$; for T_2

26. $d = \dfrac{LR_2}{R_2 + R_1}$; for R_1

27. $\dfrac{s - s_0}{v_0 + gt} = t$; for v_0

28. $\dfrac{A - P}{Pr} = t$; for P

Round your answers to four decimal places.

29. Solve for T. $\dfrac{1.98V}{1.96V_0} = 0.983 + 5.936(T - T_0)$

30. Solve for r_1. $\dfrac{1}{R} = \dfrac{1}{r_1} + \dfrac{1}{0.368} + \dfrac{1}{0.736}$

Applications

31. Map Scale On a map of Nigeria, the cities of Benin City and Onitsha are 6.5 centimeters apart. The map scale shows that 3 centimeters = 55 kilometers on the map. How far apart are the two cities?

32. Scale Drawing The lobby of the Fine Arts building at Normandale Community College was recently renovated. The scale drawing is 12 inches wide and 18 inches long. The blueprint has a scale of 3:50. Find the length and width of the lobby measured in inches. Find the length and width of the lobby measured in feet.

33. Speed Conversion A speed of 60 miles per hour (mph) is equivalent to 88 kilometers per hour (km/h). On a trip through Nova Scotia, Lora passes a sign stating that the speed limit is 80 km/h. What is the speed limit in miles per hour?

34. Currency Conversion Miguel traveled to Italy on business for a week and will now travel to Australia for a week. He wants to convert the 80 euros he now has to Australian dollars. 0.75 euros are currently equivalent to 1.2 Australian dollars. How many Australian dollars will he receive for his 80 euros?

In exercises 35–50, round your answers to the nearest hundredth, unless otherwise directed.

35. Grizzly Bear Population Thirty-five grizzly bears were captured and tagged by wildlife personnel in the Yukon and then released back into the wild. Later that same year, fifty were captured, and twenty-two had tags. Estimate the number of grizzly bears in that part of the Yukon. (Round this answer to the nearest whole number.)

36. Alligator Population Thirty alligators were captured in the Louisiana bayou (swamp) and tagged by National Park officials and then put back into the bayou. One month later, fifty alligators were captured from the same part of the bayou, eighteen of which had been tagged. Estimate to the nearest whole number how many alligators are in that part of Louisiana.

37. Composition of Russian Navy A ship in the Russian navy has a ratio of two officers for every seven seamen. The crew of the ship totals 117 people. How many officers and how many seamen are on this ship?

38. Camp Counselors A YMCA summer camp employs a total of 180 counselors each summer. The ratio of employees is three new counselors for every seven experienced counselors. How many experienced counselors are employed at the camp? How many new counselors are employed at the camp?

39. Cell Phone Employees Southwestern Cell Phones has a total of 187 employees in the marketing and sales offices in Dallas. For every four people employed in marketing, there are thirteen people employed in sales. How many people in the Dallas office work in marketing? How many people in the Dallas office work in sales?

40. Education At Elmwood University there are nine men for every five women on the faculty. If Elmwood University employs 182 faculty members, how many are men and how many are women?

▲ **41.** *Picture Framing* Rose is carefully framing an 8-inch × 10-inch family photo so that the length and width of the frame have the same ratio as the photo. She is using a piece of frame 54 inches in length. What are the length and width of the frame?

▲ **42.** *Enlarging Photographs* Becky DeWitt is a photographer at Photographics. She wants to enlarge a photograph that measures 3 inches wide and 5 inches long to an oversize photograph with the same width-to-length ratio. The perimeter of the new photograph is 115.2 inches. What are the length and the width of the new oversize photograph?

43. *E-Mail Software* A college recently installed new e-mail software on all of its computers. A poll was taken to find out how many of the faculty and staff preferred the new software. The ratio of those preferring the new software to the old software was 3:11. If there are 280 faculty and staff at the college, how many prefer the new software?

44. *Hospital Staff* The ratio of doctors to nurses at Carson City Hospital is 2:5. The hospital has a total of 119 doctors and nurses. How many are doctors? How many are nurses?

▲ **45.** *Shadow Length* A 12-foot marble statue in Italy casts a shadow that is 15 feet long. At the same time of day, a wall casts a shadow that is 8 feet long. How high is the wall?

▲ **46.** *Shadow Length* A 3.5-foot-tall child casts a shadow that is 6 feet long. At the same time of day, a tree casts a shadow that is 42 feet long. How tall is the tree?

47. *Snow Plowing* Matt can plow out all the driveways on his street in Duluth, Minnesota, with his new four-wheel-drive truck in 6 hours. Using a snow blade on a lawn tractor, his neighbor can plow out the same number of driveways in 9 hours. How long would it take them to do the work together?

48. *Mowing Lawns* The new employee at Pete's Yard Service can mow four yards in 5 hours. Pete, the owner, can do the job in 3 hours. How long would it take them to do the work together?

49. *Swimming Pool* Houghton College has just built a new Olympic-size swimming pool. One large pipe with cold water and one small pipe with hot water are used to fill the pool. When both pipes are used, the pool can be filled in 2 hours. If only the cold water pipe is used, the pool can be filled in 3 hours. How long would it take to fill the pool with just the hot water pipe?

50. *Splitting Firewood* A lumberjack and his cousin Fred can split a cord of seasoned oak firewood in 3 hours. The lumberjack can split the wood alone without any help from Fred in 4 hours. How long would it take Fred if he worked without the lumberjack?

To Think About

River Width *To find the width of a river, a hiking club laid out a triangular pattern of measurements. See the figure. Use your knowledge of similar triangles to solve exercises 51 and 52.*

▲ **51.** If any observer stands at the point shown in the figure, then $a = 2$ feet, $b = 5$ feet, and $c = 116$ feet. What is the width of the river?

▲ **52.** What is the width of the river if $a = 3$ feet, $b = 8$ feet, and $c = 297$ feet?

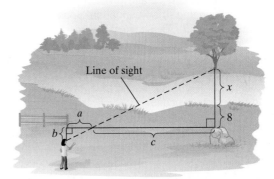

Cumulative Review

53. **[3.3.1]** Find the slope and the y-intercept of the line: $7x - 3y = 8$

54. **[3.2.1]** Find the slope of the line passing through $(-1, 2.5)$ and $(1.5, 2)$

55. **[3.3.3]** Find the equation in slope-intercept form of the line that is perpendicular to $y = \frac{1}{3}x - 4$ passing through $(0, 6)$.

56. **[4.1.6]** Solve the system. $x - 5y = -9$
$$3x + 2y = -10$$

Quick Quiz 6.5

1. Solve for H. $\dfrac{3W + 2AH}{4W - 8H} = A$

2. For every 3 students who were accepted for admission to Springfield University, there were 22 students who were not accepted. If 2400 students applied, how many were accepted for admission?

3. A window at Don Work's house is 22 inches wide and 35 inches tall. He took that window out and replaced it with a larger window that has a perimeter of 342 inches. The new window has the same proportional ratio of width to length as the old window. What is the width and what is the length of the new window?

4. **Concept Check** Explain how you would solve for H in the following equation.
$$S = \frac{2A + 3F}{B - H}$$

Developing Your Study Skills

Why Is Review Necessary?

You master a course in mathematics by learning the concepts one at a time. There are basic concepts like addition, subtraction, multiplication, and division of whole numbers, which are considered the foundation upon which all of mathematics is built. These must be mastered first. The study of mathematics is built upon this foundation, each concept supporting the next. The process is a carefully designed procedure, and no steps can be skipped. A student of mathematics needs to realize the importance of this building process to succeed.

Because new concepts depend on those previously learned, students often need to take time to review. Reviewing at the right time on the right concepts can strengthen previously learned skills and make progress possible.

Timely, periodic review of previously learned mathematical concepts is absolutely necessary in order to master new concepts. You may have forgotten a concept or grown a bit rusty in applying it. Reviewing is the answer. Make use of any review sections in your textbook, whether they are assigned or not. Look back to previous chapters whenever you have forgotten how to do something. Study the examples and practice some exercises to refresh your understanding.

Be sure that you understand and can perform the computations of each new concept. This will enable you to move successfully on to the next ones.

Remember that mathematics is a step-by-step building process. Learn one concept at a time, skipping none, and reinforce and strengthen skills with review whenever necessary.

Putting Your Skills to Work: Use Math to Save Money

GAS MONEY AND MILES PER GALLON (MPGs)

How many gallons of gas do you purchase for your car in one year? With the sharp increase in the price of gasoline, people are thinking about questions like this more than ever before. Michelle drives a 2003 SUV because she likes the room and comfort it provides. However, it has a gas-guzzling engine that only gets 16 miles per gallon.

Michelle has spent more and more money on gas each month, so she is thinking about buying another car that will get better gas mileage. She is even thinking about a hybrid model (a car with a combined gas and electric engine) so she can get the best possible mileage.

Analyzing the Options

Michelle wants to make a smart choice, so she is going to determine the cost and potential savings she would enjoy from 3 different types of cars:

Option A: A small SUV that gets 21 miles per gallon (MPG) and sells for $21,540

Option B: A hybrid model of the same SUV that gets 32 MPG and sells for $28,150

Option C: The most fuel-efficient hybrid car that gets 60 MPG and sells for $23,770

During her research Michelle learned that in the U.S., cars on average are driven 15,000 miles per year. That is just about how many miles she drives her car in a year. In the following questions, round to the nearest hundredth when necessary.

1. Determine the number of gallons of gasoline that each car (Options A, B, and C) would need if they are driven 15,000 miles in one year.

2. If the average price of gasoline is $4.15 per gallon (July 2008), how much would it cost to fuel each car for 15,000 miles?

Making the Best Choice to Save Money

The hybrid model SUV in Option B would cost Michelle $6610 more to purchase than the SUV in Option A.

3. How much gas money would Michelle save in one year with Option B compared to Option A?

4. How many years would it take for Michelle to save $6610 in gas money with Option B compared to Option A?

5. How much gas money would Michelle save after 5 years with Option B compared to Option A? After 10 years?

The incredibly fuel-efficient hybrid car in Option C would cost Michelle $2230 more to purchase than the SUV in Option A.

6. How much gas money will Michelle save in one year with Option C compared to Option A?

7. How many years would it take for Michelle to save $2230 in gas money with Option C compared to Option A?

8. How much gas money would Michelle save after 5 years with Option C compared to Option A? After 10 years?

Even though Michelle enjoyed the comfort of her SUV, she could not ignore the potential savings in gas money she saw with Option C. She decided it was time to make a smart financial decision instead of one based on comfort. She expected to have her new car for at least 5 years. At that point, Option C will have saved her nearly $10,000 in gas money!

9. Do you drive more than 15,000 miles per year or fewer? How would more miles or fewer miles per year affect this analysis?

10. How much do you pay for a gallon of gas? How would higher or lower prices for a gallon of gas affect this analysis?

Topic	Procedure	Examples
Simplifying rational expressions, p. 308.	**1.** Factor the numerator and denominator, if possible. **2.** Any factor that is common to both numerator and denominator can be divided out. This is an application of the basic rule of fractions. *Basic rule of fractions:* For any polynomials a, b, and c (where $b \neq 0$ and $c \neq 0$), $$\frac{ac}{bc} = \frac{a}{b}.$$	Simplify. $\dfrac{6x^2 - 14x + 4}{6x^2 - 20x + 6}$ $\dfrac{2(3x^2 - 7x + 2)}{2(3x^2 - 10x + 3)} = \dfrac{\cancel{2}\,\cancel{(3x-1)}(x-2)}{\cancel{2}\,\cancel{(3x-1)}(x-3)}$ $= \dfrac{x-2}{x-3}$
Multiplying rational expressions, p. 310.	**1.** Factor all numerators and denominators, if possible. **2.** Any factor that is common to a numerator of one fraction and the denominator of the same fraction or any fraction that is multiplied by it can be divided out. **3.** Write the product of the remaining factors in the numerator. Write the product of the remaining factors in the denominator. $$\frac{a}{b} \cdot \frac{c}{d} = \frac{ac}{bd}$$	Multiply. $\dfrac{x^2 - 4x}{6x - 12} \cdot \dfrac{3x^2 - 6x}{x^3 + 3x^2}$ $\dfrac{\cancel{x}(x-4)}{\underset{2}{\cancel{6}}\cancel{(x-2)}} \cdot \dfrac{\overset{1}{\cancel{3}}\,\cancel{x}\,\cancel{(x-2)}}{x^2(x+3)} = \dfrac{x-4}{2(x+3)}$ or $\dfrac{x-4}{2x+6}$
Dividing rational expressions, p. 311.	**1.** Invert the second fraction and multiply it by the first fraction. $$\frac{a}{b} \div \frac{c}{d} = \frac{a}{b} \cdot \frac{d}{c}$$ **2.** Apply the steps for multiplying rational expressions.	Divide. $\dfrac{6x^2 - 5x - 6}{24x^2 + 13x - 2} \div \dfrac{4x^2 + x - 3}{8x^2 + 7x - 1}$ $\dfrac{6x^2 - 5x - 6}{24x^2 + 13x - 2} \cdot \dfrac{8x^2 + 7x - 1}{4x^2 + x - 3}$ $= \dfrac{\cancel{(3x+2)}(2x-3)}{\cancel{(3x+2)}\cancel{(8x-1)}} \cdot \dfrac{\cancel{(8x-1)}\cancel{(x+1)}}{\cancel{(x+1)}(4x-3)} = \dfrac{2x-3}{4x-3}$
Adding rational expressions, p. 317.	**1.** If all fractions have a common denominator, add the numerators and place the result over the common denominator. $$\frac{a}{c} + \frac{b}{c} = \frac{a+b}{c}$$ **2.** If the fractions do not have a common denominator, factor the denominators (if necessary) and determine the least common denominator (LCD). **3.** Rewrite each fraction as an equivalent fraction with the LCD as the denominator. **4.** Add the numerators and place the result over the common denominator. **5.** Simplify, if possible.	Add. $\dfrac{7x}{x^2 - 9} + \dfrac{x+2}{x+3}$ $\dfrac{7x}{(x+3)(x-3)} + \dfrac{x+2}{x+3}$ LCD $= (x+3)(x-3)$ $\dfrac{7x}{(x+3)(x-3)} + \dfrac{x+2}{x+3} \cdot \dfrac{x-3}{x-3}$ $= \dfrac{7x}{(x+3)(x-3)} + \dfrac{x^2 - x - 6}{(x+3)(x-3)}$ $= \dfrac{x^2 + 6x - 6}{(x+3)(x-3)}$
Subtracting rational expressions, p. 317.	Follow the procedure for adding rational expressions through step 3. Then subtract the second numerator from the first and place the result over the common denominator. $$\frac{a}{c} - \frac{b}{c} = \frac{a-b}{c}$$ Simplify, if possible.	Subtract. $\dfrac{4x}{3x-2} - \dfrac{5x}{x+4}$ LCD $= (3x-2)(x+4)$ $\dfrac{4x}{3x-2} \cdot \dfrac{x+4}{x+4} - \dfrac{5x}{x+4} \cdot \dfrac{3x-2}{3x-2}$ $= \dfrac{4x^2 + 16x}{(3x-2)(x+4)} - \dfrac{15x^2 - 10x}{(x+4)(3x-2)}$ $= \dfrac{-11x^2 + 26x}{(3x-2)(x+4)}$ $= \dfrac{x(-11x + 26)}{(3x-2)(x+4)}$

Topic	Procedure	Examples
Simplifying a complex rational expression by Method 1, p. 324.	1. Simplify the numerator and denominator, if possible, by combining quantities to obtain one fraction in the numerator and one fraction in the denominator. 2. Divide the numerator by the denominator. (That is, multiply the numerator by the reciprocal of the denominator.) 3. Simplify the result.	Simplify by Method 1. $\dfrac{4 - \dfrac{1}{x^2}}{\dfrac{2}{x} + \dfrac{1}{x^2}}$ **Step 1:** $\dfrac{\dfrac{4x^2}{x^2} - \dfrac{1}{x^2}}{\dfrac{2x}{x^2} + \dfrac{1}{x^2}} = \dfrac{\dfrac{4x^2 - 1}{x^2}}{\dfrac{2x + 1}{x^2}}$ **Step 2:** $\dfrac{4x^2 - 1}{x^2} \cdot \dfrac{x^2}{2x + 1}$ **Step 3:** $\dfrac{\cancel{(2x + 1)}(2x - 1)}{\cancel{x^2}} \cdot \dfrac{\cancel{x^2}}{\cancel{2x + 1}} = 2x - 1$
Simplifying a complex rational expression by Method 2, p. 325.	1. Find the LCD of all the rational expressions in the numerator and the denominator. 2. Multiply the numerator and the denominator of the complex fraction by the LCD. 3. Simplify the results.	Simplify by Method 2. $\dfrac{4 - \dfrac{1}{x^2}}{\dfrac{2}{x} - \dfrac{1}{x^2}}$ **Step 1:** The LCD of the rational expressions is x^2. **Step 2:** $\dfrac{\left[4 - \dfrac{1}{x^2}\right]x^2}{\left[\dfrac{2}{x} - \dfrac{1}{x^2}\right]x^2} = \dfrac{4(x^2) - \left(\dfrac{1}{x^2}\right)(x^2)}{\left(\dfrac{2}{x}\right)(x^2) - \left(\dfrac{1}{x^2}\right)(x^2)}$ $= \dfrac{4x^2 - 1}{2x - 1}$ **Step 3:** $\dfrac{(2x + 1)\cancel{(2x - 1)}}{\cancel{(2x - 1)}} = 2x + 1$
Solving rational equations, p. 330.	1. Determine the LCD of all denominators in the equation. 2. Multiply each term in the equation by the LCD. 3. Simplify and remove parentheses. 4. Combine any like terms. 5. Solve for the variable. If the variable term drops out, there is no solution. 6. Check your answer. Be sure that the value you obtained does not make any denominator in the original equation 0. If so, there is no solution.	Solve. $\dfrac{4}{y - 1} + \dfrac{-y + 5}{3y^2 - 4y + 1} = \dfrac{9}{3y - 1}$ $\text{LCD} = (y - 1)(3y - 1)$ $\cancel{(y - 1)}(3y - 1)\left[\dfrac{4}{\cancel{y - 1}}\right]$ $+ \cancel{(y - 1)}\cancel{(3y - 1)}\left[\dfrac{-y + 5}{\cancel{(y - 1)}\cancel{(3y - 1)}}\right]$ $= (y - 1)\cancel{(3y - 1)}\left[\dfrac{9}{\cancel{3y - 1}}\right]$ $4(3y - 1) + (-y) + 5 = 9(y - 1)$ $12y - 4 - y + 5 = 9y - 9$ $11y + 1 = 9y - 9$ $11y - 9y = -9 - 1$ $2y = -10$ $y = -5$ *Check:* $\dfrac{4}{-5 - 1} + \dfrac{-(-5) + 5}{3(-5)^2 - 4(-5) + 1} \overset{?}{=} \dfrac{9}{3(-5) - 1}$ $\dfrac{4}{-6} + \dfrac{10}{96} \overset{?}{=} \dfrac{9}{-16}$ $-\dfrac{9}{16} = -\dfrac{9}{16}$ ✓

(Continued on next page)

Topic	Procedure	Examples
Solving formulas containing rational expressions for a specified variable, p. 335.	1. Multiply each term of the equation by the LCD. 2. Add a quantity to or subtract a quantity from each side of the equation so that only terms containing the desired variable are on one side of the equation while all other terms are on the other side. 3. If there are two or more unlike terms containing the desired variable, remove that variable as a common factor. 4. Divide each side of the equation by the coefficient of the desired variable. 5. Simplify, if possible.	Solve for n. $$v = c - \frac{ct}{n}$$ $$n(v) = n(c) - n\left(\frac{ct}{n}\right)$$ $$nv = nc - ct$$ $$nv - nc = -ct$$ $$n(v - c) = -ct$$ $$n = \frac{-ct}{v - c} \quad \text{or} \quad \frac{ct}{c - v}$$
Using advanced ratios to solve applied problems, p. 336.	1. Determine a given ratio in the problem for which both values are known. 2. Use variables to describe each of the quantities in the other ratio. 3. Set the two ratios equal to each other to form an equation. 4. Solve the resulting equation. Determine both quantities in the second ratio.	A new navy cruiser has a crew of 304 people. For every thirteen seamen there are three officers. How many officers and how many seamen are in the crew? The ratio of seamen to officers is: $\dfrac{13}{3}$ Let x be the number of officers. Because the crew totals 304, it follows that $304 - x$ represents the number of seamen. $\begin{array}{cc}\text{seamen} \rightarrow \\ \text{officers} \rightarrow\end{array} \dfrac{13}{3} = \dfrac{304 - x}{x} \begin{array}{c}\leftarrow \text{seamen} \\ \leftarrow \text{officers}\end{array}$ $$13x = 3(304 - x)$$ $$13x = 912 - 3x$$ $$16x = 912$$ $$x = 57$$ $$304 - x = 247$$ There are 57 officers and 247 seamen.

Chapter 6 Review Problems

Simplify.

1. $\dfrac{6x^3 - 9x^2}{12x^2 - 18x}$

2. $\dfrac{15x^4}{5x^2 - 20x}$

3. $\dfrac{26x^3 y^2}{39x y^4}$

4. $\dfrac{42a^4 b c^3}{24a^7 b}$

5. $\dfrac{2x^2 - 5x + 3}{3x^2 + 2x - 5}$

6. $\dfrac{ax + 2a - bx - 2b}{3x^2 - 12}$

7. $\dfrac{4x^2 - 1}{x^2 - 4} \cdot \dfrac{2x^2 + 4x}{4x + 2}$

8. $\dfrac{3y}{4xy - 6y^2} \cdot \dfrac{2x - 3y}{12xy}$

9. $\dfrac{y^2 + 8y - 20}{y^2 + 6y - 16} \cdot \dfrac{y^2 + 3y - 40}{y^2 + 6y - 40}$

10. $\dfrac{3x^3 y}{x^2 + 7x + 12} \cdot \dfrac{x^2 + 8x + 15}{6x y^2}$

11. $\dfrac{3x + 9}{5x - 20} \div \dfrac{3x^2 - 3x - 36}{x^2 - 8x + 16}$

12. $\dfrac{4a^3 b^2}{4x^2 - 4x - 3} \div \dfrac{8ab^4}{6x^2 - 11x + 3}$

13. $\dfrac{9y^2 - 3y - 2}{6y^2 - 13y - 5} \div \dfrac{3y^2 + 10y - 8}{2y^2 + 13y + 20}$

14. $\dfrac{4a^2 + 12a + 5}{2a^2 - 7a - 13} \div (4a^2 + 2a)$

Add or subtract the rational expressions and simplify your answers.

15. $\dfrac{x - 5}{2x + 1} - \dfrac{x + 1}{x - 2}$

16. $\dfrac{5}{4x} + \dfrac{-3}{x + 4}$

17. $\dfrac{2y - 1}{12y} - \dfrac{3y + 2}{9y}$

18. $\dfrac{4}{y + 5} + \dfrac{3y + 2}{y^2 - 25}$

19. $\dfrac{2}{y^2 - 4} + \dfrac{y}{y^2 + 4y + 4}$

20. $\dfrac{y^2 + 2y + 20}{y^2 + 8y + 12} - \dfrac{2y + 1}{y + 6}$

21. $\dfrac{a}{5 - a} - \dfrac{2}{a + 3} + \dfrac{2a^2 - 2a}{a^2 - 2a - 15}$

22. $\dfrac{2}{x^2 + 8x + 16} - \dfrac{x}{2x^2 + 9x + 4}$

23. $5b - 1 - \dfrac{b + 2}{b + 3}$

24. $\dfrac{1}{x} + \dfrac{3}{2x} + 3 + 2x$

Simplify the complex rational expressions.

25. $\dfrac{\dfrac{5}{x} + 1}{1 - \dfrac{25}{x^2}}$

26. $\dfrac{\dfrac{4}{x + 3}}{\dfrac{2}{x - 2} - \dfrac{1}{x^2 + x - 6}}$

27. $\dfrac{\dfrac{8}{y + 2} - 4}{\dfrac{2}{y + 2} - 1}$

28. $\dfrac{\dfrac{1}{a} + \dfrac{a}{a + 1}}{\dfrac{a}{a + 1} - \dfrac{1}{a}}$

29. $\dfrac{\dfrac{2}{x + 4} - \dfrac{1}{x^2 + 4x}}{\dfrac{3}{2x + 8}}$

30. $\dfrac{\dfrac{y^2}{y^2 - x^2} - 1}{x + \dfrac{xy}{x - y}}$

31. $\dfrac{\dfrac{2x + 1}{x - 1}}{1 + \dfrac{x}{x + 1}}$

32. $\dfrac{\dfrac{3}{x} - \dfrac{2}{x + 1}}{\dfrac{5}{x^2 + 5x + 4} - \dfrac{1}{x + 4}}$

Solve for the variable and check your solutions. If there is no solution, say so.

33. $\dfrac{3}{2} = 1 - \dfrac{1}{x - 1}$

34. $\dfrac{3}{7} + \dfrac{4}{x + 1} = 1$

35. $\dfrac{7}{2x - 3} + \dfrac{3}{x + 2} = \dfrac{6}{2x - 3}$

36. $\dfrac{9}{5x - 2} - \dfrac{2}{x} = \dfrac{1}{x}$

37. $\dfrac{5}{2a} = \dfrac{2}{a} - \dfrac{1}{12}$

38. $\dfrac{1}{2a} = \dfrac{2}{a} - \dfrac{3}{10}$

39. $\dfrac{1}{y} + \dfrac{1}{2y} = 2$

40. $\dfrac{5}{y^2} + \dfrac{7}{y} = \dfrac{6}{y^2}$

41. $\dfrac{3}{a + 4} + \dfrac{a + 1}{3a + 12} = \dfrac{5}{3}$

42. $\dfrac{a + 3}{2a + 12} = \dfrac{3}{2} - \dfrac{5}{a + 6}$

43. $\dfrac{3x - 23}{2x^2 - 5x - 3} + \dfrac{2}{x - 3} = \dfrac{5}{2x + 1}$

44. $\dfrac{2x - 10}{2x^2 - 5x - 3} + \dfrac{1}{x - 3} = \dfrac{3}{2x + 1}$

Solve for the variable indicated.

45. $\dfrac{N}{V} = \dfrac{m}{M + N}$; for M

46. $m = \dfrac{y - y_0}{x - x_0}$; for x

47. $\dfrac{1}{f} = \dfrac{1}{a} + \dfrac{1}{b}$; for a

48. $S = \dfrac{V_1 t + V_2 t}{2}$; for t

49. $d = \dfrac{LR_2}{R_2 + R_1}$; for R_2

50. $\dfrac{S - P}{Pr} = t$; for r

Mixed Practice

Simplify.

51. $\dfrac{x^2 - x - 42}{x^2 - 2x - 35}$

52. $\dfrac{2x^2 - 5x - 3}{x^2 - 9} \cdot \dfrac{2x^2 + 5x - 3}{2x^2 + 5x + 2}$

53. $\dfrac{-2x - 1}{x + 4} + 4x + 3$

54. $\dfrac{\dfrac{1}{x^2 - 3x + 2}}{\dfrac{3}{x - 2} - \dfrac{2}{x - 1}}$

55. Solve for x.

$\dfrac{5}{2} - \dfrac{3}{x + 1} = \dfrac{2 - x}{x + 1}$

Solve the following exercises. If necessary, round your answers to the nearest hundredth.

56. *Calculator Supply* At the beginning of the fall semester, the campus bookstore at Boston University ordered 320 scientific and graphing calculators. For every seven graphing calculators, they also ordered three scientific calculators. How many graphing calculators did they order? How many scientific calculators did they order?

57. *New Homes* Walter Johnson built a new development of 112 homes in Naperville, Illinois. For every three one-story homes, he built thirteen two-story homes. How many one-story homes did he build? How many two-story homes did he build?

▲ **58.** *Photograph Enlargement* Jill VanderWoude decided to enlarge a photograph that measures 5 inches wide and 7 inches long into a poster-size photograph with a perimeter of 168 inches. The new photograph will maintain the same width-to-length ratio. How wide will the enlarged photograph be? How long will the enlarged photograph be?

59. *Swimming Pool* How long will it take a pump to empty a 4900-gallon swimming pool if the same pump can empty a 3500-gallon swimming pool in 4 hours?

60. *Rabbit Population* In a sanctuary, a sample of one hundred wild rabbits is tagged and released by the wildlife management team. In a few weeks, after they have mixed with the general rabbit population, a sample of forty rabbits is caught, and eight have a tag. Estimate the population of rabbits in the sanctuary.

61. *Police Statistics* The ratio of officers to state troopers is 2:9. If there are 154 men and women on the force, how many are officers?

62. *Maritime Chart* The scale on a maritime sailing chart shows that 2 centimeters is equivalent to 7 nautical miles. A boat captain lays out a course on the chart that is 3.5 centimeters long. How many nautical miles will this be?

▲ **63.** *Shadow Length* A 7-foot-tall tree casts a shadow that is 6 feet long. At the same time of day, a building casts a shadow that is 156 feet long. How tall is the building?

64. *Planting Time* Each spring it takes Marianne 10 hours to plant flowers in her yard. She estimates her son would take one and a half times as long, or 15 hours, to do the job. How many hours would it take if they both worked together?

65. *Jacuzzi* If the hot water faucet at Mike's house is left on, it takes 15 minutes to fill the jacuzzi. If the cold water faucet is left on, it takes 10 minutes to fill the jacuzzi. How many minutes would it take if both faucets are left on?

Online Retail Spending The bar graph below shows the growth in online spending in the United States. Use this bar graph to answer exercises 66–69. Round all answers to the nearest billion.

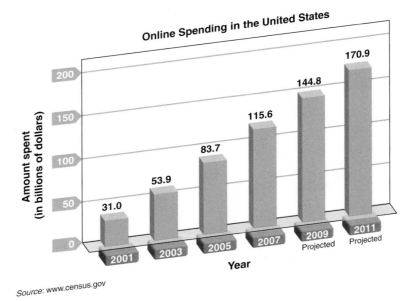

Source: www.census.gov

66. If the increase per year in online spending that occurred from 2007 to 2011 continues at the same increase per year, what would be the expected amount spent online in 2013?

67. If the ratio of the amount spent online in 2011 to the amount spent online in 2009 is proportional to the ratio of the amount spent online in 2013 to the amount spent online in 2011, what would be the expected amount spent online in 2013?

68. If the increase per year in online spending that occurred from 2001 to 2005 is double the increase in spending from 2011 to 2013, what would be the expected amount spent online in 2013?

69. If the increase per year in online spending that occurred from 2005 to 2007 is one-half as great as the increase in spending from 2011 to 2013, what would be the expected amount spent online in 2013?

How Am I Doing? Chapter 6 Test

Remember to use your Chapter Test Prep Video CD to see the worked-out solutions to the test problems you want to review.

Simplify.

1. $\dfrac{x^3 + 3x^2 + 2x}{x^3 - 2x^2 - 3x}$

2. $\dfrac{-25p^4qr^3}{45pqr^6}$

3. $\dfrac{2y^2 + 7y - 4}{y^2 + 2y - 8} \cdot \dfrac{2y^2 - 8}{3y^2 + 11y + 10}$

4. $\dfrac{4 - 2x}{3x^2 - 2x - 8} \div \dfrac{2x^2 + x - 1}{9x + 12}$

5. $\dfrac{3}{x} - \dfrac{2}{x + 1}$

6. $\dfrac{2}{x^2 + 5x + 6} + \dfrac{3x}{x^2 + 6x + 9}$

7. $\dfrac{\dfrac{4}{y + 2} - 2}{5 - \dfrac{10}{y + 2}}$

8. $\dfrac{\dfrac{1}{x} - \dfrac{3}{x + 2}}{\dfrac{2}{x^2 + 2x}}$

Solve for the variable and check your answers. If no solution exists, say so.

9. $\dfrac{7}{4} = \dfrac{x + 4}{x}$

10. $2 + \dfrac{x}{x + 4} = \dfrac{3x}{x - 4}$

11. $\dfrac{1}{2y + 4} - \dfrac{1}{6} = \dfrac{-2}{3y + 6}$

12. $2 + \dfrac{3}{x} = \dfrac{2x}{x - 1}$

13. Solve for W. $h = \dfrac{S - 2WL}{2W + 2L}$

14. Solve for h. $\dfrac{3V}{\pi h} = r^2$

15. A total of 286 employees at Kaiser Telecommunication Systems were eligible this year for a high-performance bonus. The company president announced that for every three employees who got the bonus, nineteen employees did not. If the president was correct, how many employees got the high-performance bonus? How many did not get the high-performance bonus?

▲ 16. The Newbury Elementary School had a rectangular playground that measured 500 feet wide and 850 feet long. When the school had a major addition to its property, it was decided to increase the playground to a large rectangular shape with the same proportional ratio that had a perimeter of 8100 feet. What is the width and the length of the new playground?

1. _____

2. _____

3. _____

4. _____

5. _____

6. _____

7. _____

8. _____

9. _____

10. _____

11. _____

12. _____

13. _____

14. _____

15. _____

16. _____

Cumulative Test for Chapters 1–6

Approximately one-half of this test covers the content of Chapters 1–5. The remainder covers the content of Chapter 6.

1. Simplify. $(3x^{-2}y)^2(x^4y^{-3})$

2. Solve for x. $\dfrac{3}{4}(x + 2) = \dfrac{1}{3}x - 2$

3. Graph the straight line $-6x + 2y = -12$.

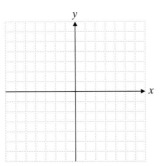

4. Find the standard form of the equation of the line parallel to $5x - 6y = 8$ that passes through $(-1, -3)$.

5. Brenda invested $7000 in two accounts at the bank. One account earns 5% simple interest. The other earns 8% simple interest. She earned $539 interest in 1 year. How much was invested at each rate?

6. Solve for x and graph. $3(2 - 6x) > 4(x + 1) + 24$

$$\xleftarrow{\quad} \overset{\;}{\underset{-3\;\;-2\;\;-1\;\;\;0\;\;\;1\;\;\;2\;\;\;3}{|\;\;\;|\;\;\;|\;\;\;|\;\;\;|\;\;\;|\;\;\;|}} \xrightarrow{\quad}$$

7. Evaluate $2x^2 - 3x - 4y^2$ when $x = -2$ and $y = 3$.

8. Solve for x. $|2x + 1| \le 8$

Factor.

9. $8x^3 - 125y^3$

10. $81x^3 - 90x^2y + 25xy^2$

Solve for x.

11. $x^2 + 20x + 36 = 0$

12. $3x^2 - 11x - 4 = 0$

Simplify.

13. $\dfrac{7x^2 - 28}{x^2 + 6x + 8}$

14. $\dfrac{2x^2 + x - 1}{2x^2 - 9x + 4} \cdot \dfrac{3x^2 - 12x}{6x + 15}$

15. $\dfrac{x^3 + 2x^2}{3x - 21} \div \dfrac{2x^3 + 5x^2 + 2x}{x - 7}$

16. $\dfrac{5}{2x - 4} - \dfrac{x + 1}{x^2 - 4}$

17. $\dfrac{\dfrac{1}{2x + 1} + 1}{4 - \dfrac{3}{4x^2 - 1}}$

18. $\dfrac{3}{x - 6} + \dfrac{4}{x + 4}$

Solve for the variable and check your answers.

19. $\dfrac{1}{2x + 3} - \dfrac{4}{4x^2 - 9} = \dfrac{3}{2x - 3}$

20. $\dfrac{2}{5x} - \dfrac{4}{x} = \dfrac{3}{10}$

21. Solve for *b*. $H = \dfrac{3b + 2x}{5 - 4b}$

22. The mayor of Chicago decided that in the coming year the city police force should be deployed in a particular ratio. He decided that for every three police officers who patrol on foot, there should be eleven police officers patrolling in squad cars. If 3234 police officers are normally on duty during the day, how many are patrolling on foot, and how many are patrolling in squad cars?

13. _____

14. _____

15. _____

16. _____

17. _____

18. _____

19. _____

20. _____

21. _____

22. _____

At what speed should a satellite travel to maintain a circular orbit around Earth at a distance of 100 miles above Earth? By how much should the speed be increased to change the orbiting distance to 125 miles above Earth? These and other questions about satellite travel require calculations using radicals. The material contained in this chapter will allow you to explore questions related to satellite travel.

Rational Exponents and Radicals

7.1   **RATIONAL EXPONENTS**   354

7.2   **RADICAL EXPRESSIONS AND FUNCTIONS**   361

7.3   **SIMPLIFYING, ADDING, AND SUBTRACTING RADICALS**   369

7.4   **MULTIPLYING AND DIVIDING RADICALS**   375

        HOW AM I DOING? SECTIONS 7.1–7.4   385

7.5   **RADICAL EQUATIONS**   386

7.6   **COMPLEX NUMBERS**   392

7.7   **VARIATION**   400

        CHAPTER 7 ORGANIZER   408

        CHAPTER 7 REVIEW PROBLEMS   411

        HOW AM I DOING? CHAPTER 7 TEST   415

        CUMULATIVE TEST FOR CHAPTERS 1–7   417

1 Simplifying Expressions with Rational Exponents

Before studying this section, you may need to review Section 1.4. For convenience, we list the rules of exponents that we learned there.

$$x^m x^n = x^{m+n} \qquad\qquad x^0 = 1$$

$$\frac{x^m}{x^n} = x^{m-n} \qquad\qquad (x^m)^n = x^{mn}$$

$$x^{-n} = \frac{1}{x^n} \qquad\qquad (xy)^n = x^n y^n$$

$$\frac{x^{-n}}{y^{-m}} = \frac{y^m}{x^n} \qquad\qquad \left(\frac{x}{y}\right)^n = \frac{x^n}{y^n}$$

To ensure that you understand these rules, study Example 1 carefully and work Practice Problem 1.

EXAMPLE 1 Simplify. $\left(\dfrac{5xy^{-3}}{2x^{-4}y}\right)^{-2}$

Solution

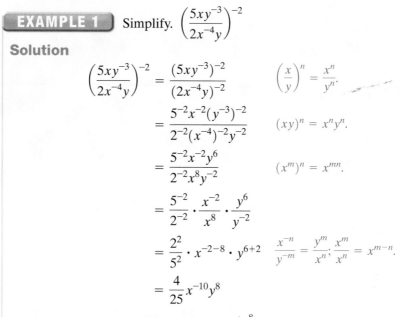

$$\left(\frac{5xy^{-3}}{2x^{-4}y}\right)^{-2} = \frac{(5xy^{-3})^{-2}}{(2x^{-4}y)^{-2}} \qquad\qquad \left(\frac{x}{y}\right)^n = \frac{x^n}{y^n}.$$

$$= \frac{5^{-2}x^{-2}(y^{-3})^{-2}}{2^{-2}(x^{-4})^{-2}y^{-2}} \qquad\qquad (xy)^n = x^n y^n.$$

$$= \frac{5^{-2}x^{-2}y^6}{2^{-2}x^8 y^{-2}} \qquad\qquad (x^m)^n = x^{mn}.$$

$$= \frac{5^{-2}}{2^{-2}} \cdot \frac{x^{-2}}{x^8} \cdot \frac{y^6}{y^{-2}}$$

$$= \frac{2^2}{5^2} \cdot x^{-2-8} \cdot y^{6+2} \qquad \frac{x^{-n}}{y^{-m}} = \frac{y^m}{x^n}; \frac{x^m}{x^n} = x^{m-n}.$$

$$= \frac{4}{25}x^{-10}y^8$$

The answer can also be written as $\dfrac{4y^8}{25x^{10}}$. Explain why.

Practice Problem 1 Simplify.

$$\left(\frac{3x^{-2}y^4}{2x^{-5}y^2}\right)^{-3}$$

SIDELIGHT: Example 1 Follow-Up Deciding when to use the rule $\dfrac{x^{-n}}{y^{-m}} = \dfrac{y^m}{x^n}$ is entirely up to you. In Example 1, we could have begun by writing

$$\left(\frac{5xy^{-3}}{2x^{-4}y}\right)^{-2} = \left(\frac{5x \cdot x^4}{2y \cdot y^3}\right)^{-2} = \left(\frac{5x^5}{2y^4}\right)^{-2}.$$

Complete the steps to simplify this expression.

Likewise, in the fourth step in Example 1, we could have written

$$\frac{5^{-2}x^{-2}y^6}{2^{-2}x^8 y^{-2}} = \frac{2^2 y^6 y^2}{5^2 x^8 x^2}.$$

Complete the steps to simplify this expression. Are the two answers the same as the answer in Example 1? Why or why not?

We generally begin to simplify a rational expression with exponents by raising a power to a power because sometimes negative powers become positive. The order in which you use the rules of exponents is up to you. Work carefully. Keep track of your exponents and where you are as you simplify the rational expression.

These rules for exponents can also be extended to include rational exponents—that is, exponents that are fractions. As you recall, rational numbers are of the form $\frac{a}{b}$, where a and b are integers and b does not equal zero. We will write fractional exponents using diagonal lines. Thus, we will write $\frac{5}{6}$ as 5/6 and $\frac{a}{b}$ as a/b throughout this chapter when writing fractional exponents. For now we restrict the base to *positive* real numbers. Later we will talk about negative bases.

EXAMPLE 2   Simplify.

(a) $(x^{2/3})^4$ **(b)** $\dfrac{x^{5/6}}{x^{1/6}}$ **(c)** $x^{2/3} \cdot x^{-1/3}$ **(d)** $5^{3/7} \cdot 5^{2/7}$

Solution   We will not write out every step or every rule of exponents that we use. You should be able to follow the solutions.

(a) $(x^{2/3})^4 = x^{(2/3)(4/1)} = x^{8/3}$

(b) $\dfrac{x^{5/6}}{x^{1/6}} = x^{5/6 - 1/6} = x^{4/6} = x^{2/3}$

(c) $x^{2/3} \cdot x^{-1/3} = x^{2/3 - 1/3} = x^{1/3}$

(d) $5^{3/7} \cdot 5^{2/7} = 5^{3/7 + 2/7} = 5^{5/7}$

Practice Problem 2   Simplify.

(a) $(x^4)^{3/8}$ **(b)** $\dfrac{x^{3/7}}{x^{2/7}}$ **(c)** $x^{-7/5} \cdot x^{4/5}$

Sometimes fractional exponents will not have the same denominator. Remember that you need to change the fractions to equivalent fractions with the same denominator when the rules of exponents require you to add or to subtract them.

EXAMPLE 3   Simplify. Express your answers with positive exponents only.

(a) $(2x^{1/2})(3x^{1/3})$ **(b)** $\dfrac{18x^{1/4}y^{-1/3}}{-6x^{-1/2}y^{1/6}}$

Solution

(a) $(2x^{1/2})(3x^{1/3}) = 6x^{1/2 + 1/3} = 6x^{3/6 + 2/6} = 6x^{5/6}$

(b) $\dfrac{18x^{1/4}y^{-1/3}}{-6x^{-1/2}y^{1/6}} = -3x^{1/4 - (-1/2)}y^{-1/3 - 1/6}$

$\qquad = -3x^{1/4 + 2/4}y^{-2/6 - 1/6}$

$\qquad = -3x^{3/4}y^{-3/6}$

$\qquad = -3x^{3/4}y^{-1/2}$

$\qquad = -\dfrac{3x^{3/4}}{y^{1/2}}$

Practice Problem 3   Simplify. Express your answers with positive exponents only.

(a) $(-3x^{1/4})(2x^{1/2})$ **(b)** $\dfrac{13x^{1/12}y^{-1/4}}{26x^{-1/3}y^{1/2}}$

EXAMPLE 4 Multiply and simplify. $-2x^{5/6}(3x^{1/2} - 4x^{-1/3})$

Solution We will need to be very careful when we add the exponents for x as we use the distributive property. Study each step of the following example. Be sure you understand each operation.

$$-2x^{5/6}(3x^{1/2} - 4x^{-1/3}) = -6x^{5/6+1/2} + 8x^{5/6-1/3}$$
$$= -6x^{5/6+3/6} + 8x^{5/6-2/6}$$
$$= -6x^{8/6} + 8x^{3/6}$$
$$= -6x^{4/3} + 8x^{1/2}$$

NOTE TO STUDENT: Fully worked-out solutions to all of the Practice Problems can be found at the back of the text starting at page SP-1

Practice Problem 4 Multiply and simplify. $-3x^{1/2}(2x^{1/4} + 3x^{-1/2})$

Sometimes we can use the rules of exponents to simplify numerical values raised to rational powers.

EXAMPLE 5 Evaluate.

(a) $(25)^{3/2}$ **(b)** $(27)^{2/3}$

Solution

(a) $(25)^{3/2} = (5^2)^{3/2} = 5^{2/1 \cdot 3/2} = 5^3 = 125$

(b) $(27)^{2/3} = (3^3)^{2/3} = 3^{3/1 \cdot 2/3} = 3^2 = 9$

Practice Problem 5 Evaluate.

(a) $(4)^{5/2}$ **(b)** $(27)^{4/3}$

② Adding Expressions with Rational Exponents

Adding expressions with rational exponents may require several steps. Sometimes this involves removing negative exponents. For example, to add $2x^{-1/2} + x^{1/2}$, we begin by writing $2x^{-1/2}$ as $\dfrac{2}{x^{1/2}}$. This is a rational expression. Recall that to add rational expressions we need to have a common denominator. Take time to look at the steps needed to write $2x^{-1/2} + x^{1/2}$ as one term.

EXAMPLE 6 Write as one fraction with positive exponents. $2x^{-1/2} + x^{1/2}$

Solution

$$2x^{-1/2} + x^{1/2} = \frac{2}{x^{1/2}} + \frac{x^{1/2} \cdot x^{1/2}}{x^{1/2}} = \frac{2}{x^{1/2}} + \frac{x^1}{x^{1/2}} = \frac{2+x}{x^{1/2}}$$

Practice Problem 6 Write as one fraction with only positive exponents.

$$3x^{1/3} + x^{-1/3}$$

③ Factoring Expressions with Rational Exponents

To factor expressions, we need to be able to recognize common factors. If the terms of the expression contain exponents, we look for the same exponential factor in each term. For example, in the expression $6x^5 + 4x^3 - 8x^2$, the common factor of each term is $2x^2$. Thus, we can factor out the common factor $2x^2$ from each term. The expression then becomes $2x^2(3x^3 + 2x - 4)$.

We do exactly the same thing when we factor expressions with rational exponents. The key is to identify the exponent of the common factor. In the expression $6x^{3/4} + 4x^{1/2} - 8x^{1/4}$, the common factor is $2x^{1/4}$. Thus, we factor the expression $6x^{3/4} + 4x^{1/2} - 8x^{1/4}$ as $2x^{1/4}(3x^{1/2} + 2x^{1/4} - 4)$. We do not always need to factor out the greatest common factor. In the following examples we simply factor out a common factor.

EXAMPLE 7 Factor out the common factor $2x$. $2x^{3/2} + 4x^{5/2}$

Solution   We rewrite the exponent of each term so that we can see that each term contains the factor $2x$ or $2x^{2/2}$.

$$2x^{3/2} + 4x^{5/2} = 2x^{2/2+1/2} + 4x^{2/2+3/2}$$

$$= 2(x^{2/2})(x^{1/2}) + 4(x^{2/2})(x^{3/2})$$

$$= 2x(x^{1/2} + 2x^{3/2})$$

Practice Problem 7 Factor out the common factor $4y$. $4y^{3/2} - 8y^{5/2}$

For convenience we list here the properties of exponents that we have discussed in this section, as well as the property $x^0 = 1$.

When x and y are **positive real numbers** and a and b are **rational numbers:**

$$x^a x^b = x^{a+b} \qquad \frac{x^a}{x^b} = x^{a-b} \qquad x^0 = 1$$

$$x^{-a} = \frac{1}{x^a} \qquad \frac{x^{-a}}{y^{-b}} = \frac{y^b}{x^a}$$

$$(x^a)^b = x^{ab} \qquad (xy)^a = x^a y^a \qquad \left(\frac{x}{y}\right)^a = \frac{x^a}{y^a}$$

Simplify. Express your answer with positive exponents.

1. $\left(\dfrac{3xy^{-1}}{z^2}\right)^4$

2. $\left(\dfrac{3xy^{-2}}{x^3}\right)^2$

3. $\left(\dfrac{4ab^{-2}}{3b}\right)^2$

4. $\left(\dfrac{-a^2b}{2b^{-1}}\right)^3$

5. $\left(\dfrac{2x^2}{y}\right)^{-3}$

6. $\left(\dfrac{4x}{y^3}\right)^{-3}$

7. $\left(\dfrac{3xy^{-2}}{y^3}\right)^{-2}$

8. $\left(\dfrac{5x^{-2}y}{x^4}\right)^{-2}$

9. $(x^{3/4})^2$

10. $(x^{5/6})^3$

11. $(y^{12})^{2/3}$

12. $(y^2)^{5/2}$

13. $\dfrac{x^{3/5}}{x^{1/5}}$

14. $\dfrac{y^{6/7}}{y^{3/7}}$

15. $\dfrac{x^{8/9}}{x^{2/9}}$

16. $\dfrac{x^{11/12}}{x^{5/12}}$

17. $\dfrac{a^2}{a^{1/4}}$

18. $\dfrac{a^3}{a^{1/5}}$

19. $x^{1/7} \cdot x^{3/7}$

20. $x^{3/5} \cdot x^{1/5}$

21. $a^{3/8} \cdot a^{1/2}$

22. $b^{2/5} \cdot b^{2/15}$

23. $y^{3/5} \cdot y^{-1/10}$

24. $y^{7/10} \cdot y^{-1/5}$

Write each expression with positive exponents.

25. $x^{-3/4}$

26. $x^{-5/6}$

27. $a^{-5/6}b^{1/3}$

28. $2a^{-1/6}b^{3/4}$

29. $6^{-1/2}$

30. $4^{-1/3}$

31. $3a^{-1/3}$

32. $-4y^{-3/4}$

Evaluate or simplify the numerical expressions.

33. $(27)^{5/3}$

34. $(16)^{3/4}$

35. $(4)^{3/2}$

36. $(9)^{3/2}$

37. $(-8)^{5/3}$

38. $(-27)^{5/3}$

39. $(-27)^{2/3}$

40. $(-64)^{2/3}$

Mixed Practice *Simplify and express your answers with positive exponents. Evaluate or simplify the numerical expressions.*

41. $(x^{1/4}y^{-1/3})(x^{3/4}y^{1/2})$

42. $(x^{2/3}y^{2/3})(x^{1/2}y^{-2/3})$

43. $(7x^{1/3}y^{1/4})(-2x^{1/4}y^{-1/6})$

44. $(8x^{-1/5}y^{1/3})(-3x^{-1/4}y^{1/6})$

45. $6^2 \cdot 6^{-2/3}$

46. $11^{1/2} \cdot 11^3$

47. $\dfrac{2x^{1/5}}{x^{-1/2}}$

48. $\dfrac{3y^{2/3}}{y^{-1/4}}$

49. $\dfrac{-20x^2y^{-1/5}}{5x^{-1/2}y}$

50. $\dfrac{12x^{-2/3}y}{-6xy^{-3/4}}$

51. $\left(\dfrac{8a^2b^6}{a^{-1}b^3}\right)^{1/3}$

52. $\left(\dfrac{16a^5b^{-2}}{a^{-1}b^{-6}}\right)^{1/2}$

53. $(-4x^{1/4}y^{5/2}z^{1/2})^2$

54. $(4x^{2/3}y^{-1/2}z^{3/5})^3$

55. $x^{2/3}(x^{4/3} - x^{1/5})$

56. $y^{-2/3}(y^{2/3} + y^{3/2})$

57. $m^{7/8}(m^{-1/2} + 2m)$

58. $m^{2/3}(m^{-1/2} + 3m)$

59. $(8)^{-1/3}$

60. $(100)^{-1/2}$

61. $(25)^{-3/2}$

62. $(16)^{-5/4}$

63. $(81)^{3/4} + (25)^{1/2}$

64. $9^{3/2} + 4^{1/2}$

Write each expression as one fraction with positive exponents.

65. $3y^{1/2} + y^{-1/2}$

66. $2y^{1/3} + y^{-2/3}$

67. $x^{-1/3} + 6^{4/3}$

68. $5^{-1/4} + x^{-1/2}$

Factor out the common factor 2a.

69. $10a^{5/4} - 4a^{8/5}$

70. $6a^{4/3} - 8a^{3/2}$

Factor out the common factor 3x.

71. $12x^{4/3} - 3x^{5/2}$

72. $18x^{5/3} - 6x^{11/8}$

To Think About

73. What is the value of a if $x^a \cdot x^{1/4} = x^{-1/8}$?

74. What is the value of b if $x^b \div x^{1/3} = x^{-1/12}$?

Applications *Radius and Volume of a Sphere* *The radius needed to create a sphere with a given volume V can be approximated by the equation* $r = 0.62(V)^{1/3}$. *Find the radius of the spheres with the following volumes.*

▲ **75.** 27 cubic meters

▲ **76.** 64 cubic meters

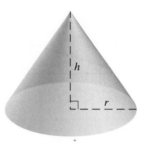

Radius and Volume of a Cone *The radius required for a cone to have a volume V and a height h is given by the equation* $r = \left(\dfrac{3V}{\pi h}\right)^{1/2}$. *Find the necessary radius to have a cone with the properties below. Use $\pi \approx 3.14$.*

▲ **77.** $V = 314$ cubic feet and $h = 12$ feet.

▲ **78.** $V = 3140$ cubic feet and $h = 30$ feet.

Cumulative Review *Solve for x.*

79. **[2.1.1]** $-4(x + 1) = \dfrac{1}{3}(3 - 2x)$

Solve for b.

80. **[2.2.1]** $A = \dfrac{h}{2}(a + b)$

Quick Quiz 7.1 Simplify.

1. $(-4x^{2/3}y^{1/4})(3x^{1/6}y^{1/2})$

2. $\dfrac{16x^4}{8x^{2/3}}$

3. $(25x^{1/4})^{3/2}$

4. **Concept Check** Explain how you would simplify the following.

$$x^{9/10} \cdot x^{-1/5}$$

① Evaluating Radical Expressions and Functions

In Section 1.3 we studied simple radical expressions called square roots. The **square root** of a number is a value that when multiplied by itself is equal to the original number. That is, since $3 \cdot 3 = 9$, 3 is a square root of 9. But $(-3) \cdot (-3) = 9$, so -3 is also a square root. We call the positive square root the **principal square root.**

The symbol $\sqrt{\ }$ is called a **radical sign.** We use it to denote positive square roots (and positive higher-order roots also). A negative square root is written $-\sqrt{\ }$. Thus, we have the following:

$$\sqrt{9} = 3 \qquad -\sqrt{9} = -3$$
$$\sqrt{64} = 8 \quad \text{(because } 8 \cdot 8 = 64\text{)}$$
$$\sqrt{121} = 11 \quad \text{(because } 11 \cdot 11 = 121\text{)}$$

Because $\sqrt{9} = \sqrt{3 \cdot 3} = \sqrt{3^2} = 3$, we can say the following:

> **DEFINITION OF SQUARE ROOT**
>
> If x is a nonnegative real number, then $\sqrt{x}$ is the *nonnegative* (or principal) *square root* of x; in other words, $\left(\sqrt{x}\right)^2 = x$.

Note that x must be *nonnegative*. Why? Suppose we want to find $\sqrt{-36}$. We must find a number that when multiplied by itself gives -36. Is there one? No, because

$$6 \cdot 6 = 36 \quad \text{and}$$
$$(-6)(-6) = 36.$$

So there is no real number that we can square to get -36.

We call $\sqrt[n]{x}$ a **radical expression.** The $\sqrt{\ }$ symbol is the radical sign, the x is the **radicand,** and the n is the **index** of the radical. When no number for n appears in the radical expression, it is understood that 2 is the index, which means that we are looking for the square root. For example, in the radical expression $\sqrt{25}$, with no number given for the index n, we take the index to be 2. Thus, $\sqrt{25}$ is the principal square root of 25.

We can extend the notion of square root to **higher-order roots,** such as cube roots, fourth roots, and so on. A **cube root** of a number is a value that when cubed is equal to the original number. The index n of the radical is 3, and the radical used is $\sqrt[3]{\ }$. Similarly, a **fourth root** of a number is a value that when raised to the fourth power is equal to the original number. The index n of the radical is 4, and the radical used is $\sqrt[4]{\ }$. Thus, we have the following:

$$\sqrt[3]{27} = 3 \quad \text{because } 3 \cdot 3 \cdot 3 = 3^3 = 27.$$
$$\sqrt[4]{81} = 3 \quad \text{because } 3 \cdot 3 \cdot 3 \cdot 3 = 3^4 = 81.$$
$$\sqrt[5]{32} = 2 \quad \text{because } 2 \cdot 2 \cdot 2 \cdot 2 \cdot 2 = 2^5 = 32.$$
$$\sqrt[3]{-64} = -4 \quad \text{because } (-4)(-4)(-4) = (-4)^3 = -64.$$
$$\sqrt[6]{729} = 3 \quad \text{because } 3 \cdot 3 \cdot 3 \cdot 3 \cdot 3 \cdot 3 = 3^6 = 729.$$

You should be able to see a pattern here.

$$\sqrt[3]{27} = \sqrt[3]{3^3} = 3$$
$$\sqrt[4]{81} = \sqrt[4]{3^4} = 3$$
$$\sqrt[5]{32} = \sqrt[5]{2^5} = 2$$
$$\sqrt[3]{-64} = \sqrt[3]{(-4)^3} = -4$$
$$\sqrt[6]{729} = \sqrt[6]{3^6} = 3$$

In these cases, we see that $\sqrt[n]{x^n} = x$. We now give the following definition.

Student Learning Objectives

After studying this section, you will be able to:

① Evaluate radical expressions and functions.

② Change radical expressions to expressions with rational exponents.

③ Change expressions with rational exponents to radical expressions.

④ Evaluate higher-order radicals containing variable radicands that represent any real number (including negative real numbers).

DEFINITION OF HIGHER-ORDER ROOTS

1. If x is a *nonnegative* real number, then $\sqrt[n]{x}$ is a nonnegative nth root and has the property that

$$\left(\sqrt[n]{x}\right)^n = x.$$

2. If x is a *negative* real number, then
   (a) When n is an *odd integer*, $\sqrt[n]{x}$ has the property that $\left(\sqrt[n]{x}\right)^n = x$.
   (b) When n is an *even integer*, $\sqrt[n]{x}$ is *not* a real number.

EXAMPLE 1 If possible, find the root of each negative number. If there is no real number root, say so.

(a) $\sqrt[3]{-216}$ (b) $\sqrt[5]{-32}$ (c) $\sqrt[4]{-16}$ (d) $\sqrt[6]{-64}$

Solution

(a) $\sqrt[3]{-216} = \sqrt[3]{(-6)^3} = -6$ because $(-6)^3 = -216$.

(b) $\sqrt[5]{-32} = \sqrt[5]{(-2)^5} = -2$ because $(-2)^5 = -32$.

(c) $\sqrt[4]{-16}$ is not a real number because n is even and x is negative.

(d) $\sqrt[6]{-64}$ is not a real number because n is even and x is negative.

Practice Problem 1 If possible, find the roots. If there is no real number root, say so.

(a) $\sqrt[3]{216}$ (b) $\sqrt[5]{32}$ (c) $\sqrt[3]{-8}$ (d) $\sqrt[4]{-81}$

Because the symbol $\sqrt{x}$ represents exactly one real number for all real numbers x that are nonnegative, we can use it to define the **square root function** $f(x) = \sqrt{x}$.

This function has a domain of all real numbers x that are greater than or equal to zero.

EXAMPLE 2 Find the indicated function values of the function $f(x) = \sqrt{2x + 4}$. Round your answers to the nearest tenth when necessary.

(a) $f(-2)$ (b) $f(6)$ (c) $f(3)$

Solution

(a) $f(-2) = \sqrt{2(-2) + 4} = \sqrt{-4 + 4} = \sqrt{0} = 0$ The square root of zero is zero.

(b) $f(6) = \sqrt{2(6) + 4} = \sqrt{12 + 4} = \sqrt{16} = 4$

(c) $f(3) = \sqrt{2(3) + 4} = \sqrt{6 + 4} = \sqrt{10} \approx 3.2$ We use a calculator or a square root table to approximate $\sqrt{10}$.

NOTE TO STUDENT: Fully worked-out solutions to all of the Practice Problems can be found at the back of the text starting at page SP-1

Practice Problem 2 Find the indicated values of the function $f(x) = \sqrt{4x - 3}$. Round your answers to the nearest tenth when necessary.

(a) $f(3)$ (b) $f(4)$ (c) $f(7)$

EXAMPLE 3 Find the domain of the function. $f(x) = \sqrt{3x - 6}$

Solution We know that the expression $3x - 6$ must be nonnegative. That is, $3x - 6 \geq 0$.

$$3x - 6 \geq 0$$
$$3x \geq 6$$
$$x \geq 2$$

Thus, the domain is all real numbers x where $x \geq 2$.

Practice Problem 3 Find the domain of the function. $f(x) = \sqrt{0.5x + 2}$

EXAMPLE 4 Graph the function $f(x) = \sqrt{x + 2}$. Use the values $f(-2)$, $f(-1), f(0), f(1), f(2)$, and $f(7)$. Round your answers to the nearest tenth when necessary.

Solution We show the table of values here.

x	f(x)
−2	0
−1	1
0	1.4
1	1.7
2	2
7	3

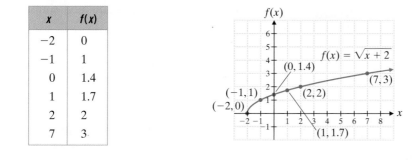

Practice Problem 4 Graph the function $f(x) = \sqrt{3x - 9}$. Use the values $f(3), f(4), f(5)$, and $f(6)$.

 Changing Radical Expressions to Expressions with Rational Exponents

Now we want to extend our definition of roots to rational exponents. By the laws of exponents we know that

$$x^{1/2} \cdot x^{1/2} = x^{1/2+1/2} = x^1 = x.$$

Since $x^{1/2}x^{1/2} = x$, $x^{1/2}$ must be a square root of x. That is, $x^{1/2} = \sqrt{x}$. Is this true? By the definition of square root, $(\sqrt{x})^2 = x$. Does $(x^{1/2})^2 = x$? Using the law of exponents we have

$$(x^{1/2})^2 = x^{(1/2)(2)} = x^1 = x.$$

We conclude that

$$x^{1/2} = \sqrt{x}.$$

In the same way we can write the following:

$$x^{1/3} \cdot x^{1/3} \cdot x^{1/3} = x \qquad x^{1/3} = \sqrt[3]{x}$$
$$x^{1/4} \cdot x^{1/4} \cdot x^{1/4} \cdot x^{1/4} = x \qquad x^{1/4} = \sqrt[4]{x}$$
$$\vdots \qquad\qquad\qquad \vdots$$
$$\underbrace{x^{1/n} \cdot x^{1/n} \cdot \ \cdots \ \cdot x^{1/n}}_{n \text{ factors}} = x \qquad x^{1/n} = \sqrt[n]{x}$$

Therefore, we are ready to define fractional exponents in general.

PRACTICE PROBLEM 4

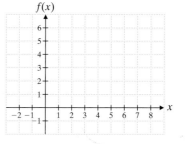

DEFINITION

If n is a positive integer and x is a nonnegative real number, then

$$x^{1/n} = \sqrt[n]{x}.$$

EXAMPLE 5 Change to rational exponents and simplify. Assume that all variables are nonnegative real numbers.

(a) $\sqrt[4]{x^4}$

(b) $\sqrt[5]{(32)^5}$

Solution

(a) $\sqrt[4]{x^4} = (x^4)^{1/4} = x^{4/4} = x^1 = x$

(b) $\sqrt[5]{(32)^5} = (32^5)^{1/5} = 32^{5/5} = 32^1 = 32$

Practice Problem 5 Change to rational exponents and simplify. Assume that all variables are nonnegative real numbers.

(a) $\sqrt[3]{x^3}$

(b) $\sqrt[4]{y^4}$

EXAMPLE 6 Replace all radicals with rational exponents.

(a) $\sqrt[3]{x^2}$

(b) $\left(\sqrt[5]{w}\right)^7$

Solution

(a) $\sqrt[3]{x^2} = (x^2)^{1/3} = x^{2/3}$

(b) $\left(\sqrt[5]{w}\right)^7 = (w^{1/5})^7 = w^{7/5}$

Practice Problem 6 Replace all radicals with rational exponents.

(a) $\sqrt[4]{x^3}$

(b) $\sqrt[5]{(xy)^7}$

EXAMPLE 7 Evaluate or simplify. Assume that all variables are nonnegative.

(a) $\sqrt[5]{32x^{10}}$

(b) $\sqrt[3]{125x^9}$

(c) $(16x^4)^{3/4}$

Solution

(a) $\sqrt[5]{32x^{10}} = (2^5 x^{10})^{1/5} = 2x^2$

(b) $\sqrt[3]{125x^9} = [5^3 x^9]^{1/3} = 5x^3$

(c) $(16x^4)^{3/4} = (2^4 x^4)^{3/4} = 2^3 x^3 = 8x^3$

Practice Problem 7 Evaluate or simplify. Assume that all variables are nonnegative.

(a) $\sqrt[4]{81x^{12}}$

(b) $\sqrt[3]{27x^6}$

(c) $(32x^5)^{3/5}$

③ Changing Expressions with Rational Exponents to Radical Expressions

Sometimes we need to change an expression with rational exponents to a radical expression. This is especially helpful because the value of the radical form of an expression is sometimes more recognizable. For example, because of our experience with radicals, we know that $\sqrt{25} = 5$. It is not as easy to see that $25^{1/2} = 5$. Therefore, we simplify expressions with rational exponents by first rewriting them as radical expressions. Recall that

$$x^{1/n} = \sqrt[n]{x}.$$

Again, using the laws of exponents, we know that

$$x^{m/n} = (x^m)^{1/n} = (x^{1/n})^m,$$

when x is nonnegative. We can make the following general definition.

DEFINITION

For positive integers m and n and any real number x for which $x^{1/n}$ is defined,

$$x^{m/n} = \left(\sqrt[n]{x}\right)^m = \sqrt[n]{x^m}.$$

If it is also true that $x \neq 0$, then

$$x^{-m/n} = \frac{1}{x^{m/n}} = \frac{1}{\left(\sqrt[n]{x}\right)^m} = \frac{1}{\sqrt[n]{x^m}}.$$

EXAMPLE 8   Change to radical form.

(a) $(xy)^{5/7}$ **(b)** $w^{-2/3}$ **(c)** $3x^{3/4}$ **(d)** $(3x)^{3/4}$

Solution

(a) $(xy)^{5/7} = \sqrt[7]{(xy)^5} = \sqrt[7]{x^5 y^5}$ or $(xy)^{5/7} = \left(\sqrt[7]{xy}\right)^5$

(b) $w^{-2/3} = \dfrac{1}{w^{2/3}} = \dfrac{1}{\sqrt[3]{w^2}}$ or $w^{-2/3} = \dfrac{1}{w^{2/3}} = \dfrac{1}{\left(\sqrt[3]{w}\right)^2}$

(c) $3x^{3/4} = 3\sqrt[4]{x^3}$ or $3x^{3/4} = 3\left(\sqrt[4]{x}\right)^3$

(d) $(3x)^{3/4} = \sqrt[4]{(3x)^3} = \sqrt[4]{27x^3}$ or $(3x)^{3/4} = \left(\sqrt[4]{3x}\right)^3$

Practice Problem 8   Change to radical form.

(a) $(xy)^{3/4}$ **(b)** $y^{-1/3}$ **(c)** $(2x)^{4/5}$ **(d)** $2x^{4/5}$

Graphing Calculator

Rational Exponents

To evaluate Example 9(a) on a graphing calculator we use

125 ∧ (2 ÷ 3)
enter .

Note the need to include the parentheses around the quantity $\frac{2}{3}$. Try each part of Example 9 on your graphing calculator.

EXAMPLE 9   Change to radical form and evaluate.

(a) $125^{2/3}$ **(b)** $(-16)^{5/2}$ **(c)** $144^{-1/2}$

Solution

(a) $125^{2/3} = \left(\sqrt[3]{125}\right)^2 = (5)^2 = 25$

(b) $(-16)^{5/2} = \left(\sqrt{-16}\right)^5$; however, $\sqrt{-16}$ is not a real number. Thus, $(-16)^{5/2}$ is not a real number.

(c) $144^{-1/2} = \dfrac{1}{144^{1/2}} = \dfrac{1}{\sqrt{144}} = \dfrac{1}{12}$

NOTE TO STUDENT: Fully worked-out solutions to all of the Practice Problems can be found at the back of the text starting at page SP-1

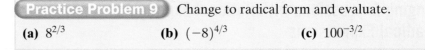

Practice Problem 9 Change to radical form and evaluate.

(a) $8^{2/3}$ **(b)** $(-8)^{4/3}$ **(c)** $100^{-3/2}$

4 Evaluating Higher-Order Radicals Containing Variable Radicands That Represent Any Real Number (Including Negative Real Numbers)

We now give a definition of higher-order radicals that works for all radicals, no matter what their signs are.

DEFINITION

For all real numbers x (including negative real numbers),

$$\sqrt[n]{x^n} = |x| \quad \text{when } n \text{ is an } even \text{ positive integer, and}$$
$$\sqrt[n]{x^n} = x \quad \text{when } n \text{ is an } odd \text{ positive integer.}$$

EXAMPLE 10 Evaluate; x may be any real number.

(a) $\sqrt[3]{(-2)^3}$ **(b)** $\sqrt[4]{(-2)^4}$ **(c)** $\sqrt[5]{x^5}$ **(d)** $\sqrt[6]{x^6}$

Solution

(a) $\sqrt[3]{(-2)^3} = -2$ because the index is odd .

(b) $\sqrt[4]{(-2)^4} = |-2| = 2$ because the index is even .

(c) $\sqrt[5]{x^5} = x$ because the index is odd .

(d) $\sqrt[6]{x^6} = |x|$ because the index is even .

Practice Problem 10 Evaluate; y and w may be any real numbers.

(a) $\sqrt[5]{(-3)^5}$ **(b)** $\sqrt[4]{(-5)^4}$ **(c)** $\sqrt[4]{w^4}$ **(d)** $\sqrt[7]{y^7}$

EXAMPLE 11 Simplify. Assume that x and y may be any real numbers.

(a) $\sqrt{49x^2}$ **(b)** $\sqrt[4]{81y^{16}}$ **(c)** $\sqrt[3]{27x^6y^9}$

Solution

(a) We observe that the index is an even positive number. We will need the absolute value. $\sqrt{49x^2} = 7|x|$

(b) Again, we need the absolute value. $\sqrt[4]{81y^{16}} = 3|y^4|$
Since we know that $3y^4$ is positive (anything to the fourth power will be positive), we can write $3|y^4|$ without the absolute value symbol. Thus, $\sqrt[4]{81y^{16}} = 3y^4$.

(c) The index is an odd integer. The absolute value is never needed in such a case.
$\sqrt[3]{27x^6y^9} = \sqrt[3]{3^3(x^2)^3(y^3)^3} = 3x^2y^3$

Practice Problem 11 Simplify. Assume that x and y may be any real numbers.

(a) $\sqrt{36x^2}$ **(b)** $\sqrt[4]{16y^8}$ **(c)** $\sqrt[3]{125x^3y^6}$

Verbal and Writing Skills

1. In a simple sentence, explain what a square root is.

2. In a simple sentence, explain what a cube root is.

3. Give an example to show why the cube root of a negative number is a negative number.

4. Give an example to show why it is not possible to find a real number that is the square root of a negative number.

Evaluate if possible.

5. $\sqrt{100}$

6. $\sqrt{64}$

7. $\sqrt{16} + \sqrt{81}$

8. $\sqrt{25} + \sqrt{49}$

9. $-\sqrt{\dfrac{1}{9}}$

10. $-\sqrt{\dfrac{4}{25}}$

11. $\sqrt{-36}$

12. $\sqrt{-49}$

13. $\sqrt{0.04}$

14. $\sqrt{0.16}$

For the given function, find the indicated function values. Find the domain of each function. Round your answers to the nearest tenth when necessary. You may use a table of square roots or a calculator as needed.

15. $f(x) = \sqrt{3x + 21};\quad f(0), f(1), f(5), f(-4)$

16. $f(x) = \sqrt{10x + 5};\quad f(0), f(1), f(2), f(3)$

17. $f(x) = \sqrt{0.5x - 5};\quad f(10), f(12), f(18), f(20)$

18. $f(x) = \sqrt{2.5x - 1};\quad f(2), f(4), f(6), f(8)$

Graph each of the following functions. Plot at least four points for each function.

19. $f(x) = \sqrt{x - 1}$

20. $f(x) = \sqrt{x - 3}$

21. $f(x) = \sqrt{3x + 9}$

22. $f(x) = \sqrt{2x + 4}$

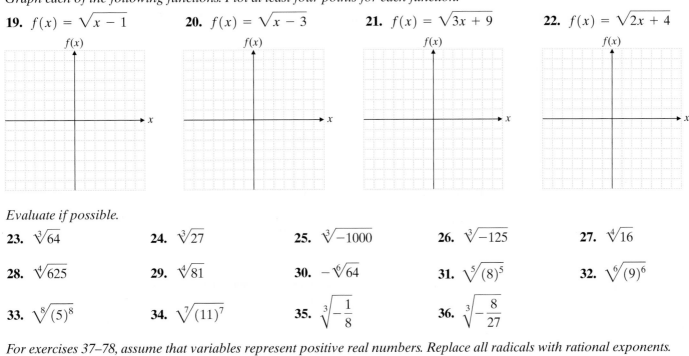

Evaluate if possible.

23. $\sqrt[3]{64}$

24. $\sqrt[3]{27}$

25. $\sqrt[3]{-1000}$

26. $\sqrt[3]{-125}$

27. $\sqrt[4]{16}$

28. $\sqrt[4]{625}$

29. $\sqrt[4]{81}$

30. $-\sqrt[6]{64}$

31. $\sqrt[5]{(8)^5}$

32. $\sqrt[6]{(9)^6}$

33. $\sqrt[8]{(5)^8}$

34. $\sqrt[7]{(11)^7}$

35. $\sqrt[3]{-\dfrac{1}{8}}$

36. $\sqrt[3]{-\dfrac{8}{27}}$

For exercises 37–78, assume that variables represent positive real numbers. Replace all radicals with rational exponents.

37. $\sqrt[3]{y}$

38. $\sqrt{a}$

39. $\sqrt[5]{m^3}$

40. $\sqrt[6]{b^5}$

41. $\sqrt[4]{2a}$

42. $\sqrt[5]{4b}$

43. $\sqrt[7]{(a + b)^3}$

44. $\sqrt[9]{(a - b)^5}$

45. $\sqrt{\sqrt[3]{x}}$

46. $\sqrt[5]{\sqrt{y}}$

47. $\left(\sqrt[6]{3x}\right)^5$

48. $\left(\sqrt[5]{2x}\right)^3$

Simplify.

49. $\sqrt[6]{(12)^6}$

50. $\sqrt[5]{(-11)^5}$

51. $\sqrt[3]{x^{12}y^3}$

52. $\sqrt[3]{x^9y^{18}}$

53. $\sqrt{36x^8y^4}$

54. $\sqrt{49x^2y^8}$

55. $\sqrt[4]{16a^8b^4}$

56. $\sqrt[4]{81a^{12}b^{20}}$

Change to radical form.

57. $y^{4/7}$

58. $x^{5/6}$

59. $7^{-2/3}$

60. $5^{-3/5}$

61. $(a + 5b)^{3/4}$

62. $(3a + b)^{2/7}$

63. $(-x)^{3/5}$

64. $(-y)^{5/7}$

65. $(2xy)^{3/5}$

66. $(3ab)^{2/7}$

Mixed Practice *Evaluate or simplify.*

67. $9^{3/2}$

68. $8^{2/3}$

69. $\left(\dfrac{4}{25}\right)^{1/2}$

70. $\left(\dfrac{16}{81}\right)^{3/4}$

71. $\left(\dfrac{1}{8}\right)^{-1/3}$

72. $\left(\dfrac{25}{9}\right)^{-1/2}$

73. $(64x^4)^{-1/2}$

74. $(81y^{10})^{-1/2}$

75. $\sqrt{121x^4}$

76. $\sqrt{49x^8}$

77. $\sqrt{144a^6b^{24}}$

78. $\sqrt{25a^{14}b^{18}}$

Simplify. Assume that the variables represent any positive or negative real number.

79. $\sqrt{25x^2}$

80. $\sqrt{100x^2}$

81. $\sqrt[3]{-8x^6}$

82. $\sqrt[3]{-27x^9}$

83. $\sqrt[4]{x^8y^{16}}$

84. $\sqrt[4]{x^{16}y^{40}}$

85. $\sqrt[4]{a^{12}b^4}$

86. $\sqrt[4]{a^4b^{20}}$

87. $\sqrt{25x^{12}y^4}$

88. $\sqrt{36x^{16}y^8}$

Cumulative Review

89. **[2.2.1]** Solve for x. $-5x + 2y = 6$

90. **[2.2.1]** Solve for y. $x = \dfrac{2}{3}y + 4$

Quick Quiz 7.2 Simplify. Assume that variables represent positive real numbers.

1. $\left(\dfrac{4}{25}\right)^{3/2}$

2. $\sqrt[3]{-64}$

3. $\sqrt{121x^{10}y^{12}}$

4. **Concept Check** Explain how you would simplify the following.
$$\sqrt[4]{81x^8y^{16}}$$

7.3 SIMPLIFYING, ADDING, AND SUBTRACTING RADICALS

① Simplifying a Radical by Using the Product Rule

When we simplify a radical, we want to get an equivalent expression with the smallest possible quantity in the radicand. We can use the product rule for radicals to simplify radicals.

Student Learning Objectives

After studying this section, you will be able to:

① Simplify a radical by using the product rule.

② Add and subtract like radical terms.

PRODUCT RULE FOR RADICALS

For all nonnegative real numbers a and b and positive integers n,

$$\sqrt[n]{a}\,\sqrt[n]{b} = \sqrt[n]{ab}.$$

You should be able to derive the product rule from your knowledge of the laws of exponents. We have

$$\sqrt[n]{a}\,\sqrt[n]{b} = a^{1/n}\,b^{1/n} = (ab)^{1/n} = \sqrt[n]{ab}.$$

Throughout the remainder of this chapter, assume that all variables in any radicand represent nonnegative numbers, unless a specific statement is made to the contrary.

EXAMPLE 1 Simplify. $\sqrt{32}$

Solution 1: $\sqrt{32} = \sqrt{16 \cdot 2} = \sqrt{16}\sqrt{2} = 4\sqrt{2}$

Solution 2: $\sqrt{32} = \sqrt{4 \cdot 8} = \sqrt{4}\sqrt{8} = 2\sqrt{8} = 2\sqrt{4 \cdot 2} = 2\sqrt{4}\sqrt{2} = 4\sqrt{2}$

Although we obtained the same answer both times, the first solution is much shorter. You should try to use the largest factor that is a perfect square when you use the product rule.

Practice Problem 1 Simplify. $\sqrt{20}$

EXAMPLE 2 Simplify. $\sqrt{48}$

Solution $$\sqrt{48} = \sqrt{16}\sqrt{3} = 4\sqrt{3}$$

Practice Problem 2 Simplify. $\sqrt{27}$

EXAMPLE 3 Simplify.

(a) $\sqrt[3]{16}$ (b) $\sqrt[3]{-81}$

Solution

(a) $\sqrt[3]{16} = \sqrt[3]{8}\sqrt[3]{2} = 2\sqrt[3]{2}$

(b) $\sqrt[3]{-81} = \sqrt[3]{-27}\sqrt[3]{3} = -3\sqrt[3]{3}$

Practice Problem 3 Simplify.

(a) $\sqrt[3]{24}$ (b) $\sqrt[3]{-108}$

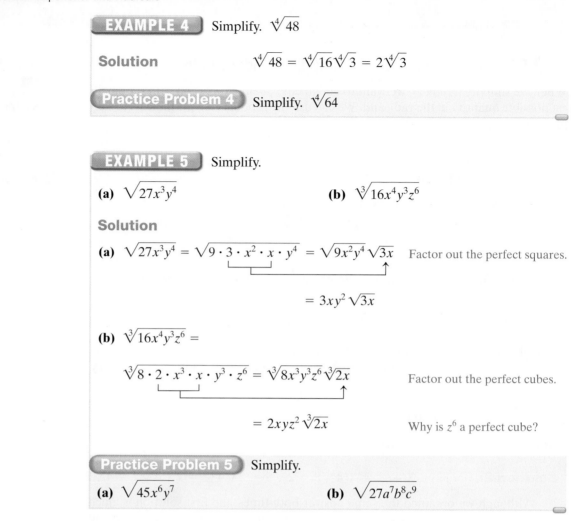

EXAMPLE 4   Simplify. $\sqrt[4]{48}$

Solution $\sqrt[4]{48} = \sqrt[4]{16}\sqrt[4]{3} = 2\sqrt[4]{3}$

Practice Problem 4   Simplify. $\sqrt[4]{64}$

EXAMPLE 5   Simplify.

(a) $\sqrt{27x^3y^4}$ **(b)** $\sqrt[3]{16x^4y^3z^6}$

Solution

(a) $\sqrt{27x^3y^4} = \sqrt{9 \cdot 3 \cdot x^2 \cdot x \cdot y^4} = \sqrt{9x^2y^4}\sqrt{3x}$ Factor out the perfect squares.

$$= 3xy^2\sqrt{3x}$$

(b) $\sqrt[3]{16x^4y^3z^6} =$

$\sqrt[3]{8 \cdot 2 \cdot x^3 \cdot x \cdot y^3 \cdot z^6} = \sqrt[3]{8x^3y^3z^6}\sqrt[3]{2x}$ Factor out the perfect cubes.

$$= 2xyz^2\sqrt[3]{2x}$$ Why is z^6 a perfect cube?

Practice Problem 5   Simplify.

(a) $\sqrt{45x^6y^7}$ **(b)** $\sqrt{27a^7b^8c^9}$

② Adding and Subtracting Like Radical Terms

Only like radicals can be added or subtracted. Two radicals are **like radicals** if they have the same radicand and the same index. $2\sqrt{5}$ and $3\sqrt{5}$ are like radicals. $2\sqrt{5}$ and $2\sqrt{3}$ are not like radicals; $2\sqrt{5}$ and $2\sqrt[3]{5}$ are not like radicals. When we combine radicals, we combine like terms by using the distributive property.

EXAMPLE 6   Combine. $2\sqrt{5} + 3\sqrt{5} - 4\sqrt{5}$

Solution $2\sqrt{5} + 3\sqrt{5} - 4\sqrt{5} = (2 + 3 - 4)\sqrt{5} = 1\sqrt{5} = \sqrt{5}$

Practice Problem 6   Combine. $19\sqrt{xy} + 5\sqrt{xy} - 10\sqrt{xy}$

Sometimes when you simplify radicands, you may find you have like radicals.

EXAMPLE 7 Combine. $5\sqrt{3} - \sqrt{27} + 2\sqrt{48}$

Solution

$$5\sqrt{3} - \sqrt{27} + 2\sqrt{48} = 5\sqrt{3} - \sqrt{9}\sqrt{3} + 2\sqrt{16}\sqrt{3}$$
$$= 5\sqrt{3} - 3\sqrt{3} + 2(4)\sqrt{3}$$
$$= 5\sqrt{3} - 3\sqrt{3} + 8\sqrt{3}$$
$$= 10\sqrt{3}$$

Practice Problem 7 Combine.

$$4\sqrt{2} - 5\sqrt{50} - 3\sqrt{98}$$

EXAMPLE 8 Combine. $6\sqrt{x} + 4\sqrt{12x} - \sqrt{75x} + 3\sqrt{x}$

Solution

$$6\sqrt{x} + 4\sqrt{12x} - \sqrt{75x} + 3\sqrt{x} = 6\sqrt{x} + 4\sqrt{4}\sqrt{3x} - \sqrt{25}\sqrt{3x} + 3\sqrt{x}$$
$$= 6\sqrt{x} + 8\sqrt{3x} - 5\sqrt{3x} + 3\sqrt{x}$$
$$= 6\sqrt{x} + 3\sqrt{x} + 8\sqrt{3x} - 5\sqrt{3x}$$
$$= 9\sqrt{x} + 3\sqrt{3x}$$

Practice Problem 8 Combine.

$$4\sqrt{2x} + \sqrt{18x} - 2\sqrt{125x} - 6\sqrt{20x}$$

EXAMPLE 9 Combine. $2\sqrt[3]{81x^3y^4} + 3xy\sqrt[3]{24y}$

Solution

$$2\sqrt[3]{81x^3y^4} + 3xy\sqrt[3]{24y} = 2\sqrt[3]{27x^3y^3}\sqrt[3]{3y} + 3xy\sqrt[3]{8}\sqrt[3]{3y}$$
$$= 2(3xy)\sqrt[3]{3y} + 3xy(2)\sqrt[3]{3y}$$
$$= 6xy\sqrt[3]{3y} + 6xy\sqrt[3]{3y}$$
$$= 12xy\sqrt[3]{3y}$$

Practice Problem 9 Combine.

$$3x\sqrt[3]{54x^4} - 3\sqrt[3]{16x^7}$$

Simplify. Assume that all variables are nonnegative real numbers.

1. $\sqrt{8}$ **2.** $\sqrt{12}$ **3.** $\sqrt{18}$ **4.** $\sqrt{75}$

5. $\sqrt{28}$ **6.** $\sqrt{54}$ **7.** $\sqrt{50}$

8. $\sqrt{60}$ **9.** $\sqrt{9x^3}$ **10.** $\sqrt{16x^5}$

11. $\sqrt{40a^6b^7}$ **12.** $\sqrt{45a^3b^8}$ **13.** $\sqrt{90x^3yz^4}$

14. $\sqrt{24xy^8z^3}$ **15.** $\sqrt[3]{8}$ **16.** $\sqrt[3]{125}$

17. $\sqrt[3]{40}$ **18.** $\sqrt[3]{128}$ **19.** $\sqrt[3]{54a^2}$

20. $\sqrt[3]{32n}$ **21.** $\sqrt[3]{27a^5b^9}$ **22.** $\sqrt[3]{125a^6b^2}$

23. $\sqrt[3]{24x^6y^{11}}$ **24.** $\sqrt[3]{40x^7y^{26}}$ **25.** $\sqrt[4]{81kp^{23}}$

26. $\sqrt[4]{16k^{12}p^{18}}$ **27.** $\sqrt[5]{-32x^5y^6}$ **28.** $\sqrt[5]{-243x^4y^{10}}$

To Think About

29. $\sqrt[4]{1792} = a\sqrt[4]{7}$. What is the value of a?   **30.** $\sqrt[3]{3072} = b\sqrt[3]{6}$. What is the value of b?

Combine.

31. $4\sqrt{5} + 8\sqrt{5}$ **32.** $3\sqrt{13} + 7\sqrt{13}$ **33.** $4\sqrt{3} + \sqrt{7} - 5\sqrt{7}$

34. $2\sqrt{6} + \sqrt{2} - 5\sqrt{6}$ **35.** $3\sqrt{32} - \sqrt{2}$ **36.** $\sqrt{90} - \sqrt{10}$

37. $4\sqrt{12} + \sqrt{27}$ **38.** $5\sqrt{75} + \sqrt{48}$ **39.** $\sqrt{8} + \sqrt{50} - 2\sqrt{72}$

40. $\sqrt{45} + \sqrt{80} - 3\sqrt{20}$

41. $\sqrt{48} - 2\sqrt{27} + \sqrt{12}$

42. $-4\sqrt{18} + \sqrt{98} - \sqrt{32}$

43. $-5\sqrt{45} + 6\sqrt{20} + 3\sqrt{5}$

44. $-7\sqrt{10} + 4\sqrt{40} - 8\sqrt{90}$

Combine. Assume that all variables represent nonnegative real numbers.

45. $3\sqrt{48x} - 2\sqrt{12x}$

46. $5\sqrt{27x} - 4\sqrt{75x}$

47. $5\sqrt{2x} + 2\sqrt{18x} + 2\sqrt{32x}$

48. $4\sqrt{50x} + 3\sqrt{2x} + \sqrt{72x}$

49. $\sqrt{44} - 3\sqrt{63x} + 4\sqrt{28x}$

50. $\sqrt{63x} - \sqrt{54x} + \sqrt{24x}$

51. $\sqrt{200x^3} - x\sqrt{32x}$
$4x\sqrt{2x}$

52. $\sqrt{75a^3} + a\sqrt{12a}$

53. $\sqrt[3]{16} + 3\sqrt[3]{54}$

54. $\sqrt[3]{128} - 4\sqrt[3]{16}$

55. $4\sqrt[3]{x^4y^3} - 3\sqrt[3]{xy^5}$

56. $2y\sqrt[3]{27x^3} - 3\sqrt[3]{125x^3y^3}$

To Think About

57. Use a calculator to show that
$\sqrt{48} + \sqrt{27} + \sqrt{75} = 12\sqrt{3}$.

58. Use a calculator to show that
$\sqrt{98} + \sqrt{50} + \sqrt{128} = 20\sqrt{2}$.

Applications *Electric Current* *We can approximate the amount of current in amps I (amperes) drawn by an appliance in the home using the formula*

$$I = \sqrt{\frac{P}{R}},$$

where P is the power measured in watts and R is the resistance measured in ohms. In exercises 59 and 60, round your answers to three decimal places.

59. What is the current I if $P = 500$ watts and $R = 10$ ohms?

60. What is the current I if $P = 640$ and $R = 8$?

Period of a Pendulum The **period** of a pendulum is the amount of time it takes the pendulum to make one complete swing back and forth. If the length of the pendulum L is measured in feet, then its period T measured in seconds is given by the formula

$$T = 2\pi\sqrt{\frac{L}{32}}.$$

Use $\pi \approx 3.14$ for exercises 61–62.

61. Find the period of a pendulum if its length is 8 feet.

62. How much longer is the period of a pendulum consisting of a person swinging on a 128-foot rope than that of a person swinging on a 32-foot rope?

Cumulative Review *Factor completely.*

63. **[5.7.1]** $16x^3 - 56x^2y + 49xy^2$

64. **[5.7.1]** $81x^2y - 25y$

FDA Recommendations for Phosphorus The FDA recommends that an adult's minimum daily intake of the mineral phosphorus be 1 gram. A small serving of scallops (six average-size scallops) has 0.2 gram of phosphorus, while one small serving of skim milk (1 cup) has 0.25 gram of phosphorus.

65. **[4.3.1]** If the number of servings of scallops and the number of servings of skim milk totals 4.5 servings, how many of each would you need to meet the minimum daily requirement of phosphorus?

66. **[1.2.2]** If you eat only scallops, how many servings would you need to obtain the minimum daily requirement of phosphorus? If you drink only skim milk, how many servings would you need to obtain the minimum daily requirement of phosphorus?

Quick Quiz 7.3 Simplify.

1. $\sqrt{120x^7y^8}$

2. $\sqrt[3]{16x^{15}y^{10}}$

3. Combine.

$$2\sqrt{75} + 3\sqrt{48} - 4\sqrt{27}$$

4. **Concept Check** Explain how you would simplify the following.

$$\sqrt[4]{16x^{13}y^{16}}$$

① Multiplying Radical Expressions

We use the product rule for radicals to multiply radical expressions. Recall that $\sqrt[n]{a}\sqrt[n]{b} = \sqrt[n]{ab}$.

EXAMPLE 1 Multiply. $(3\sqrt{2})(5\sqrt{11x})$

Solution $(3\sqrt{2})(5\sqrt{11x}) = (3)(5)\sqrt{2 \cdot 11x} = 15\sqrt{22x}$

Practice Problem 1 Multiply. $(-4\sqrt{2})(-3\sqrt{13x})$

EXAMPLE 2 Multiply. $\sqrt{6x}(\sqrt{3} + \sqrt{2x} + \sqrt{5})$

Solution

$$\sqrt{6x}(\sqrt{3} + \sqrt{2x} + \sqrt{5}) = (\sqrt{6x})(\sqrt{3}) + (\sqrt{6x})(\sqrt{2x}) + (\sqrt{6x})(\sqrt{5})$$
$$= \sqrt{18x} + \sqrt{12x^2} + \sqrt{30x}$$
$$= \sqrt{9}\sqrt{2x} + \sqrt{4x^2}\sqrt{3} + \sqrt{30x}$$
$$= 3\sqrt{2x} + 2x\sqrt{3} + \sqrt{30x}$$

Practice Problem 2 Multiply. $\sqrt{2x}(\sqrt{5} + 2\sqrt{3x} + \sqrt{8})$

To multiply two binomials containing radicals, we can use the distributive property. Most students find that the FOIL method is helpful in remembering how to find the four products.

EXAMPLE 3 Multiply. $(\sqrt{2} + 3\sqrt{5})(2\sqrt{2} - \sqrt{5})$

Solution By FOIL:

$$(\sqrt{2} + 3\sqrt{5})(2\sqrt{2} - \sqrt{5}) = 2\sqrt{4} - \sqrt{10} + 6\sqrt{10} - 3\sqrt{25}$$
$$= 4 + 5\sqrt{10} - 15$$
$$= -11 + 5\sqrt{10}$$

By the distributive property:

$$(\sqrt{2} + 3\sqrt{5})(2\sqrt{2} - \sqrt{5}) = (\sqrt{2} + 3\sqrt{5})(2\sqrt{2}) - (\sqrt{2} + 3\sqrt{5})(\sqrt{5})$$
$$= (\sqrt{2})(2\sqrt{2}) + (3\sqrt{5})(2\sqrt{2}) - (\sqrt{2})(\sqrt{5}) - (3\sqrt{5})(\sqrt{5})$$
$$= 2\sqrt{4} + 6\sqrt{10} - \sqrt{10} - 3\sqrt{25}$$
$$= 4 + 5\sqrt{10} - 15$$
$$= -11 + 5\sqrt{10}$$

Practice Problem 3 Multiply. $(\sqrt{7} + 4\sqrt{2})(2\sqrt{7} - 3\sqrt{2})$

EXAMPLE 4 Multiply. $(7 - 3\sqrt{2})(4 - \sqrt{3})$

Solution

$$(7 - 3\sqrt{2})(4 - \sqrt{3}) = 28 - 7\sqrt{3} - 12\sqrt{2} + 3\sqrt{6}$$

Practice Problem 4 Multiply. $(2 - 5\sqrt{5})(3 - 2\sqrt{2})$

EXAMPLE 5 Multiply. $\left(\sqrt{7} + \sqrt{3x}\right)^2$

Solution

Method 1: We can use the FOIL method or the distributive property.

$$\left(\sqrt{7} + \sqrt{3x}\right)\left(\sqrt{7} + \sqrt{3x}\right) = \sqrt{49} + \sqrt{21x} + \sqrt{21x} + \sqrt{9x^2}$$

$$= 7 + 2\sqrt{21x} + 3x$$

Method 2: We could also use the Chapter 5 formula.

$$(a + b)^2 = a^2 + 2ab + b^2,$$

where $a = \sqrt{7}$ and $b = \sqrt{3x}$. Then

$$\left(\sqrt{7} + \sqrt{3x}\right)^2 = \left(\sqrt{7}\right)^2 + 2\sqrt{7}\sqrt{3x} + \left(\sqrt{3x}\right)^2$$

$$= 7 + 2\sqrt{21x} + 3x$$

Practice Problem 5 Multiply, using the approach that seems easiest to you. $\left(\sqrt{5x} + \sqrt{10}\right)^2$

EXAMPLE 6 Multiply.

(a) $\sqrt[3]{3x}\left(\sqrt[3]{x^2} + 3\sqrt[3]{4y}\right)$ **(b)** $\left(\sqrt[3]{2y} + \sqrt[3]{4}\right)\left(2\sqrt[3]{4y^2} - 3\sqrt[3]{2}\right)$

Solution

(a) $\sqrt[3]{3x}\left(\sqrt[3]{x^2} + 3\sqrt[3]{4y}\right) = \left(\sqrt[3]{3x}\right)\left(\sqrt[3]{x^2}\right) + 3\left(\sqrt[3]{3x}\right)\left(\sqrt[3]{4y}\right)$

$$= \sqrt[3]{3x^3} + 3\sqrt[3]{12xy}$$

$$= x\sqrt[3]{3} + 3\sqrt[3]{12xy}$$

(b) $\left(\sqrt[3]{2y} + \sqrt[3]{4}\right)\left(2\sqrt[3]{4y^2} - 3\sqrt[3]{2}\right) = 2\sqrt[3]{8y^3} - 3\sqrt[3]{4y} + 2\sqrt[3]{16y^2} - 3\sqrt[3]{8}$

$$= 2(2y) - 3\sqrt[3]{4y} + 2\sqrt[3]{8}\sqrt[3]{2y^2} - 3(2)$$

$$= 4y - 3\sqrt[3]{4y} + 4\sqrt[3]{2y^2} - 6$$

Practice Problem 6 Multiply.

(a) $\sqrt[3]{2x}\left(\sqrt[3]{4x^2} + 3\sqrt[3]{y}\right)$

(b) $\left(\sqrt[3]{7} + \sqrt[3]{x^2}\right)\left(2\sqrt[3]{49} - \sqrt[3]{x}\right)$

2 Dividing Radical Expressions

We can use the laws of exponents to develop a rule for dividing two radicals.

$$\sqrt[n]{\frac{a}{b}} = \left(\frac{a}{b}\right)^{1/n} = \frac{a^{1/n}}{b^{1/n}} = \frac{\sqrt[n]{a}}{\sqrt[n]{b}}$$

This quotient rule is very useful. We now state it more formally.

QUOTIENT RULE FOR RADICALS

For all nonnegative real numbers a, all positive real numbers b, and positive integers n,

$$\frac{\sqrt[n]{a}}{\sqrt[n]{b}} = \sqrt[n]{\frac{a}{b}}.$$

Sometimes it will be best to change $\sqrt[n]{\dfrac{a}{b}}$ to $\dfrac{\sqrt[n]{a}}{\sqrt[n]{b}}$, whereas at other times it will be best to change $\dfrac{\sqrt[n]{a}}{\sqrt[n]{b}}$ to $\sqrt[n]{\dfrac{a}{b}}$. To use the quotient rule for radicals, you need to have good number sense. You should know your squares up to 15^2 and your cubes up to 5^3.

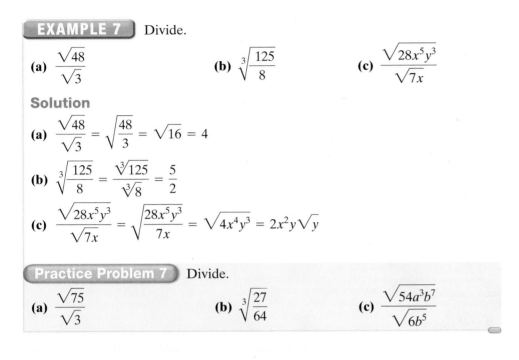

EXAMPLE 7   Divide.

(a) $\dfrac{\sqrt{48}}{\sqrt{3}}$ **(b)** $\sqrt[3]{\dfrac{125}{8}}$ **(c)** $\dfrac{\sqrt{28x^5y^3}}{\sqrt{7x}}$

Solution

(a) $\dfrac{\sqrt{48}}{\sqrt{3}} = \sqrt{\dfrac{48}{3}} = \sqrt{16} = 4$

(b) $\sqrt[3]{\dfrac{125}{8}} = \dfrac{\sqrt[3]{125}}{\sqrt[3]{8}} = \dfrac{5}{2}$

(c) $\dfrac{\sqrt{28x^5y^3}}{\sqrt{7x}} = \sqrt{\dfrac{28x^5y^3}{7x}} = \sqrt{4x^4y^3} = 2x^2y\sqrt{y}$

Practice Problem 7   Divide.

(a) $\dfrac{\sqrt{75}}{\sqrt{3}}$ **(b)** $\sqrt[3]{\dfrac{27}{64}}$ **(c)** $\dfrac{\sqrt{54a^3b^7}}{\sqrt{6b^5}}$

③ Simplifying Radical Expressions by Rationalizing the Denominator

Recall that to simplify a radical we want to get the smallest possible quantity in the radicand. Whenever possible, we find the square root of a perfect square. Thus, to simplify $\sqrt{\dfrac{7}{16}}$ we have

$$\sqrt{\dfrac{7}{16}} = \dfrac{\sqrt{7}}{\sqrt{16}} = \dfrac{\sqrt{7}}{4}.$$

Notice that the denominator does not contain a square root. The expression $\dfrac{\sqrt{7}}{4}$ is in simplest form.

Let's look at $\sqrt{\dfrac{16}{7}}$. We have

$$\sqrt{\dfrac{16}{7}} = \dfrac{\sqrt{16}}{\sqrt{7}} = \dfrac{4}{\sqrt{7}}.$$

Notice that the denominator contains a square root. If an expression contains a square root in the denominator, it is not considered to be simplified. How can we rewrite $\dfrac{4}{\sqrt{7}}$ as an equivalent expression that does not contain the $\sqrt{7}$ in the denominator? Since $\sqrt{7}\sqrt{7} = 7$, we can multiply the numerator and the denominator by the radical in the denominator.

$$\dfrac{4}{\sqrt{7}} \cdot \dfrac{\sqrt{7}}{\sqrt{7}} = \dfrac{4\sqrt{7}}{\sqrt{49}} = \dfrac{4\sqrt{7}}{7}$$

This expression is considered to be in simplest form. We call this process rationalizing the denominator.

Rationalizing the denominator is the process of transforming a fraction with one or more radicals in the denominator into an equivalent fraction without a radical in the denominator.

EXAMPLE 8 Simplify by rationalizing the denominator. $\dfrac{3}{\sqrt{2}}$

Solution

$$\frac{3}{\sqrt{2}} = \frac{3}{\sqrt{2}} \cdot \frac{\sqrt{2}}{\sqrt{2}} \qquad \text{Since } \frac{\sqrt{2}}{\sqrt{2}} = 1.$$

$$= \frac{3\sqrt{2}}{\sqrt{4}} \qquad \text{Product rule for radicals.}$$

$$= \frac{3\sqrt{2}}{2}$$

Practice Problem 8 Simplify by rationalizing the denominator.

$$\frac{7}{\sqrt{3}}$$

We can rationalize the denominator either before or after we simplify the denominator.

EXAMPLE 9 Simplify. $\dfrac{3}{\sqrt{12x}}$

Solution

Method 1: First we simplify the radical in the denominator, and then we multiply in order to rationalize the denominator.

$$\frac{3}{\sqrt{12x}} = \frac{3}{\sqrt{4}\sqrt{3x}} = \frac{3}{2\sqrt{3x}} \cdot \frac{\sqrt{3x}}{\sqrt{3x}} = \frac{3\sqrt{3x}}{2(3x)} = \frac{\sqrt{3x}}{2x}$$

Method 2: We can multiply numerator and denominator by a value that will make the radicand in the denominator a perfect square (i.e., rationalize the denominator).

$$\frac{3}{\sqrt{12x}} = \frac{3}{\sqrt{12x}} \cdot \frac{\sqrt{3x}}{\sqrt{3x}}$$

$$= \frac{3\sqrt{3x}}{\sqrt{36x^2}} \qquad \text{Since } \sqrt{12x}\sqrt{3x} = \sqrt{36x^2}.$$

$$= \frac{3\sqrt{3x}}{6x} = \frac{\sqrt{3x}}{2x}$$

Practice Problem 9 Simplify.

$$\frac{8}{\sqrt{20x}}$$

If a radicand contains a fraction, it is not considered to be simplified. We can use the quotient rule for radicals and then rationalize the denominator to simplify the radical. We have already rationalized denominators when they contain square roots. Now we will rationalize denominators when they contain radical expressions that are cube roots or higher-order roots.

EXAMPLE 10   Simplify. $\sqrt[3]{\dfrac{2}{3x^2}}$

Solution

Method 1: $\sqrt[3]{\dfrac{2}{3x^2}} = \dfrac{\sqrt[3]{2}}{\sqrt[3]{3x^2}}$ Quotient rule for radicals.

$$= \dfrac{\sqrt[3]{2}}{\sqrt[3]{3x^2}} \cdot \dfrac{\sqrt[3]{9x}}{\sqrt[3]{9x}}$$ Multiply the numerator and denominator by an appropriate value so that the new radicand in the denominator will be a perfect cube.

$$= \dfrac{\sqrt[3]{18x}}{\sqrt[3]{27x^3}}$$ Observe that we can evaluate the cube root in the denominator.

$$= \dfrac{\sqrt[3]{18x}}{3x}$$

Method 2: $\sqrt[3]{\dfrac{2}{3x^2}} = \sqrt[3]{\dfrac{2}{3x^2} \cdot \dfrac{9x}{9x}}$

$$= \sqrt[3]{\dfrac{18x}{27x^3}}$$

$$= \dfrac{\sqrt[3]{18x}}{\sqrt[3]{27x^3}}$$

$$= \dfrac{\sqrt[3]{18x}}{3x}$$

Practice Problem 10   Simplify.

$$\sqrt[3]{\dfrac{6}{5x}}$$

If the denominator of a radical expression contains a sum or difference with radicals, we multiply the numerator and denominator by the *conjugate* of the denominator. For example, the conjugate of $x + \sqrt{y}$ is $x - \sqrt{y}$; similarly, the conjugate of $x - \sqrt{y}$ is $x + \sqrt{y}$. What is the conjugate of $3 + \sqrt{2}$? It is $3 - \sqrt{2}$. How about $\sqrt{11} + \sqrt{xyz}$? It is $\sqrt{11} - \sqrt{xyz}$.

CONJUGATES

The expressions $a + b$ and $a - b$, where a and b represent any algebraic term, are called **conjugates.** Each expression is the conjugate of the other expression.

Multiplying by conjugates is simply an application of the formula

$$(a + b)(a - b) = a^2 - b^2.$$

For example,

$$\left(\sqrt{x} + \sqrt{y}\right)\left(\sqrt{x} - \sqrt{y}\right) = \left(\sqrt{x}\right)^2 - \left(\sqrt{y}\right)^2 = x - y.$$

EXAMPLE 11 Simplify. $\dfrac{5}{3 + \sqrt{2}}$

Solution

$$\dfrac{5}{3 + \sqrt{2}} = \dfrac{5}{3 + \sqrt{2}} \cdot \dfrac{3 - \sqrt{2}}{3 - \sqrt{2}}$$ Multiply the numerator and denominator by the conjugate of $3 + \sqrt{2}$.

$$= \dfrac{15 - 5\sqrt{2}}{3^2 - \left(\sqrt{2}\right)^2}$$

$$= \dfrac{15 - 5\sqrt{2}}{9 - 2} = \dfrac{15 - 5\sqrt{2}}{7}$$

NOTE TO STUDENT: *Fully worked-out solutions to all of the Practice Problems can be found at the back of the text starting at page SP-1*

Practice Problem 11 Simplify.

$$\dfrac{4}{2 + \sqrt{5}}$$

EXAMPLE 12 Simplify. $\dfrac{\sqrt{7} + \sqrt{3}}{\sqrt{7} - \sqrt{3}}$

Solution The conjugate of $\sqrt{7} - \sqrt{3}$ is $\sqrt{7} + \sqrt{3}$.

$$\dfrac{\sqrt{7} + \sqrt{3}}{\sqrt{7} - \sqrt{3}} \cdot \dfrac{\sqrt{7} + \sqrt{3}}{\sqrt{7} + \sqrt{3}} = \dfrac{\sqrt{49} + 2\sqrt{21} + \sqrt{9}}{\left(\sqrt{7}\right)^2 - \left(\sqrt{3}\right)^2}$$

$$= \dfrac{7 + 2\sqrt{21} + 3}{7 - 3}$$

$$= \dfrac{10 + 2\sqrt{21}}{4}$$

$$= \dfrac{2\left(5 + \sqrt{21}\right)}{2 \cdot 2}$$

$$= \dfrac{\cancel{2}\left(5 + \sqrt{21}\right)}{\cancel{2} \cdot 2}$$

$$= \dfrac{5 + \sqrt{21}}{2}$$

Practice Problem 12 Simplify.

$$\dfrac{\sqrt{11} + \sqrt{5}}{\sqrt{11} - \sqrt{5}}$$

Multiply and simplify. Assume that all variables represent nonnegative numbers.

1. $\sqrt{5}\sqrt{7}$

2. $\sqrt{11}\sqrt{6}$

3. $(5\sqrt{2})(-6\sqrt{5})$

4. $(-7\sqrt{3})(-4\sqrt{10})$

5. $(3\sqrt{10})(-4\sqrt{2})$

6. $(-5\sqrt{6})(2\sqrt{3})$

7. $(-3\sqrt{y})(\sqrt{5x})$

8. $(\sqrt{2x})(-7\sqrt{3y})$

9. $(3x\sqrt{2x})(-2\sqrt{10xy})$

10. $(4\sqrt{3a})(a\sqrt{6ab})$

11. $5\sqrt{a}(3\sqrt{b} - 5)$

12. $-\sqrt{x}(5\sqrt{y} + 3)$

13. $-3\sqrt{a}(\sqrt{2b} + 2\sqrt{5})$

14. $4\sqrt{x}(\sqrt{3y} - 3\sqrt{6})$

15. $-\sqrt{a}(\sqrt{a} - 2\sqrt{b})$

16. $4\sqrt{xy}(3\sqrt{x} - 10\sqrt{xy})$

17. $7\sqrt{x}(2\sqrt{3} - 5\sqrt{x})$

18. $3\sqrt{y}(4\sqrt{6} + 11\sqrt{y})$

19. $(3 - \sqrt{2})(8 + \sqrt{2})$

20. $(\sqrt{5} + 4)(\sqrt{5} - 1)$

21. $(2\sqrt{3} + \sqrt{2})(2\sqrt{3} - 4\sqrt{2})$

22. $(3\sqrt{3} + \sqrt{5})(\sqrt{3} - 2\sqrt{5})$

23. $(\sqrt{7} + 4\sqrt{5x})(2\sqrt{7} + 3\sqrt{5x})$

24. $(\sqrt{6} + 3\sqrt{3y})(5\sqrt{6} + 2\sqrt{3y})$

25. $(\sqrt{3} + 2\sqrt{2})(\sqrt{5} + \sqrt{3})$

26. $(3\sqrt{5} + \sqrt{3})(\sqrt{2} + 2\sqrt{5})$

27. $(\sqrt{5} - 2\sqrt{6})^2$

28. $(\sqrt{3} + 4\sqrt{7})^2$

29. $(9 - 2\sqrt{b})^2$

30. $(4\sqrt{a} + 5)^2$

31. $(\sqrt{3x + 4} + 3)^2$

32. $(\sqrt{2x + 1} - 2)^2$

33. $(\sqrt[3]{x^2})(3\sqrt[3]{4x} - 4\sqrt[3]{x^5})$

34. $(2\sqrt[3]{x})(\sqrt[3]{4x^2} - \sqrt[3]{14x})$

35. $(\sqrt[3]{3} + \sqrt[3]{2})(\sqrt[3]{9} - \sqrt[3]{4})$

36. $(\sqrt[3]{4} - \sqrt[3]{6})(\sqrt[3]{2} + \sqrt[3]{36})$

Divide and simplify. Assume that all variables represent positive numbers.

37. $\sqrt{\dfrac{49}{25}}$

38. $\sqrt{\dfrac{16}{36}}$

39. $\sqrt{\dfrac{12x}{49y^6}}$

40. $\sqrt{\dfrac{27a^4}{64x^2}}$

41. $\sqrt[3]{\dfrac{8x^5y^6}{27}}$

42. $\sqrt[3]{\dfrac{125a^3b^4}{64}}$

43. $\dfrac{\sqrt[3]{5y^8}}{\sqrt[3]{27x^3}}$

44. $\dfrac{\sqrt[3]{9y^{10}}}{\sqrt[3]{64x^6}}$

Simplify by rationalizing the denominator.

45. $\dfrac{3}{\sqrt{2}}$

46. $\dfrac{5}{\sqrt{7}}$

47. $\sqrt{\dfrac{4}{3}}$

48. $\sqrt{\dfrac{25}{2}}$

49. $\dfrac{1}{\sqrt{5y}}$

50. $\dfrac{1}{\sqrt{3x}}$

51. $\dfrac{\sqrt{14a}}{\sqrt{2y}}$

52. $\dfrac{\sqrt{15a}}{\sqrt{3b}}$

53. $\dfrac{\sqrt{2}}{\sqrt{6x}}$

54. $\dfrac{\sqrt{5y}}{\sqrt{10x}}$

55. $\dfrac{x}{\sqrt{5}-\sqrt{2}}$

56. $\dfrac{y}{\sqrt{7}+\sqrt{3}}$

57. $\dfrac{2y}{\sqrt{6}+\sqrt{5}}$

58. $\dfrac{3x}{\sqrt{10}-\sqrt{2}}$

59. $\dfrac{\sqrt{y}}{\sqrt{6}+\sqrt{2y}}$

60. $\dfrac{\sqrt{y}}{\sqrt{3y}+\sqrt{5}}$

61. $\dfrac{\sqrt{5}+\sqrt{3}}{\sqrt{5}-\sqrt{3}}$

62. $\dfrac{\sqrt{11}-\sqrt{5}}{\sqrt{11}+\sqrt{5}}$

63. $\dfrac{\sqrt{3x}-2\sqrt{y}}{\sqrt{3x}+\sqrt{y}}$

64. $\dfrac{\sqrt{x}+\sqrt{y}}{\sqrt{x}-2\sqrt{y}}$

Mixed Practice *Simplify each of the following.*

65. $2\sqrt{32}-\sqrt{72}+3\sqrt{18}$

66. $\sqrt{45}-2\sqrt{125}-3\sqrt{20}$

67. $\left(3\sqrt{2}-5\sqrt{3}\right)\left(\sqrt{2}+2\sqrt{3}\right)$

68. $\left(5\sqrt{6}-3\sqrt{2}\right)\left(\sqrt{6}+2\sqrt{2}\right)$

69. $\dfrac{9}{\sqrt{8x}}$

70. $\dfrac{5}{\sqrt{12x}}$

71. $\dfrac{\sqrt{5}+1}{\sqrt{5}+2}$

72. $\dfrac{\sqrt{2}-1}{2\sqrt{2}+1}$

To Think About

73. A student rationalized the denominator of $\dfrac{\sqrt{6}}{2\sqrt{3}-\sqrt{2}}$ and obtained $\dfrac{\sqrt{3}+3\sqrt{2}}{5}$. Find a decimal approximation of each expression. Are the decimals equal? Did the student do the work correctly?

74. A student rationalized the denominator of $\dfrac{\sqrt{5}}{\sqrt{5}+\sqrt{3}}$ and obtained $\dfrac{5-\sqrt{15}}{2}$. Find a decimal approximation of each expression. Are the decimals equal? Did the student do the work correctly?

In calculus, students are sometimes required to rationalize the numerator of an expression. In this case the numerator will not have a radical in the answer. Rationalize the numerator in each of the following:

75. $\dfrac{\sqrt{2}+3\sqrt{5}}{4}$

76. $\dfrac{\sqrt{6}-3\sqrt{3}}{7}$

Applications

Fertilizer Costs *The cost of fertilizing a lawn is $0.25 per square foot. Find the cost to fertilize each of the triangular lawns in exercises 77 and 78. Round your answers to the nearest cent.*

▲ **77.** The base of the triangle is $\left(10+\sqrt{15}\right)$ feet, and the altitude is $\sqrt{60}$ feet.

▲ **78.** The base of the triangle is $\left(8+\sqrt{11}\right)$ feet, and the altitude is $\sqrt{44}$ feet.

▲ **79.** ***Pacemaker Control Panel*** A medical doctor has designed a pacemaker that has a rectangular control panel. This rectangle has a width of $\left(\sqrt{x}+3\right)$ millimeters and a length of $\left(\sqrt{x}+5\right)$ millimeters. Find the area in square millimeters of this rectangle.

▲ **80.** ***FBI Listening Device*** An FBI agent has designed a secret listening device that has a rectangular base. The rectangle has a width of $\left(\sqrt{x}+7\right)$ centimeters and a length of $\left(\sqrt{x}+11\right)$ centimeters. Find the area in square centimeters of this rectangle.

Cumulative Review

Solve the system for x and y.

81. [4.1.6] $\begin{aligned} -2x+3y &= 21 \\ 3x+2y &= 1 \end{aligned}$

Solve the system for x, y, and z.

82. [4.2.2] $\begin{aligned} 2x+3y-\ z &= 8 \\ -x+2y+3z &= -14 \\ 3x-\ y-\ z &= 10 \end{aligned}$

Price of Engagement Rings Tiffany & Co., one of the most famous diamond merchants, offers engagement rings in a variety of price ranges. These data are displayed in the bar graph to the right. The data indicate the price of all diamond rings sold during a recent year.

83. **[1.2.2]** What percent of the rings sold for $23,000 or less?

84. **[1.2.2]** If Tiffany & Co. sold 85,000 diamond engagement rings last year, what is the number of rings that cost more than $5000?

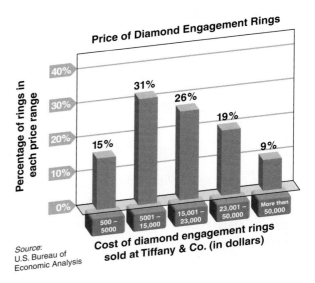

Source:
U.S. Bureau of
Economic Analysis

Quick Quiz 7.4

1. Multiply and simplify. $\left(2\sqrt{3} - \sqrt{5}\right)\left(3\sqrt{3} + 2\sqrt{5}\right)$

Rationalize the denominator.

2. $\dfrac{9}{\sqrt{3x}}$

3. $\dfrac{1 + 2\sqrt{5}}{4 - \sqrt{5}}$

4. **Concept Check** Explain how you would rationalize the denominator of the following.

$$\frac{5\sqrt{3} - 3\sqrt{2}}{3\sqrt{2} - 2\sqrt{3}}$$

How are you doing with your homework assignments in Sections 7.1 to 7.4? Do you feel you have mastered the material so far? Do you understand the concepts you have covered? Before you go further in the textbook, take some time to do each of the following problems.

Leave all answers with positive exponents. All variables represent positive real numbers.

7.1

1. Multiply and simplify your answer. $(-3x^{1/4}y^{1/2})(-2x^{-1/2}y^{1/3})$

Simplify.

2. $(-4x^{-1/4}y^{1/3})^3$

3. $\dfrac{-18x^{-2}y^2}{-3x^{-5}y^{1/3}}$

4. $\left(\dfrac{27x^2y^{-5}}{x^{-4}y^4}\right)^{2/3}$

7.2

Evaluate.

5. $27^{-4/3}$

6. $\sqrt[5]{-243}$

7. $\sqrt{169} + \sqrt[3]{-64}$

8. $\sqrt{64a^8y^{16}}$

9. $\sqrt[3]{27a^{12}b^6c^{15}}$

10. Replace the radical with a rational exponent. $\left(\sqrt[6]{4x}\right)^5$

7.3

11. Simplify. $\sqrt[4]{16x^{20}y^{28}}$

12. Simplify. $\sqrt[3]{32x^8y^{15}}$

13. Combine. $\sqrt{44} - 2\sqrt{99} + 7\sqrt{11}$

14. Combine where possible. $3\sqrt{48y^3} - 2\sqrt[3]{16} + 3\sqrt[3]{54} - 5y\sqrt{12y}$

7.4

15. Multiply and simplify. $\left(5\sqrt{2} - 3\sqrt{5}\right)\left(\sqrt{8} + 3\sqrt{5}\right)$

16. Rationalize the denominator and simplify your answer. $\dfrac{5}{\sqrt{18x}}$

17. Rationalize the denominator and simplify your answer. $\dfrac{\sqrt{2} + \sqrt{3}}{\sqrt{2} - \sqrt{3}}$

Now turn to page SA-18 for the answers to each of these problems. Each answer also includes a reference to the objective in which the problem is first taught. If you missed any of these problems, you should stop and review the Examples and Practice Problems in the referenced objective. A little review now will help you master the material in the upcoming sections of the text.

1. _____

2. _____

3. _____

4. _____

5. _____

6. _____

7. _____

8. _____

9. _____

10. _____

11. _____

12. _____

13. _____

14. _____

15. _____

16. _____

17. _____

1 Solving a Radical Equation by Squaring Each Side Once

A **radical equation** is an equation with a variable in one or more of the radicals. $3\sqrt{x} = 8$ and $\sqrt{3x - 1} = 5$ are radical equations. We solve radical equations by raising each side of the equation to the appropriate power. In other words, we square both sides if the radicals are square roots, cube both sides if the radicals are cube roots, and so on. Once we have done this, solving for the unknown becomes routine.

Sometimes after we square each side, we obtain a quadratic equation. In this case we collect all terms on one side and use the zero factor method that we developed in Section 5.8. After solving the equation, *always* check your answers to see whether extraneous solutions have been introduced.

We will now generalize this rule because it is very useful in higher-level mathematics courses.

RAISING EACH SIDE OF AN EQUATION TO A POWER

If $y = x$, then $y^n = x^n$ for all natural numbers n.

EXAMPLE 1 Solve. $\sqrt{2x + 9} = x + 3$

Solution

$$\left(\sqrt{2x + 9}\right)^2 = (x + 3)^2 \qquad \text{Square each side.}$$

$$2x + 9 = x^2 + 6x + 9 \qquad \text{Simplify.}$$

$$0 = x^2 + 4x \qquad \text{Collect all terms on one side.}$$

$$0 = x(x + 4) \qquad \text{Factor.}$$

$$x = 0 \quad \text{or} \quad x + 4 = 0 \qquad \text{Set each factor equal to zero.}$$

$$x = 0 \qquad\qquad x = -4 \qquad \text{Solve for } x.$$

Check:

For $x = 0$: $\sqrt{2(0) + 9} \overset{?}{=} 0 + 3$ 　　 For $x = -4$: $\sqrt{2(-4) + 9} \overset{?}{=} -4 + 3$

$\sqrt{9} \overset{?}{=} 3$ 　　　　　　　　　　　　$\sqrt{1} \overset{?}{=} -1$

$3 = 3$ ✓ 　　　　　　　　　　　　　　$1 \neq -1$

Therefore, 0 is the only solution to this equation.

Practice Problem 1 Solve and check your solution(s). $\sqrt{3x - 8} = x - 2$

As you begin to solve more complicated radical equations, it is important to make sure that one radical expression is alone on one side of the equation. This is often referred to as **isolating the radical term.**

EXAMPLE 2 Solve. $\sqrt{10x + 5} - 1 = 2x$

Solution

$$\sqrt{10x + 5} = 2x + 1 \qquad \text{Isolate the radical term.}$$

$$\left(\sqrt{10x + 5}\right)^2 = (2x + 1)^2 \qquad \text{Square each side.}$$

$$10x + 5 = 4x^2 + 4x + 1 \qquad \text{Simplify.}$$

$$0 = 4x^2 - 6x - 4 \qquad \text{Collect all terms on one side.}$$

$$0 = 2(2x^2 - 3x - 2) \qquad \text{Factor out the common factor.}$$

$$0 = 2(2x + 1)(x - 2) \qquad \text{Factor completely.}$$

$$2x + 1 = 0 \quad \text{or} \quad x - 2 = 0 \qquad \text{Set each factor equal to zero.}$$

$$2x = -1 \qquad\qquad x = 2 \qquad \text{Solve for } x.$$

$$x = -\frac{1}{2}$$

Check:

$$x = -\frac{1}{2}: \sqrt{10\left(-\frac{1}{2}\right) + 5} - 1 \overset{?}{=} 2\left(-\frac{1}{2}\right) \qquad x = 2: \sqrt{10(2) + 5} - 1 \overset{?}{=} 2(2)$$

$$\sqrt{-5 + 5} - 1 \overset{?}{=} -1 \qquad\qquad \sqrt{25} - 1 \overset{?}{=} 4$$

$$\sqrt{0} - 1 \overset{?}{=} -1 \qquad\qquad 5 - 1 \overset{?}{=} 4$$

$$-1 = -1 \ \checkmark \qquad\qquad 4 = 4 \ \checkmark$$

Both answers check, so $-\dfrac{1}{2}$ and 2 are roots of the equation.

Practice Problem 2 Solve and check your solution(s). $\sqrt{x + 4} = x + 4$

② Solving a Radical Equation by Squaring Each Side Twice

In some exercises, we must square each side twice in order to remove all the radicals. It is important to isolate at least one radical before squaring each side.

EXAMPLE 3 Solve. $\sqrt{5x + 1} - \sqrt{3x} = 1$

Solution

$$\sqrt{5x + 1} = 1 + \sqrt{3x} \qquad\qquad \text{Isolate one of the radicals.}$$

$$\left(\sqrt{5x + 1}\right)^2 = \left(1 + \sqrt{3x}\right)^2 \qquad \text{Square each side.}$$

$$5x + 1 = \left(1 + \sqrt{3x}\right)\left(1 + \sqrt{3x}\right)$$

$$5x + 1 = 1 + 2\sqrt{3x} + 3x$$

$$2x = 2\sqrt{3x} \qquad\qquad \text{Isolate the remaining radical.}$$

$$x = \sqrt{3x} \qquad\qquad \text{Divide each side by 2.}$$

$$(x)^2 = \left(\sqrt{3x}\right)^2 \qquad\qquad \text{Square each side.}$$

$$x^2 = 3x$$

$$x^2 - 3x = 0 \qquad\qquad \text{Collect all terms on one side.}$$

$$x(x - 3) = 0 \qquad\qquad \text{Factor.}$$

$$x = 0 \quad \text{or} \quad x - 3 = 0 \qquad \text{Solve for } x.$$

$$x = 3$$

Check:

$$x = 0: \ \sqrt{5(0) + 1} - \sqrt{3(0)} \overset{?}{=} 1 \qquad x = 3: \ \sqrt{5(3) + 1} - \sqrt{3(3)} \overset{?}{=} 1$$

$$\sqrt{1} - \sqrt{0} \overset{?}{=} 1 \qquad\qquad \sqrt{16} - \sqrt{9} \overset{?}{=} 1$$

$$1 = 1 \ \checkmark \qquad\qquad 1 = 1 \ \checkmark$$

Both answers check. The solutions are 0 and 3.

Practice Problem 3 Solve and check your solution(s).

$$\sqrt{2x + 5} - 2\sqrt{2x} = 1$$

We will now formalize the procedure for solving radical equations.

PROCEDURE FOR SOLVING RADICAL EQUATIONS

1. Perform algebraic operations to obtain one radical by itself on one side of the equation.
2. If the equation contains square roots, square each side of the equation. Otherwise, raise each side to the appropriate power for third- and higher-order roots.
3. Simplify, if possible.
4. If the equation still contains a radical, repeat steps 1 to 3.
5. Collect all terms on one side of the equation.
6. Solve the resulting equation.
7. Check all apparent solutions. Solutions to radical equations must be verified.

EXAMPLE 4 Solve. $\sqrt{2y + 5} - \sqrt{y - 1} = \sqrt{y + 2}$

Solution

$$\left(\sqrt{2y + 5} - \sqrt{y - 1}\right)^2 = \left(\sqrt{y + 2}\right)^2$$

$$\left(\sqrt{2y + 5} - \sqrt{y - 1}\right)\left(\sqrt{2y + 5} - \sqrt{y - 1}\right) = y + 2$$

$$2y + 5 - 2\sqrt{(y - 1)(2y + 5)} + y - 1 = y + 2$$

$$-2\sqrt{(y - 1)(2y + 5)} = -2y - 2$$

$$\sqrt{(y - 1)(2y + 5)} = y + 1 \qquad \text{Divide each side by } -2.$$

$$\left(\sqrt{2y^2 + 3y - 5}\right)^2 = (y + 1)^2 \qquad \text{Square each side.}$$

$$2y^2 + 3y - 5 = y^2 + 2y + 1$$

$$y^2 + y - 6 = 0 \qquad \text{Collect all terms on one side.}$$

$$(y + 3)(y - 2) = 0$$

$$y = -3 \quad \text{or} \quad y = 2$$

Check: Verify that 2 is a valid solution but -3 is not a valid solution.

Practice Problem 4 Solve and check your solution(s).

$$\sqrt{y - 1} + \sqrt{y - 4} = \sqrt{4y - 11}$$

Verbal and Writing Skills

1. Before squaring each side of a radical equation, what step should be taken first?

2. Why do we have to check the solutions when we solve radical equations?

Solve each radical equation. Check your solution(s).

3. $\sqrt{8x + 1} = 5$

4. $\sqrt{5x - 4} = 6$

5. $\sqrt{7x - 3} - 2 = 0$

6. $1 = 5 - \sqrt{9x - 2}$

7. $y + 1 = \sqrt{5y - 1}$

8. $\sqrt{y + 10} = y - 2$

9. $2x = \sqrt{11x + 3}$

10. $3x = \sqrt{15x - 4}$

11. $2 = 5 + \sqrt{2x + 1}$

12. $12 + \sqrt{4x + 5} = 7$

13. $y - \sqrt{y - 3} = 5$

14. $\sqrt{2y - 4} + 2 = y$

15. $\sqrt{y + 1} - 1 = y$

16. $5 + \sqrt{2y + 5} = y$

17. $x - 2\sqrt{x - 3} = 3$

18. $2\sqrt{4x + 1} + 5 = x + 9$

19. $\sqrt{3x^2 - x} = x$

20. $\sqrt{5x^2 - 3x} = 2x$

21. $\sqrt[3]{2x + 3} = 2$

22. $\sqrt[3]{3x + 7} = 4$

23. $\sqrt[3]{4x - 1} = 3$

24. $\sqrt[3]{3 - 5x} = 2$

Solve each radical equation. This will usually involve squaring each side twice. Check your solutions.

25. $\sqrt{x + 4} = 1 + \sqrt{x - 3}$

26. $\sqrt{x + 6} + 1 = \sqrt{x + 15}$

27. $\sqrt{7x + 1} = 1 + \sqrt{5x}$

28. $2\sqrt{x + 4} = 1 + \sqrt{2x + 9}$

29. $\sqrt{x + 6} = 1 + \sqrt{x + 2}$

30. $\sqrt{3x + 1} - \sqrt{x - 4} = 3$

31. $\sqrt{6x + 6} = 1 + \sqrt{4x + 5}$ **32.** $\sqrt{8x + 17} = \sqrt{2x + 8} + 3$ **33.** $\sqrt{2x + 9} - \sqrt{x + 1} = 2$

34. $\sqrt{2x + 6} = \sqrt{7 - 2x} + 1$ **35.** $\sqrt{4x + 6} = \sqrt{x + 1} - \sqrt{x + 5}$ **36.** $\sqrt{3x + 4} + \sqrt{x + 5} = \sqrt{7 - 2x}$

37. $2\sqrt{x} - \sqrt{x - 5} = \sqrt{2x - 2}$ **38.** $\sqrt{3 - 2\sqrt{x}} = \sqrt{x}$

Optional Graphing Calculator Problems

Solve for x. Round your answer to four decimal places.

39. $x = \sqrt{5.326x - 1.983}$ **40.** $\sqrt[3]{5.62x + 9.93} = 1.47$

Applications

41. *Length of Skid Marks* When a car traveling on wet pavement at a speed V in miles per hour stops suddenly, it will produce skid marks of length S feet according to the formula $V = 2\sqrt{3S}$.

   (a) Solve the equation for S.

   (b) Use your result from **(a)** to find the length of the skid mark S if the car is traveling at 30 miles per hour.

42. *Flight Data Recorder* The volume V of a steel container inside a flight data recorder is defined by the equation

$$x = \sqrt{\frac{V}{5}},$$

where x is the sum of the length and the width of the container in inches and the height of the container is 5 inches.

   (a) Solve the equation for V.

   (b) Use the result from **(a)** to find the volume of the container whose length and width total 3.5 inches.

Stopping Distance Recently an experiment was conducted relating the speed a car is traveling and the stopping distance. In this experiment, a car is traveling on dry pavement at a constant rate of speed. From the instant that a driver recognizes the need to stop, the number of feet it takes for him to stop the car is recorded. For example, for a driver traveling at 50 miles per hour, it requires a stopping distance of 190 feet. In general, the stopping distance x in feet is related to the speed of the car y in miles per hour by the equation

$$0.11y + 1.25 = \sqrt{3.7625 + 0.22x}.$$

(Source: National Highway Traffic Safety Administration)

43. Solve this equation for x.

44. Use your answer from exercise 43 to find what the stopping distance x would have been for a car traveling at $y = 60$ miles per hour.

To Think About

45. The solution to the equation
$$\sqrt{x^2 - 3x + c} = x - 2$$
is $x = 3$. What is the value of c?

46. The solution to the equation
$$\sqrt{x + a} - \sqrt{x} = -4$$
is $x = 25$. What is the value of a?

Cumulative Review *Simplify.*

47. **[1.4.4]** $(4^3 x^6)^{2/3}$

48. **[1.4.1]** $(2^{-3} x^{-6})^{1/3}$

49. **[7.2.2]** $\sqrt[3]{-216 x^6 y^9}$

50. **[7.2.2]** $\sqrt[5]{-32 x^{15} y^5}$

51. **[4.3.1]** *Mississippi Queen* The *Mississippi Queen* is a steamboat that travels up and down the Mississippi River at a cruising speed of 12 miles per hour in still water. After traveling 4 hours downstream with the current, it takes 5 hours to get upstream against the current and return to its original starting point. What is the speed of the Mississippi River's current?

52. **[4.3.1]** *Veterinarian Costs* Concord Veterinary Clinic saw a total of 28 dogs and cats in one day for a standard annual checkup. The checkup cost for a cat is $55, and the checkup cost for a dog is $68. The total amount charged that day for checkups was $1748. How many dogs were examined? How many cats?

Quick Quiz 7.5 Solve and check your solutions.

1. $\sqrt{5x - 4} = x$

2. $x = 3 - \sqrt{2x - 3}$

3. $4 - \sqrt{x - 4} = \sqrt{2x - 1}$

4. **Concept Check** When you try to solve the equation $2 + \sqrt{x + 10} = x$, you obtain the values $x = -1$ and $x = 6$. Explain how you would determine if either of these values is a solution of the radical equation.

Student Learning Objectives

After studying this section, you will be able to:

1. Simplify expressions involving complex numbers.

2. Add and subtract complex numbers.

3. Multiply complex numbers.

4. Evaluate complex numbers of the form i^n.

5. Divide two complex numbers.

1 Simplifying Expressions Involving Complex Numbers

Until now we have not been able to solve an equation such as $x^2 = -4$ because there is no *real* number that satisfies this equation. However, this equation *does* have a nonreal solution. This solution is an *imaginary number*.

We define a new number:

$$i = \sqrt{-1} \text{ or } i^2 = -1$$

Now let us use the product rule

$$\sqrt{-a} = \sqrt{-1}\sqrt{a} \quad \text{and see if it is valid.}$$

If $x^2 = -4$, $x = \sqrt{-4}$. Then $\sqrt{-4} = \sqrt{4(-1)} = \sqrt{4}\sqrt{-1} = \sqrt{4} \cdot i = 2i$.

Thus, one solution to the equation $x^2 = -4$ is $2i$. Let's check it.

$$x^2 = -4$$
$$(2i)^2 \stackrel{?}{=} -4$$
$$4i^2 \stackrel{?}{=} -4$$
$$4(-1) \stackrel{?}{=} -4$$
$$-4 = -4 \checkmark$$

The value $-2i$ is also a solution. You should verify this.

Now we formalize our definitions and give some examples of imaginary numbers.

DEFINITION OF IMAGINARY NUMBERS

The **imaginary number i** is defined as follows:

$$i = \sqrt{-1} \quad \text{and} \quad i^2 = -1.$$

The set of imaginary numbers consists of numbers of the form bi, where b is a real number and $b \neq 0$.

DEFINITION

For all positive real numbers a,

$$\sqrt{-a} = \sqrt{-1}\sqrt{a} = i\sqrt{a}.$$

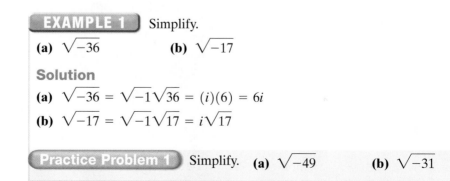

EXAMPLE 1 Simplify.

(a) $\sqrt{-36}$ **(b)** $\sqrt{-17}$

Solution

(a) $\sqrt{-36} = \sqrt{-1}\sqrt{36} = (i)(6) = 6i$

(b) $\sqrt{-17} = \sqrt{-1}\sqrt{17} = i\sqrt{17}$

Practice Problem 1 Simplify. **(a)** $\sqrt{-49}$ **(b)** $\sqrt{-31}$

NOTE TO STUDENT: Fully worked-out solutions to all of the Practice Problems can be found at the back of the text starting at page SP-1

To avoid confusing $\sqrt{17}i$ with $\sqrt{17i}$, we write the i before the radical. That is, we write $i\sqrt{17}$.

EXAMPLE 2 Simplify. $\sqrt{-45}$

Solution

$$\sqrt{-45} = \sqrt{-1}\sqrt{45} = i\sqrt{45} = i\sqrt{9}\sqrt{5} = 3i\sqrt{5}$$

Practice Problem 2 Simplify. $\sqrt{-98}$

The rule $\sqrt{a}\sqrt{b} = \sqrt{ab}$ requires that $a \geq 0$ and $b \geq 0$. Therefore, we cannot use our product rule when the radicands are negative unless we first use the definition of $\sqrt{-1}$. Notice that

$$\sqrt{-1} \cdot \sqrt{-1} = i \cdot i = i^2 = -1.$$

EXAMPLE 3 Multiply. $\sqrt{-16} \cdot \sqrt{-25}$

Solution First we must use the definition $\sqrt{-1} = i$. Thus, we have the following:

$$(\sqrt{-16})(\sqrt{-25}) = (i\sqrt{16})(i\sqrt{25})$$
$$= i^2(4)(5)$$
$$= -1(20) \qquad i^2 = -1.$$
$$= -20$$

Practice Problem 3 Multiply. $\sqrt{-8} \cdot \sqrt{-2}$

Now we formally define a complex number.

DEFINITION OF COMPLEX NUMBER
A number that can be written in the form $a + bi$, where a and b are real numbers, is a **complex number.** We say that a is the **real part** and bi is the **imaginary part.**

Under this definition, every real number is also a complex number. For example, the real number 5 can be written as $5 + 0i$. Therefore, 5 is a complex number. In a similar fashion, the imaginary number $2i$ can be written as $0 + 2i$. So $2i$ is a complex number. Thus, the set of complex numbers includes the set of real numbers and the set of imaginary numbers.

DEFINITION
Two complex numbers $a + bi$ and $c + di$ are equal if and only if $a = c$ and $b = d$.

This definition means that two complex numbers are equal if and only if their real parts are equal *and* their imaginary parts are equal.

Graphing Calculator

Complex Numbers

Some graphing calculators, such as the TI-84, have a complex number mode. If your graphing calculator has this capability, you will be able to use it to do complex number operations. First you must use the [Mode] command to transfer selection from "Real" to "Complex" or "$a + bi$." To verify your status, try to find $\sqrt{-7}$ on your graphing calculator. If you obtain an approximate answer of "2.645751311 i," then your calculator is operating in the complex number mode. If you obtain "ERROR: NONREAL ANSWER," then your calculator is not operating in the complex number mode.

EXAMPLE 4   Find the real numbers x and y if $x + 3i\sqrt{7} = -2 + yi$.

Solution   By our definition, the real parts must be equal, so x must be -2; the imaginary parts must also be equal, so y must be $3\sqrt{7}$.

Practice Problem 4   Find the real numbers x and y if

$$-7 + 2yi\sqrt{3} = x + 6i\sqrt{3}.$$

2 Adding and Subtracting Complex Numbers

ADDING AND SUBTRACTING COMPLEX NUMBERS

For all real numbers $a, b, c,$ and $d,$

$$(a + bi) + (c + di) = (a + c) + (b + d)i \quad \text{and}$$
$$(a + bi) - (c + di) = (a - c) + (b - d)i.$$

In other words, to combine complex numbers we add (or subtract) the real parts, and we add (or subtract) the imaginary parts.

EXAMPLE 5   Subtract. $(6 - 2i) - (3 - 5i)$

Solution

$$(6 - 2i) - (3 - 5i) = (6 - 3) + [-2 - (-5)]i = 3 + (-2 + 5)i = 3 + 3i$$

Practice Problem 5   Subtract. $(3 - 4i) - (-2 - 18i)$

3 Multiplying Complex Numbers

As we might expect, the procedure for multiplying complex numbers is similar to the procedure for multiplying polynomials. We will see that the complex numbers obey the associative, commutative, and distributive properties.

EXAMPLE 6   Multiply. $(7 - 6i)(2 + 3i)$

Solution   Use FOIL.

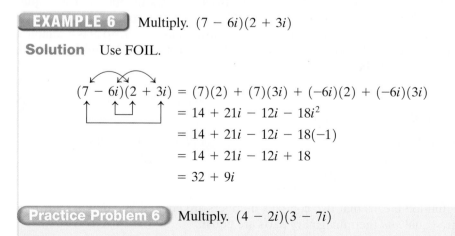

$$(7 - 6i)(2 + 3i) = (7)(2) + (7)(3i) + (-6i)(2) + (-6i)(3i)$$
$$= 14 + 21i - 12i - 18i^2$$
$$= 14 + 21i - 12i - 18(-1)$$
$$= 14 + 21i - 12i + 18$$
$$= 32 + 9i$$

Practice Problem 6   Multiply. $(4 - 2i)(3 - 7i)$

Verbal and Writing Skills

1. Does $x^2 = -9$ have a real number solution? Why or why not?

2. Describe a complex number and give an example.

3. Are the complex numbers $2 + 3i$ and $3 + 2i$ equal? Why or why not?

4. Describe in your own words how to add or subtract complex numbers.

Simplify. Express in terms of i.

5. $\sqrt{-25}$

6. $\sqrt{-100}$

7. $\sqrt{-50}$

8. $\sqrt{-48}$

9. $\sqrt{-\dfrac{25}{4}}$

10. $\sqrt{-\dfrac{49}{64}}$

11. $-\sqrt{-81}$

12. $-\sqrt{-36}$

13. $2 + \sqrt{-3}$

14. $5 + \sqrt{-7}$

15. $-2.8 + \sqrt{-16}$

16. $\dfrac{3}{2} + \sqrt{-121}$

17. $-3 + \sqrt{-24}$

18. $-6 - \sqrt{-32}$

19. $\left(\sqrt{-5}\right)\left(\sqrt{-2}\right)$

20. $\left(\sqrt{-6}\right)\left(\sqrt{-5}\right)$

21. $\left(\sqrt{-36}\right)\left(\sqrt{-4}\right)$

22. $\left(\sqrt{-25}\right)\left(\sqrt{-9}\right)$

Find the real numbers x and y.

23. $x - 3i = 5 + yi$

24. $x - 6i = 7 + yi$

25. $1.3 - 2.5yi = x - 5i$

26. $3.4 - 0.8i = 2x - yi$

27. $23 + yi = 17 - x + 3i$

28. $2 + x - 11i = 19 + yi$

Perform the addition or subtraction.

29. $(1 + 8i) + (-6 + 3i)$

30. $(-8 - 5i) + (12 - i)$

31. $\left(-\dfrac{3}{2} + \dfrac{1}{2}i\right) + \left(\dfrac{5}{2} - \dfrac{3}{2}i\right)$

32. $\left(\dfrac{2}{3} - \dfrac{1}{3}i\right) + \left(\dfrac{10}{3} + \dfrac{4}{3}i\right)$

33. $(2.8 - 0.7i) - (1.6 - 2.8i)$

34. $(5.4 + 4.1i) - (4.8 + 2.6i)$

Multiply and simplify your answers. Place in i notation before doing any other operations.

35. $(2i)(7i)$

36. $(6i)(3i)$

37. $(-7i)(6i)$

38. $(i)(-3i)$

39. $(2 + 3i)(2 - i)$

40. $(4 - 6i)(2 + i)$

41. $9i - 3(-2 + i)$

42. $15i - 5(4 + i)$

43. $2i(5i - 6)$

44. $4i(7 - 2i)$

45. $\left(\dfrac{1}{2} + i\right)^2$

46. $\left(\dfrac{1}{3} - i\right)^2$

47. $\left(i\sqrt{3}\right)\left(i\sqrt{7}\right)$

48. $\left(i\sqrt{2}\right)\left(i\sqrt{6}\right)$

49. $\left(3 + \sqrt{-2}\right)\left(4 + \sqrt{-5}\right)$

50. $\left(2 + \sqrt{-3}\right)\left(6 + \sqrt{-2}\right)$

Evaluate.

51. i^{17}

52. i^{21}

53. i^{24}

54. i^{16}

55. i^{46}

56. i^{83}

57. i^{37}

58. i^{10}

59. $i^{30} + i^{28}$

60. $i^{26} + i^{24}$

61. $i^{100} - i^7$

62. $3i^{64} - 2i^{11}$

Divide.

63. $\dfrac{2 + i}{3 - i}$

64. $\dfrac{4 + 2i}{2 - i}$

65. $\dfrac{2i}{3 + 3i}$

66. $\dfrac{-3i}{2 + 5i}$

67. $\dfrac{5 - 2i}{6i}$

68. $\dfrac{7 + 10i}{3i}$

69. $\dfrac{2}{i}$

70. $\dfrac{-5}{i}$

71. $\dfrac{7}{5 - 6i}$

72. $\dfrac{3}{4 + 2i}$

73. $\dfrac{5 - 2i}{3 + 2i}$

74. $\dfrac{6 + 3i}{6 - 3i}$

Mixed Practice *Simplify.*

75. $\sqrt{-98}$

76. $\sqrt{-72}$

77. $(8 - 5i) - (-1 + 3i)$

78. $(-4 + 6i) - (3 - 4i)$

79. $(3i - 1)(5i - 3)$

80. $(4i + 5)(2i - 5)$

81. $\dfrac{2 - 3i}{2 + i}$

82. $\dfrac{4 - 3i}{5 + 2i}$

Applications *The impedance Z in an alternating current circuit (like the one used in your home and in your classroom) is given by the formula* $Z = \dfrac{V}{I}$, *where V is the voltage and I is the current.*

83. Find the value of Z if $V = 3 + 2i$ and $I = 3i$.

84. Find the value of Z if $V = 4 + 2i$ and $I = -3i$.

Cumulative Review

85. **[2.4.1]** *Factory Production* A grape juice factory produces juice in three different types of containers. $x + 3$ hours per week are spent on producing juice in glass bottles. $2x - 5$ hours per week are spent on producing juice in cans. $4x + 2$ hours per week are spent on producing juice in plastic bottles. If the factory operates 105 hours per week, how much time is spent producing juice in each type of container?

86. **[1.2.2]** *Donation of Computers* Citizens Bank has decided to donate its older personal computers to the Boston Public Schools. Each computer donated is worth $120 in tax-deductible dollars to the bank. In addition, the computer company supplying the bank with its new computers gives a 7% rebate to any customer donating used computers to schools. If sixty new computers are purchased at a list price of $1850 each and sixty older computers are donated to the Boston Public Schools, what is the net cost to the bank for this purchase?

Quick Quiz 7.6 *Simplify.*

1. $(6 - 7i)(3 + 2i)$

2. $\dfrac{4 + 3i}{1 - 2i}$

3. i^{33}

4. **Concept Check** Explain how you would simplify the following.
$$(3 + 5i)^2$$

Student Learning Objectives

After studying this section, you will be able to:

① Solve problems requiring the use of direct variation.

② Solve problems requiring the use of inverse variation.

③ Solve problems requiring the use of joint or combined variation.

NOTE TO STUDENT: Fully worked-out solutions to all of the Practice Problems can be found at the back of the text starting at page SP-1

① Solving Problems Using Direct Variation

Many times in daily life we observe how a change in one quantity produces a change in another. If we order one large pepperoni pizza, we pay $8.95. If we order two large pepperoni pizzas, we pay $17.90. For three large pepperoni pizzas, it is $26.85. The change in the number of pizzas we order results in a corresponding change in the price we pay.

Notice that the price we pay for each pizza stays the same. That is, each pizza costs $8.95. The number of pizzas changes, and the corresponding price of the order changes. From our experience with functions and with equations, we see that the cost of the order is $y = \$8.95x$, where the price y depends on the number of pizzas x. We see that the variable y is a constant multiple of x. The two variables are said to *vary directly*. That is, y varies directly with x. We write a general equation that represents this idea as follows: $y = kx$.

When we solve problems using direct variation, we usually are not given the value of the constant of variation k. This is something that we must find. Usually all we are given is a point of reference. That is, we are given the value of y for a specific value of x. Using this information, we can find k.

EXAMPLE 1 The time of a pendulum's period varies directly with the square root of its length. If the pendulum is 1 foot long when the time is 0.2 second, find the time if the length is 4 feet.

Solution Let t = the time and L = the length.
We then have the equation

$$t = k\sqrt{L}.$$

We can evaluate k by substituting $L = 1$ and $t = 0.2$ into the equation.

$$t = k\sqrt{L}$$
$$0.2 = k\left(\sqrt{1}\right)$$
$$0.2 = k \qquad \text{Because } \sqrt{1} = 1.$$

Now we know the value of k and can write the equation more completely.

$$t = 0.2\sqrt{L}$$

When $L = 4$, we have the following:

$$t = 0.2\sqrt{4}$$
$$t = (0.2)(2)$$
$$t = 0.4 \text{ second}$$

Practice Problem 1 The maximum speed of a racing car varies directly with the square root of the horsepower of the engine. If the maximum speed of a car with 256 horsepower is 128 miles per hour, what is the maximum speed of a car with 225 horsepower?

② Solving Problems Using Inverse Variation

In some cases when one variable increases, another variable decreases. For example, as the amount of money you earn each year increases, the percentage of your income that you get to keep after taxes decreases. If one variable is a constant multiple of the reciprocal of the other, the two variables are said to *vary inversely*.

EXAMPLE 2 If y varies inversely with x and $y = 12$ when $x = 5$, find the value of y when $x = 14$.

Solution If y varies inversely with x, we can write the equation $y = \dfrac{k}{x}$. We can find the value of k by substituting the values $y = 12$ and $x = 5$.

$$12 = \frac{k}{5}$$
$$60 = k$$

We can now write the equation

$$y = \frac{60}{x}.$$

To find the value of y when $x = 14$, we substitute 14 for x in the equation.

$$y = \frac{60}{14}$$
$$y = \frac{30}{7}$$

Practice Problem 2 If y varies inversely with x and $y = 45$ when $x = 16$, find the value of y when $x = 36$.

EXAMPLE 3 The amount of light from a light source varies inversely with the square of the distance to the light source. If an object receives 6.25 lumens when the light source is 8 meters away, how much light will the object receive if the light source is 4 meters away?

Solution Let L = the amount of light and d = the distance to the light source.

Since the amount of light varies inversely with the *square of the distance* to the light source, we have

$$L = \frac{k}{d^2}.$$

Substituting the known values of $L = 6.25$ and $d = 8$, we can find the value of k.

$$6.25 = \frac{k}{8^2}$$

$$6.25 = \frac{k}{64}$$

$$400 = k$$

We are now able to write a more specific equation,

$$L = \frac{400}{d^2}.$$

We will use this to find L when $d = 4$ meters.

$$L = \frac{400}{4^2}$$

$$L = \frac{400}{16}$$

$$L = 25 \text{ lumens}$$

Check: Does this answer seem reasonable? Would we expect to have more light if we move closer to the light source? ✓

NOTE TO STUDENT: *Fully worked-out solutions to all of the Practice Problems can be found at the back of the text starting at page SP-1*

Practice Problem 3 The weight that can safely be supported on top of a cylindrical column varies inversely with the square of its height. If a 7.5-ft column can support 2 tons, how much weight can a 3-ft column support?

③ Solving Problems Using Joint or Combined Variation

Sometimes a quantity depends on the variation of two or more variables. This is called joint or **combined variation.**

EXAMPLE 4 y varies directly with x and z and inversely with d^2. When $x = 7$, $z = 3$, and $d = 4$, the value of y is 20. Find the value of y when $x = 5$, $z = 6$, and $d = 2$.

Solution We can write the equation

$$y = \frac{kxz}{d^2}.$$

To find the value of k, we substitute into the equation $y = 20$, $x = 7$, $z = 3$, and $d = 4$.

$$20 = \frac{k(7)(3)}{4^2}$$

$$20 = \frac{21k}{16}$$

$$320 = 21k$$

$$\frac{320}{21} = k$$

Now we substitute $\frac{320}{21}$ for k into our original equation.

$$y = \frac{\frac{320}{21}xz}{d^2} \quad \text{or} \quad y = \frac{320xz}{21d^2}$$

We use this equation to find y for the known values of x, z, and d. We want to find y when $x = 5$, $z = 6$, and $d = 2$.

$$y = \frac{320(5)(6)}{21(2)^2} = \frac{9600}{84}$$

$$y = \frac{800}{7}$$

Practice Problem 4 y varies directly with z and w^2 and inversely with x. $y = 20$ when $z = 3$, $w = 5$, and $x = 4$. Find y when $z = 4$, $w = 6$, and $x = 2$.

Many applied problems involve joint variation. For example, a cylindrical concrete column has a safe load capacity that varies directly with the diameter raised to the fourth power and inversely with the square of its length.

Therefore, if d = diameter and l = length, the equation would be of the form

$$y = \frac{kd^4}{l^2}.$$

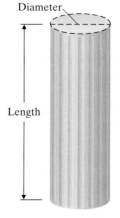

Diameter

Length

Verbal and Writing Skills

1. Give an example in everyday life of direct variation and write an equation as a mathematical model.

2. The general equation $y = kx$ means that y varies _____ with x. k is called the _____ of variation.

3. If y varies inversely with x, we write the equation

_____ .

4. Write a mathematical model for the following situation: The strength of a rectangular beam varies directly with its width and the square of its depth.

Round all answers to the nearest tenth unless otherwise directed.

5. If y varies directly with x and $y = 15$ when $x = 40$, find y when $x = 64$.

6. If y varies directly with x and $y = 30$ when $x = 18$, find y when $x = 120$.

7. *Pressure on a Submarine* A marine biology submarine was searching the waters for blue whales at 50 feet below the surface, where it experienced a pressure of 21 pounds per square inch (psi). If the pressure of water on a submerged object varies directly with its distance beneath the surface, how much pressure would the submarine experience if it had to dive to 170 feet?

8. *Spring Stretching* The distance a spring stretches varies directly with the weight of the object hung on the spring. If a 10-pound weight stretches a spring 6 inches, how far will a 35-pound weight stretch this spring?

9. *Stopping Distance* A car's stopping distance varies directly with the square of its speed. A car that is traveling 30 miles per hour can stop in 40 feet. What distance will it take to stop if it is traveling 60 miles per hour?

10. *Time of Fall in Gravitation* When an object is dropped, the distance it falls in feet varies directly with the square of the duration of the fall in seconds. An apple that falls from a tree falls 1 foot in $\frac{1}{4}$ second. How far will it fall in 1 second? How far will it fall in 2 seconds?

11. If y varies inversely with the square of x, and $y = 10$ when $x = 2$, find y when $x = 0.5$.

12. If y varies inversely with the square root of x, and $y = 2.4$ when $x = 0.09$, find y when $x = 0.2$.

13. *Gasoline Prices* Last summer the price of gasoline changed frequently. One station owner noticed that the number of gallons he sold each day seemed to vary inversely with the price per gallon. If he sold 2800 gallons when the price was $4.10, how many gallons could he expect to sell if the price fell to $3.90? Round your answer to the nearest gallon.

14. *Weight of an Object* The weight of an object on Earth's surface varies inversely with the square of its distance from the center of Earth. An object weighs 1000 pounds on Earth's surface. This is approximately 4000 miles from the center of Earth. How much would this object weigh 4500 miles from the center of Earth?

15. *Beach Cleanup* Every year on Earth Day, a group of volunteers pick up garbage at Hidden Falls Park. The time it takes to clean the beach varies inversely with the number of people picking up garbage. Last year, 39 volunteers took 6 hours to clean the park. If 60 volunteers come to pick up garbage this year, how long will it take to clean the park?

16. *Electric Current* If the voltage in an electric circuit is kept at the same level, the current varies inversely with the resistance. The current measures 40 amperes when the resistance is 270 ohms. Find the current when the resistance is 100 ohms.

17. *Support Beam* The weight that can be safely supported by a 2- by 6-inch support beam varies inversely with its length. A builder finds that a support beam that is 8 feet long will support 900 pounds. Find the weight that can be safely supported by a beam that is 18 feet long.

18. *Satellite Orbit Speed* The speed that is required to maintain a satellite in a circular orbit around Earth varies directly with the square root of the distance of the satellite from the center of Earth. We will assume that the radius of Earth is approximately 4000 miles. A satellite that is 100 miles above the surface of Earth is orbiting at approximately 18,000 miles per hour. What speed would be necessary for the satellite to orbit 500 miles above the surface of Earth? Round to the nearest mile per hour.

19. **Whirlpool Tub** The amount of time it takes to fill a whirlpool tub is inversely proportional to the square of the radius of the pipe used to fill it. If a pipe of radius 2.5 inches can fill the tub in 6 minutes, how long will it take the tub to fill if a pipe of radius 3.5 inches is used?

20. **Wind Generator** The force on a blade of a wind generator varies jointly with the product of the blade's area and the square of the wind velocity. The force of the wind is 20 pounds when the area is 3 square feet and the velocity is 30 feet per second. Find the force when the area is increased to 5 square feet and the velocity is reduced to 25 feet per second.

Cumulative Review *Solve each of the following equations or word problems.*

21. **[5.8.1]** $3x^2 - 8x + 4 = 0$

22. **[5.8.1]** $4x^2 = -28x + 32$

23. **[1.2.2]** **Sales Tax** In Champaign, Illinois, the sales tax is 6.25%. Donny bought an amplifier for his stereo that cost $488.75 after tax. What was the original price of the amplifier?

24. **[1.2.2]** **Tennis Courts** It takes 7.5 gallons of white paint to properly paint lines on three tennis courts. How much paint is needed to paint twenty-two tennis courts?

Quick Quiz 7.7

1. If y varies inversely with x and $y = 9$ when $x = 3$, find the value of y when $x = 6$.

2. Suppose y varies directly with x and inversely with the square of z. $y = 6$ when $x = 3$ and $z = 5$. Find y when $x = 6$ and $z = 10$.

3. The distance a pickup truck requires to stop varies directly with the square of the speed. A new Ford pickup truck traveling on dry pavement can stop in 80 feet when traveling at 50 miles per hour. What distance will the truck take to stop if it is traveling at 65 miles per hour?

4. **Concept Check** If y varies directly with the square root of x and $y = 50$ when $x = 5$, explain how you would find the constant of variation k.

Putting Your Skills to Work: Use Math to Save Money

BALANCE YOUR FINANCES

A Personal Question for You

How often do you balance your checkbook? Once a day, once a week, never? One of the first steps in saving money is to determine your current spending trends. The first step in that process is learning to balance your finances. Consider the story of Teresa.

Keeping a Record of Deposits

Teresa balanced her checkbook once a month when she received her bank statement. Below is a table that records the deposits Teresa made for the month of May. The beginning balance for May was $300.50.

Date	Deposits
May 1	$200.00
May 3	$150.50
May 10	$120.25
May 25	$50.00
May 28	$25.00

Keeping a Record of Checks

Below is a table that records each check that Teresa wrote for the month of May.

Date	Check Number	Checks
May 2	102	$238.50
May 6	103	$75.00
May 12	104	$200.00
May 28	105	$28.56
May 30	106	$36.00

Finding the Facts

1. What is the total amount of her deposits?

2. What is the total amount of her checks?

3. Based on the given information, will Teresa be able to cover all her checks?

4. All checks written before or on May 25 have cleared. What was her balance on May 25?

5. What should Teresa assume her balance is at the beginning of June? (Hint: She should assume all the checks and deposits for May will clear.)

6. If Teresa continues her spending habits, what will happen?

Applying This Lesson to Your Life

Do you know your monthly income and how much you spend each month? Take that first step toward using math to help you save. Balance your checkbook and organize your deposits and expenses. Find out how much you earn and spend each month.

Topic	Procedure	Examples
Multiplication of variables with rational exponents, p. 355.	$x^m x^n = x^{m+n}$	$(3x^{1/5})(-2x^{3/5}) = -6x^{4/5}$
Division of variables with rational exponents, p. 355.	$\dfrac{x^m}{x^n} = x^{m-n}, \qquad x \neq 0$	$\dfrac{-16x^{3/20}}{24x^{5/20}} = -\dfrac{2x^{-1/10}}{3}$
Removing negative exponents, p. 355.	$x^{-n} = \dfrac{1}{x^n}, \qquad x \text{ and } y \neq 0$ $\dfrac{x^{-n}}{y^{-m}} = \dfrac{y^m}{x^n}$	Write with positive exponents. $3x^{-4} = \dfrac{3}{x^4}$ $\dfrac{2x^{-6}}{5y^{-8}} = \dfrac{2y^8}{5x^6}$ $4^{-2} = \dfrac{1}{4^2} = \dfrac{1}{16}$
Zero exponent, p. 357.	$x^0 = 1 \quad (\text{if } x \neq 0)$	$(3x^{1/2})^0 = 1$
Raising a variable with an exponent to a power, p. 354.	$(x^m)^n = x^{mn}$ $(xy)^n = x^n y^n$ $\left(\dfrac{x}{y}\right)^n = \dfrac{x^n}{y^n}, \quad y \neq 0$	$(x^{-1/2})^{-2/3} = x^{1/3}$ $(3x^{-2}y^{-1/2})^{2/3} = 3^{2/3}x^{-4/3}y^{-1/3}$ $\left(\dfrac{4x^{-2}}{3^{-1}y^{-1/2}}\right)^{1/4} = \dfrac{4^{1/4}x^{-1/2}}{3^{-1/4}y^{-1/8}}$
Multiplication of expressions with rational exponents, p. 356.	Add exponents whenever expressions with the same base are multiplied.	$x^{2/3}(x^{1/3} - x^{1/4}) = x^{3/3} - x^{2/3+1/4} = x - x^{11/12}$
Higher-order roots, p. 361.	If x is a nonnegative real number, $\sqrt[n]{x}$ is a nonnegative nth root and has the property that $\left(\sqrt[n]{x}\right)^n = x.$ If x is a negative real number, and n is an odd integer, then $\sqrt[n]{x}$ has the property that $(\sqrt[n]{x})^n = x$. If x is a negative real number, and n is an even integer, $\sqrt[n]{x}$ is not a real number.	$\sqrt[3]{27} = 3$ because $3^3 = 27$. $\sqrt[5]{-32} = -2$ because $(-2)^5 = -32$. $\sqrt[4]{-16}$ is *not* a real number.
Rational exponents and radicals, p. 365.	For positive integers m and n and any real number x for which $x^{1/n}$ is defined, $x^{m/n} = \left(\sqrt[n]{x}\right)^m = \sqrt[n]{x^m}.$ If it is also true that $x \neq 0$, then $x^{-m/n} = \dfrac{1}{x^{m/n}} = \dfrac{1}{\left(\sqrt[n]{x}\right)^m} = \dfrac{1}{\sqrt[n]{x^m}}.$	Write as a radical. $x^{3/7} = \sqrt[7]{x^3}$, $3^{1/5} = \sqrt[5]{3}$ Write as an expression with a fractional exponent. $\sqrt[3]{w^4} = w^{4/3}$ Evaluate. $25^{3/2} = \left(\sqrt{25}\right)^3 = (5)^3 = 125$
Higher-order roots and absolute value, p. 366.	$\sqrt[n]{x^n} = |x|$ when n is an even positive integer. $\sqrt[n]{x^n} = x$ when n is an odd positive integer.	$\sqrt[6]{x^6} = |x|$ $\sqrt[5]{x^5} = x$
Evaluation of higher-order roots, p. 366.	Use exponent notation.	$\sqrt[5]{-32x^{15}} = \sqrt[5]{(-2)^5 x^{15}}$ $= [(-2)^5 x^{15}]^{1/5} = (-2)^1 x^3 = -2x^3$

Topic	Procedure	Examples
Simplification of radicals with the product rule, p. 369.	For nonnegative real numbers a and b and positive integers n, $$\sqrt[n]{a}\ \sqrt[n]{b} = \sqrt[n]{ab}.$$	Simplify when $x \geq 0$, $y \geq 0$. $$\sqrt{75x^3} = \sqrt{25x^2}\sqrt{3x}$$ $$= 5x\sqrt{3x}$$ $$\sqrt[3]{16x^5y^6} = \sqrt[3]{8x^3y^6}\sqrt[3]{2x^2}$$ $$= 2xy^2\sqrt[3]{2x^2}$$
Combining radicals, p. 370.	Simplify radicals and combine them if they have the same index and the same radicand.	Combine. $$2\sqrt{50} - 3\sqrt{98} = 2\sqrt{25}\sqrt{2} - 3\sqrt{49}\sqrt{2}$$ $$= 2(5)\sqrt{2} - 3(7)\sqrt{2}$$ $$= 10\sqrt{2} - 21\sqrt{2} = -11\sqrt{2}$$
Multiplying radicals, p. 375.	1. Multiply coefficients outside the radical and then multiply the radicands. 2. Simplify your answer.	$$(2\sqrt{3})(4\sqrt{5}) = 8\sqrt{15}$$ $$2\sqrt{6}(\sqrt{2} - 3\sqrt{12}) = 2\sqrt{12} - 6\sqrt{72}$$ $$= 2\sqrt{4}\sqrt{3} - 6\sqrt{36}\sqrt{2}$$ $$= 4\sqrt{3} - 36\sqrt{2}$$ $(\sqrt{2} + \sqrt{3})(2\sqrt{2} - \sqrt{3})$ Use the FOIL method. $$= 2\sqrt{4} - \sqrt{6} + 2\sqrt{6} - \sqrt{9}$$ $$= 4 + \sqrt{6} - 3$$ $$= 1 + \sqrt{6}$$
Simplifying quotients of radicals with the quotient rule, p. 377.	For nonnegative real numbers a, positive real numbers b, and positive integers n, $$\sqrt[n]{\frac{a}{b}} = \frac{\sqrt[n]{a}}{\sqrt[n]{b}}.$$	$$\sqrt[3]{\frac{5}{27}} = \frac{\sqrt[3]{5}}{\sqrt[3]{27}} = \frac{\sqrt[3]{5}}{3}$$
Rationalizing denominators, p. 378.	Multiply numerator and denominator by a value that eliminates the radical in the denominator.	$$\frac{2}{\sqrt{7}} = \frac{2}{\sqrt{7}} \cdot \frac{\sqrt{7}}{\sqrt{7}} = \frac{2\sqrt{7}}{7}$$ $$\frac{3}{\sqrt{5} + \sqrt{2}} = \frac{3}{\sqrt{5} + \sqrt{2}} \cdot \frac{\sqrt{5} - \sqrt{2}}{\sqrt{5} - \sqrt{2}}$$ $$= \frac{3\sqrt{5} - 3\sqrt{2}}{(\sqrt{5})^2 - (\sqrt{2})^2} = \frac{3\sqrt{5} - 3\sqrt{2}}{5 - 2}$$ $$= \frac{3\sqrt{5} - 3\sqrt{2}}{3} = \sqrt{5} - \sqrt{2}$$
Solving radical equations, p. 386.	1. Perform algebraic operations to obtain one radical by itself on one side of the equation. 2. If the equation contains square roots, square each side of the equation. Otherwise, raise each side to the appropriate power for third- and higher-order roots. 3. Simplify, if possible. 4. If the equation still contains a radical, repeat steps 1 to 3. 5. Collect all terms on one side of the equation. 6. Solve the resulting equation. 7. Check all apparent solutions. Solutions to radical equations must be verified.	Solve. $$x = \sqrt{2x + 9} - 3$$ $$x + 3 = \sqrt{2x + 9}$$ $$(x + 3)^2 = (\sqrt{2x + 9})^2$$ $$x^2 + 6x + 9 = 2x + 9$$ $$x^2 + 6x - 2x + 9 - 9 = 0$$ $$x^2 + 4x = 0$$ $$x(x + 4) = 0$$ $$x = 0 \quad \text{or} \quad x = -4$$ *Check:* $x = 0$: $0 \overset{?}{=} \sqrt{2(0) + 9} - 3$ $$0 \overset{?}{=} \sqrt{9} - 3$$ $$0 = 3 - 3 \quad \checkmark$$ $x = -4$: $-4 \overset{?}{=} \sqrt{2(-4) + 9} - 3$ $$-4 \overset{?}{=} \sqrt{1} - 3$$ $$-4 \neq -2$$ The only solution is 0.

(Continued on next page)

Topic	Procedure	Examples
Simplifying imaginary numbers, p. 392.	Use $i = \sqrt{-1}$, $i^2 = -1$, and $\sqrt{-a} = \sqrt{-1}\sqrt{a}$.	$\sqrt{-16} = \sqrt{-1}\sqrt{16} = 4i$ $\sqrt{-18} = \sqrt{-1}\sqrt{18} = i\sqrt{9}\sqrt{2} = 3i\sqrt{2}$
Adding and subtracting complex numbers, p. 394.	Combine real parts and imaginary parts separately.	$(5 + 6i) + (2 - 4i) = 7 + 2i$ $(-8 + 3i) - (4 - 2i) = -8 + 3i - 4 + 2i$ $\qquad\qquad\qquad\qquad\quad = -12 + 5i$
Multiplying complex numbers, p. 394.	Use the FOIL method and $i^2 = -1$.	$(5 - 6i)(2 - 4i) = 10 - 20i - 12i + 24i^2$ $\qquad\qquad\quad = 10 - 32i + 24(-1)$ $\qquad\qquad\quad = 10 - 32i - 24$ $\qquad\qquad\quad = -14 - 32i$
Dividing complex numbers, p. 396.	Multiply the numerator and denominator by the conjugate of the denominator.	$\dfrac{5 + 2i}{4 - i} = \dfrac{5 + 2i}{4 - i} \cdot \dfrac{4 + i}{4 + i} = \dfrac{20 + 5i + 8i + 2i^2}{4^2 - i^2}$ $\qquad = \dfrac{20 + 13i + 2(-1)}{16 - (-1)}$ $\qquad = \dfrac{20 + 13i - 2}{16 + 1}$ $\qquad = \dfrac{18 + 13i}{17}$ or $\dfrac{18}{17} + \dfrac{13}{17}i$
Raising i to a power, p. 395.	$i^1 = i$ $i^2 = -1$ $i^3 = -i$ $i^4 = 1$	Evaluate. $\qquad\quad i^{27} = i^{24} \cdot i^3$ $\qquad\qquad = (i^4)^6 \cdot i^3$ $\qquad\qquad = (1)^6(-i)$ $\qquad\qquad = -i$
Direct variation, p. 400.	If y varies directly with x, there is a constant of variation k such that $y = kx$. After k is determined, other values of y or x can easily be computed.	y varies directly with x. When $x = 2$, $y = 7$. $\qquad y = kx$ $\qquad 7 = k(2)$ Substitute. $\qquad k = \dfrac{7}{2}$ Solve. $\qquad y = \dfrac{7}{2}x$ What is y when $x = 18$? $\qquad y = \dfrac{7}{2}x = \dfrac{7}{2} \cdot 18 = 63$
Inverse variation, p. 401.	If y varies inversely with x, the constant k is such that $$y = \frac{k}{x}.$$	y varies inversely with x. When x is 5, y is 12. What is y when x is 30? $\qquad y = \dfrac{k}{x}$ $\qquad 12 = \dfrac{k}{5}$ Substitute. $\qquad k = 60$ Solve. $\qquad y = \dfrac{60}{x}$ Substitute. When $x = 30$, $y = \dfrac{60}{30} = 2$.

Chapter 7 Review Problems

In all exercises assume that the variables represent positive real numbers unless otherwise stated. Simplify using only positive exponents in your answers.

1. $(3xy^{1/2})(5x^2y^{-3})$

2. $\dfrac{3x^{2/3}}{6x^{1/6}}$

3. $(16a^6b^5)^{1/2}$

4. $3^{1/2} \cdot 3^{1/6}$

5. $(2a^{1/3}b^{1/4})(-3a^{1/2}b^{1/2})$

6. $\dfrac{6x^{2/3}y^{1/10}}{12x^{1/6}y^{-1/5}}$

7. $(2x^{-1/5}y^{1/10}z^{4/5})^{-5}$

8. $\left(\dfrac{49a^3b^6}{a^{-7}b^4}\right)^{1/2}$

9. $\dfrac{(x^{3/4}y^{2/5})^{1/2}}{x^{-1/8}}$

10. $\left(\dfrac{8a^4}{a^{-2}}\right)^{1/3}$

11. $(4^{5/3})^{6/5}$

12. Combine as one fraction containing only positive exponents. $2x^{1/3} + x^{-2/3}$

13. Factor out the common factor $3x$ from $6x^{3/2} - 9x^{1/2}$.

In exercises 14–47, assume that all variables represent nonnegative real numbers.

Evaluate, if possible.

14. $-\sqrt{16}$

15. $\sqrt[5]{-32}$

16. $\sqrt[6]{-20}$

17. $-\sqrt{\dfrac{1}{25}}$

18. $\sqrt{0.04}$

19. $\sqrt[4]{-256}$

20. $\sqrt[3]{-\dfrac{1}{8}}$

21. $\sqrt[3]{\dfrac{27}{64}}$

22. $64^{2/3}$

23. $125^{4/3}$

Simplify.

24. $\sqrt{49x^4y^{10}z^2}$

25. $\sqrt[3]{64a^{12}b^{30}}$

26. $\sqrt[3]{-8a^{12}b^{15}c^{21}}$

27. $\sqrt{49x^{22}y^2}$

Replace radicals with rational exponents.

28. $\sqrt[5]{a^2}$

29. $\sqrt[4]{y^3}$

30. $\sqrt{2b}$

31. $\sqrt[3]{5a}$

32. $\left(\sqrt[5]{xy}\right)^7$

Change to radical form.

33. $m^{1/2}$

34. $n^{1/4}$

35. $y^{3/5}$

36. $(3z)^{2/3}$

37. $(2x)^{3/7}$

Evaluate or simplify.

38. $16^{3/4}$

39. $64^{5/6}$

40. $(-27)^{2/3}$

41. $(-8)^{1/3}$

42. $\left(\dfrac{1}{9}\right)^{1/2}$

43. $(0.49)^{1/2}$

44. $\left(\dfrac{1}{16}\right)^{-1/4}$

45. $\left(\dfrac{1}{36}\right)^{-1/2}$

46. $(25a^2b^4)^{3/2}$

47. $(4a^6b^2)^{5/2}$

Combine where possible.

48. $\sqrt{50} + 2\sqrt{32} - \sqrt{8}$

49. $\sqrt{28} - 4\sqrt{7} + 5\sqrt{63}$

50. $2\sqrt{12} - \sqrt{48} + 5\sqrt{75}$

51. $\sqrt{125x^3} + x\sqrt{45x}$

52. $2\sqrt{32x} - 5x\sqrt{2} + \sqrt{18x}$

53. $3\sqrt[3]{16} - 4\sqrt[3]{54}$

Multiply and simplify.

54. $\left(5\sqrt{12}\right)\left(3\sqrt{6}\right)$

55. $\left(-2\sqrt{15}\right)\left(4x\sqrt{3}\right)$

56. $3\sqrt{x}\left(2\sqrt{8x} - 3\sqrt{48}\right)$

57. $\sqrt{3a}\left(4 - \sqrt{21a}\right)$

58. $-\sqrt{5xy}\left(\sqrt{3x} - \sqrt{5y}\right)$

59. $2\sqrt{7b}\left(\sqrt{ab} - b\sqrt{3bc}\right)$

60. $\left(5\sqrt{2} + \sqrt{3}\right)\left(\sqrt{2} - 2\sqrt{3}\right)$

61. $\left(5\sqrt{6} - 2\sqrt{2}\right)\left(\sqrt{6} - \sqrt{2}\right)$

62. $\left(2\sqrt{5} - 3\sqrt{6}\right)^2$

63. $\left(\sqrt[3]{2x} + \sqrt[3]{6}\right)\left(\sqrt[3]{4x^2} - \sqrt[3]{y}\right)$

64. Let $f(x) = \sqrt{4x + 16}$.
   (a) Find $f(12)$.
   (b) What is the domain of $f(x)$?

65. Let $f(x) = \sqrt{36 - 3x}$.
   (a) Find $f(9)$.
   (b) What is the domain of $f(x)$?

66. Let $f(x) = \sqrt{\dfrac{3}{4}x - \dfrac{1}{2}}$.
   (a) Find $f(1)$.
   (b) What is the domain of $f(x)$?

Rationalize the denominator and simplify the expression.

67. $\sqrt{\dfrac{6y^2}{x}}$

68. $\dfrac{3}{\sqrt{5y}}$

69. $\dfrac{3\sqrt{7x}}{\sqrt{21x}}$

70. $\dfrac{2}{\sqrt{6}-\sqrt{5}}$

71. $\dfrac{\sqrt{x}}{3\sqrt{x}+\sqrt{y}}$

72. $\dfrac{\sqrt{5}}{\sqrt{7}-3}$

73. $\dfrac{2\sqrt{3}+\sqrt{6}}{\sqrt{3}+2\sqrt{6}}$

74. $\dfrac{5\sqrt{2}-\sqrt{3}}{\sqrt{6}-\sqrt{3}}$

75. $\dfrac{3\sqrt{x}+\sqrt{y}}{\sqrt{x}-\sqrt{y}}$

76. $\dfrac{2xy}{\sqrt[3]{16xy^5}}$

77. Simplify. $\sqrt{-16}+\sqrt{-45}$

78. Find x and y. $2x-3i+5=yi-2+\sqrt{6}$

Simplify by performing the operation indicated.

79. $(-12-6i)+(3-5i)$

80. $(2-i)-(12-3i)$

81. $(5-2i)(3+3i)$

82. $(6-2i)^2$

83. $2i(3+4i)$

84. $3-4(2+i)$

85. Evaluate. i^{34}

86. Evaluate. i^{65}

Divide.

87. $\dfrac{7-2i}{3+4i}$

88. $\dfrac{5-2i}{1-3i}$

89. $\dfrac{4-3i}{5i}$

90. $\dfrac{12}{3-5i}$

91. $\dfrac{10-4i}{2+5i}$

Solve and check your solution(s).

92. $\sqrt{3x - 2} = 5$

93. $\sqrt[3]{3x - 1} = 2$

94. $\sqrt{2x + 1} = 2x - 5$

95. $1 + \sqrt{3x + 1} = x$

96. $\sqrt{3x + 1} - \sqrt{2x - 1} = 1$

97. $\sqrt{7x + 2} = \sqrt{x + 3} + \sqrt{2x - 1}$

Round all answers to the nearest tenth.

98. If y varies directly with x and $y = 11$ when $x = 4$, find the value of y when $x = 6$.

99. If y varies directly with x and $y = 5$ when $x = 20$, find the value of y when $x = 50$.

100. *Saturated Fat Intake* The maximum amount of saturated fat a person should consume varies directly with the number of calories consumed. A person who consumes 2000 calories per day should have a maximum of 18 grams of saturated fat per day. For a person who consumes 2500 calories, what is the maximum amount of saturated fat he or she should consume?

101. *Time of Falling Object* The time it takes a falling object to drop a given distance varies directly with the square root of the distance traveled. A steel ball takes 2 seconds to drop a distance of 64 feet. How many seconds will it take to drop a distance of 196 feet?

102. If y varies inversely with x and $y = 8$ when $x = 3$, find the value of y when $x = 48$.

103. *Volume of a Gas* The volume of a gas varies inversely with the pressure of the gas on its container. If a pressure of 24 pounds per square inch corresponds to a volume of 70 cubic inches, what pressure corresponds to a volume of 100 cubic inches?

104. Suppose that y varies directly with x and inversely with the square of z. $y = 20$ when $x = 10$ and $z = 5$. Find y when $x = 8$ and $z = 2$.

▲ **105.** *Capacity of a Cylinder* The capacity of a cylinder varies directly with the height and the square of the radius. A cylinder with a radius of 3 centimeters and a height of 5 centimeters has a capacity of 50 cubic centimeters. What is the capacity of a cylinder with a height of 9 centimeters and a radius of 4 centimeters?

Remember to use your Chapter Test Prep Video CD to see the worked-out solutions to the test problems you want to review.

Simplify.

1. $(2x^{1/2}y^{1/3})(-3x^{1/3}y^{1/6})$

2. $\dfrac{7x^3}{4x^{3/4}}$

3. $(8x^{1/3})^{3/2}$

4. Evaluate. $\left(\dfrac{4}{9}\right)^{3/2}$

5. Evaluate. $\sqrt[5]{-32}$

Evaluate.

6. $8^{-2/3}$

7. $16^{5/4}$

Simplify. Assume that all variables are nonnegative.

8. $\sqrt{75a^4b^9}$

9. $\sqrt{49a^4b^{10}}$

10. $\sqrt[3]{54m^3n^5}$

Combine where possible.

11. $3\sqrt{24} - \sqrt{18} + \sqrt{50}$

12. $\sqrt{40x} - \sqrt{27x} + 2\sqrt{12x}$

Multiply and simplify.

13. $\left(-3\sqrt{2y}\right)\left(5\sqrt{10xy}\right)$

14. $2\sqrt{3}\left(3\sqrt{6} - 5\sqrt{2}\right)$

15. $\left(5\sqrt{3} - \sqrt{6}\right)\left(2\sqrt{3} + 3\sqrt{6}\right)$

1. _____

2. _____

3. _____

4. _____

5. _____

6. _____

7. _____

8. _____

9. _____

10. _____

11. _____

12. _____

13. _____

14. _____

15. _____

16. _____

17. _____

18. _____

19. _____

20. _____

21. _____

22. _____

23. _____

24. _____

25. _____

26. _____

27. _____

28. _____

29. _____

30. _____

Rationalize the denominator.

16. $\dfrac{30}{\sqrt{5x}}$

17. $\sqrt{\dfrac{xy}{3}}$

18. $\dfrac{1 + 2\sqrt{3}}{3 - \sqrt{3}}$

Solve and check your solution(s).

19. $\sqrt{3x - 2} = x$

20. $5 + \sqrt{x + 15} = x$

21. $5 - \sqrt{x - 2} = \sqrt{x + 3}$

Simplify by using the properties of complex numbers.

22. $(8 + 2i) - 3(2 - 4i)$

23. $i^{18} + \sqrt{-16}$

24. $(3 - 2i)(4 + 3i)$

25. $\dfrac{2 + 5i}{1 - 3i}$

26. $(6 + 3i)^2$

27. i^{43}

28. If y varies inversely with x and $y = 9$ when $x = 2$, find the value of y when $x = 6$.

29. Suppose y varies directly with x and inversely with the square of z. When $x = 8$ and $z = 4$, then $y = 3$. Find y when $x = 5$ and $z = 6$.

30. A car's stopping distance varies directly with the square of its speed. A car traveling on pavement can stop in 30 feet when traveling at 30 miles per hour. What distance will the car take to stop if it is traveling at 50 miles per hour?

Cumulative Test for Chapters 1–7

Approximately one half of this test covers the content of Chapters 1–6. The remainder covers the content of Chapter 7.

1. Identify what property of real numbers is illustrated by the equation $7 + (2 + 3) = (7 + 2) + 3$.

2. Remove the parentheses and combine like terms. $-3a(2ab - a^3) + b(ab^2 + 4a^2)$

3. Simplify. $7(12 - 14)^3 - 7 + 3 \div (-3)$

4. Solve for x. $y = -\dfrac{2}{5}x + 3$

5. Graph $3x - 5y = 15$.

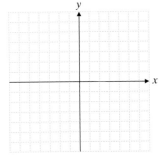

6. Factor completely. $16x^2 - 16x - 12$

7. Solve the system for x, y, and z.
$$x + 4y - z = 10$$
$$3x + 2y + z = 4$$
$$2x - 3y + 2z = -7$$

8. Combine. $\dfrac{7x}{x^2 - 2x - 15} - \dfrac{2}{x - 5}$

▲9. The length of a rectangle is 3 meters longer than twice its width. The perimeter of the rectangle is 48 meters. Find the dimensions of the rectangle.

10. Solve for a. $2ax - 3b = y - 5ax$

Simplify.

11. $\dfrac{3x^3 y^{-2}}{9x^{-1/2} y^{5/2}}$

12. $(3x^{-1/2} y^2)^{-1/3}$

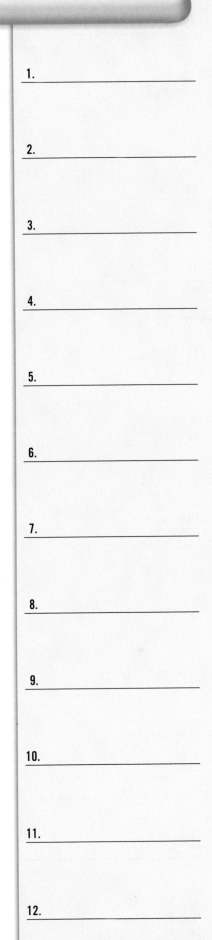

1. _____

2. _____

3. _____

4. _____

5. _____

6. _____

7. _____

8. _____

9. _____

10. _____

11. _____

12. _____

417

13. _____

13. Evaluate. $64^{-1/3}$

14. _____

14. Simplify. $\sqrt[3]{40x^5y^9}$

15. _____

15. Combine. $\sqrt{80x} + 2\sqrt{45x} - 3\sqrt{20x}$

16. _____

16. Multiply and simplify. $\left(2\sqrt{3} - 5\sqrt{2}\right)\left(\sqrt{3} + 4\sqrt{2}\right)$

17. _____

17. Rationalize the denominator. $\dfrac{\sqrt{3} + 2}{2\sqrt{3} - 5}$

18. _____

18. Simplify. $i^{21} + \sqrt{-16} + \sqrt{-49}$

19. _____

19. Simplify. $(3 - 4i)^2$

20. _____

20. Simplify. $\dfrac{1 + 2i}{1 - 3i}$

Solve for x and check your solutions.

21. $x - 4 = \sqrt{3x + 6}$

21. _____

22. $1 + \sqrt{x + 1} = \sqrt{x + 2}$

22. _____

23. If y varies directly with the square of x and $y = 12$ when $x = 2$, find the value of y if $x = 5$.

23. _____

24. The amount of light provided by a lightbulb varies inversely with the square of the distance from the lightbulb. A lightbulb provides 120 lumens at a distance of 10 feet from the light. How many lumens are provided if the distance from the light is 15 feet?

24. _____

One of the most exciting events in baseball is the home run. The path of the ball can be predicted by a quadratic function. The mathematics of this chapter provides players, coaches, and managers information that allows players to improve in their ability to hit home runs.

Quadratic Equations and Inequalities

8.1   **QUADRATIC EQUATIONS**   420

8.2   **THE QUADRATIC FORMULA AND SOLUTIONS TO QUADRATIC EQUATIONS**   427

8.3   **EQUATIONS THAT CAN BE TRANSFORMED INTO QUADRATIC FORM**   436

      HOW AM I DOING? SECTIONS 8.1–8.3   443

8.4   **FORMULAS AND APPLICATIONS**   444

8.5   **QUADRATIC FUNCTIONS**   455

8.6   **QUADRATIC INEQUALITIES IN ONE VARIABLE**   464

      CHAPTER 8 ORGANIZER   472

      CHAPTER 8 REVIEW PROBLEMS   475

      HOW AM I DOING? CHAPTER 8 TEST   479

      CUMULATIVE TEST FOR CHAPTERS 1–8   481

Student Learning Objectives

After studying this section, you will be able to:

1. Solve quadratic equations by the square root property.

2. Solve quadratic equations by completing the square.

① Solving Quadratic Equations by the Square Root Property

Recall that an equation written in the form $ax^2 + bx + c = 0$, where a, b, and c are real numbers and $a \neq 0$, is called a **quadratic equation.** Recall also that we call this the **standard form** of a quadratic equation. We have previously solved quadratic equations using the zero factor property. This has allowed us to factor the left side of an equation such as $x^2 - 7x + 12 = 0$ and obtain $(x - 3)(x - 4) = 0$ and then solve to find that $x = 3$ and $x = 4$. In this chapter we develop new methods of solving quadratic equations. The first method is often called the **square root property.**

THE SQUARE ROOT PROPERTY

If $x^2 = a$, then $x = \pm\sqrt{a}$ for all real numbers a.

The notation $\pm\sqrt{a}$ is a shorthand way of writing "$+\sqrt{a}$ or $-\sqrt{a}$." The symbol $\pm$ is read "plus or minus." We can justify this property by using the zero factor property. If we write $x^2 = a$ in the form $x^2 - a = 0$, we can factor it to obtain $\left(x + \sqrt{a}\right)\left(x - \sqrt{a}\right) = 0$ and thus, $x = -\sqrt{a}$ or $x = +\sqrt{a}$. This can be written more compactly as $x = \pm\sqrt{a}$.

EXAMPLE 1 Solve and check. $x^2 - 36 = 0$

Solution If we add 36 to each side, we have $x^2 = 36$.

$$x = \pm\sqrt{36}$$
$$x = \pm 6$$

Thus, the two roots are 6 and -6.

Check:

$(6)^2 - 36 \overset{?}{=} 0$	$(-6)^2 - 36 \overset{?}{=} 0$
$36 - 36 \overset{?}{=} 0$	$36 - 36 \overset{?}{=} 0$
$0 = 0$ ✓	$0 = 0$ ✓

Practice Problem 1 Solve and check. $x^2 - 121 = 0$

EXAMPLE 2 Solve. $x^2 = 48$

Solution
$$x = \pm\sqrt{48} = \pm\sqrt{16 \cdot 3}$$
$$x = \pm 4\sqrt{3}$$

The roots are $4\sqrt{3}$ and $-4\sqrt{3}$.

Practice Problem 2 Solve. $x^2 = 18$

EXAMPLE 3 Solve and check. $3x^2 + 2 = 77$

Solution
$$3x^2 + 2 = 77$$
$$3x^2 = 75$$
$$x^2 = 25$$
$$x = \pm\sqrt{25}$$
$$x = \pm 5$$

The roots are 5 and -5.

Check:

$$3(5)^2 + 2 \overset{?}{=} 77$$
$$3(25) + 2 \overset{?}{=} 77$$
$$75 + 2 \overset{?}{=} 77$$
$$77 = 77 \checkmark$$

$$3(-5)^2 + 2 \overset{?}{=} 77$$
$$3(25) + 2 \overset{?}{=} 77$$
$$75 + 2 \overset{?}{=} 77$$
$$77 = 77 \checkmark$$

Practice Problem 3 Solve. $5x^2 + 1 = 46$

Sometimes we obtain roots that are complex numbers.

EXAMPLE 4 Solve and check. $4x^2 = -16$

Solution

$$x^2 = -4$$
$$x = \pm\sqrt{-4}$$
$$x = \pm 2i \qquad \text{Simplify using } \sqrt{-1} = i.$$

The roots are $2i$ and $-2i$.

Check:

$$4(2i)^2 \overset{?}{=} -16$$
$$4(4i^2) \overset{?}{=} -16$$
$$4(-4) \overset{?}{=} -16$$
$$-16 = -16 \checkmark$$

$$4(-2i)^2 \overset{?}{=} -16$$
$$4(4i^2) \overset{?}{=} -16$$
$$4(-4) \overset{?}{=} -16$$
$$-16 = -16 \checkmark$$

Practice Problem 4 Solve and check. $3x^2 = -27$

EXAMPLE 5 Solve. $(4x - 1)^2 = 5$

Solution

$$4x - 1 = \pm\sqrt{5}$$
$$4x = 1 \pm \sqrt{5}$$
$$x = \frac{1 \pm \sqrt{5}}{4}$$

The roots are $\dfrac{1 + \sqrt{5}}{4}$ and $\dfrac{1 - \sqrt{5}}{4}$.

Practice Problem 5 Solve. $(2x + 3)^2 = 7$

② Solving Quadratic Equations by Completing the Square

Often, a quadratic equation cannot be factored (or it may be difficult to factor). So we use another method of solving the equation, called **completing the square.** When we complete the square, we are changing the polynomial to a perfect square trinomial. The form of the equation then becomes $(x + d)^2 = e$.

We already know that

$$(x + d)^2 = x^2 + 2dx + d^2 .$$

Notice three things about the quadratic equation on the right-hand side.

1. The coefficient of the quadratic term (x^2) is 1.
2. The coefficient of the linear (x) term is $2d$.
3. The constant term (d^2) is the square of *half* the coefficient of the linear term.

For example, in the trinomial $x^2 + 6x + 9$, the coefficient of the linear term is 6 and the constant term is $\left(\dfrac{6}{2}\right)^2 = (3)^2 = 9$.

For the trinomial $x^2 - 10x + 25$, the coefficient of the linear term is -10 and the constant term is $\left(\dfrac{-10}{2}\right)^2 = (-5)^2 = 25$.

What number n makes the trinomial $x^2 + 12x + n$ a perfect square?

$$n = \left(\frac{12}{2}\right)^2 = 6^2 = 36$$

Hence, the trinomial $x^2 + 12x + 36$ is a perfect square trinomial and can be written as $(x + 6)^2$.

Now let's solve some equations.

EXAMPLE 6   Solve by completing the square and check. $x^2 + 6x + 1 = 0$

Solution

Step 1 First we rewrite the equation in the form $ax^2 + bx = c$ by subtracting 1 from each side of the equation. Thus, we obtain

$$x^2 + 6x = -1.$$

Step 2 We want to complete the square of $x^2 + 6x$. That is, we want to add a constant term to $x^2 + 6x$ so that we get a perfect square trinomial. We do this by taking half the coefficient of x and squaring it.

$$\left(\frac{6}{2}\right)^2 = 3^2 = 9$$

Adding 9 to $x^2 + 6x$ gives the perfect square trinomial $x^2 + 6x + 9$, which we factor as $(x + 3)^2$. But we cannot add 9 to the left side of our equation unless we also add 9 to the right side. (Why?) We now have

$$x^2 + 6x + 9 = -1 + 9$$

Step 3 Now we factor.

$$(x + 3)^2 = 8$$

Step 4 We now use the square root property.

$$x + 3 = \pm\sqrt{8}$$
$$x + 3 = \pm 2\sqrt{2}$$

Step 5 Next we solve for x by subtracting 3 from each side of the equation.

$$x = -3 \pm 2\sqrt{2}$$

The roots are $-3 + 2\sqrt{2}$ and $-3 - 2\sqrt{2}$.

Step 6 We *must* check our solution in the *original* equation (not the perfect square trinomial we constructed).

$$x^2 + 6x + 1 = 0 \qquad\qquad\qquad x^2 + 6x + 1 = 0$$
$$\left(-3 + 2\sqrt{2}\right)^2 + 6\left(-3 + 2\sqrt{2}\right) + 1 \overset{?}{=} 0 \qquad \left(-3 - 2\sqrt{2}\right)^2 + 6\left(-3 - 2\sqrt{2}\right) + 1 \overset{?}{=} 0$$
$$9 - 12\sqrt{2} + 8 - 18 + 12\sqrt{2} + 1 \overset{?}{=} 0 \qquad 9 + 12\sqrt{2} + 8 - 18 - 12\sqrt{2} + 1 \overset{?}{=} 0$$
$$18 - 18 - 12\sqrt{2} + 12\sqrt{2} \overset{?}{=} 0 \qquad\qquad 18 - 18 + 12\sqrt{2} - 12\sqrt{2} \overset{?}{=} 0$$
$$0 = 0 \ \checkmark \qquad\qquad\qquad\qquad\qquad 0 = 0 \ \checkmark$$

Practice Problem 6   Solve by completing the square. $x^2 + 8x + 3 = 0$

Let us summarize for future reference the six steps we use to solve a quadratic equation by completing the square. Notice that step 2 is used when the coefficient of x^2 is not 1. We will see this step performed in Example 7.